CAM SYNTHESIS

SOLID MECHANICS AND ITS APPLICATIONS
Volume 26

Series Editor: G.M.L. GLADWELL

Solid Mechanics Division, Faculty of Engineering
University of Waterloo
Waterloo, Ontario, Canada N2L 3G1

Aims and Scope of the Series

The fundamental questions arising in mechanics are: *Why?*, *How?*, and *How much?* The aim of this series is to provide lucid accounts written by authoritative researchers giving vision and insight in answering these questions on the subject of mechanics as it relates to solids.

The scope of the series covers the entire spectrum of solid mechanics. Thus it includes the foundation of mechanics; variational formulations; computational mechanics; statics, kinematics and dynamics of rigid and elastic bodies; vibrations of solids and structures; dynamical systems and chaos; the theories of elasticity, plasticity and viscoelasticity; composite materials; rods, beams, shells and membranes; structural control and stability; soils, rocks and geomechanics; fracture; tribology; experimental mechanics; biomechanics and machine design.

The median level of presentation is the first year graduate student. Some texts are monographs defining the current state of the field; others are accessible to final year undergraduates; but essentially the emphasis is on readability and clarity.

For a list of related mechanics titles, see final pages.

Cam Synthesis

by

M. A. GONZALEZ-PALACIOS
Universidad Iberoamericana Plantel León,
Departamento de Ingeniería y Ciencias Aplicadas,
León, Guanajuato, Mexico

and

J. ANGELES
McGill University,
Department of Mechanical Engineering &
McGill Research Centre for Intelligent Machines,
Montreal, Quebec, Canada

SPRINGER SCIENCE+BUSINESS MEDIA, B.V.

A C.I.P. Catalogue record for this book is available from the Library of Congress.

ISBN 978-94-010-4835-4 ISBN 978-94-011-1890-3 (eBook)
DOI 10.1007/978-94-011-1890-3

Printed on acid-free paper

To the memory of my father
To my mother
To Betty
Max Antonio González-Palacios

To Patricia, Bruno and Romain
Jorge Angeles

"It would be foolish to suggest
that drawing and theoretical knowledge
at any stage replaced the craftsman's skill
at working with materials."

Francis Pugh (Derek, 1987)

FOREWORD

While the synthesis of planar cam mechanisms has come of age, that of spatial mechanisms is still in its infancy. A milestone in the development of methods of synthesis of planar cam mechanisms was the introduction of envelopes, a similar breakthrough as applied to spatial mechanisms, being yet to be reported. To be sure, attempts in this direction are recorded in the literature, as pointed out in Chapter 1. This state of affairs motivated the work summarized in this book.

The objectives of the book are manifold. We include first a unified approach to cam synthesis, whereby planar, spherical and spatial cam mechanisms are synthesized within a single theoretical framework. We thus start by studying the synthesis of the contact surfaces of spatial cam mechanisms composed of three links, namely, a frame, a cam and a follower. By properly setting the mechanism parameters to values that render the cam and follower axes concurrent, we obtain spherical mechanisms. Likewise, by setting the mechanism parameters so as to render these axes parallel, we end up with planar mechanisms. As a matter of fact, planar mechanisms are derived as a special case of their spherical counterparts. In synthesizing all these mechanisms, we aim at producing contact surfaces of minimum sliding, thereby minimizing the power losses due to dry friction.

Once the study of three-link cam mechanisms has been accomplished, we introduce a fourth element, the roller. Here, the roller is a quite general element that moves with both translation and rotation with respect to cam and follower. The roller thus introduced is, in fact, a generalized roller, for it *i*) not only rolls, but also slides in the most general spatial case, and *ii*) has a hyperboloid shape of revolution in the spatial case, a conical shape in the spherical case and a cylindrical shape in the planar case. Practical issues pertaining to the mechanical design of such a roller are still to be investigated, but the underlying theoretical bases are fully discussed here.

Moreover, the approach adopted at the outset has allowed us to synthesize novel complex mechanisms for the realization of intermittent motion. These mechanisms, known as indexing mechanisms, turn out to be a composition of the two above-mentioned types, namely, the three (no roller)- and four (with roller)-link mechanisms. What we obtained with this combination is a pure-rolling indexing cam mechanism with positive action, free of springs.

We believe that the approach and tools introduced here can assist the designer in producing a much broader spectrum of mechanisms than those motivating this study. We include, for example, in Chapter 7, an example pertaining to the design of a shaking mechanism, which is readily synthesized with the aid of USYCAMS (pronounced 'you see cams'), the software package that we developed to implement all the synthesis procedures described herein. This software, running on Silicon Graphics Inc. IRIS machines, is available at the Internet address `cams@mcrcim.mcgill.ca`, at no cost.

The overall philosophy in producing this book has been to exploit sound mathematical and kinematical tools of analysis and synthesis that could be used only with the available current software and hardware. The latter refers to powerful UNIX workstations that offer the capability of rendering solid models on a graphics monitor at the touch of a key or at the move of a mouse. Modern workstations of this kind thus offer the possibility of rendering animations of multi-body mechanical systems with a great degree of realism. We decided to base our software development on these workstations and not on more common, low-end hardware, because *a*) the McRCIM (McGill research Centre for Intelligent Machines) network is UNIX™ based, and *b*) the prices of graphics workstations are going down very rapidly, while their capabilities are going up at the same pace. We believe that workstations will become the norm in mechanical engineering design in the foreseeable future.

Current software available includes Matlab™, IMSL™ and the like for numerical applications (i.e. *number crunching*), with very easy-to-use routines for interpolation and matrix computations, two very common tasks in mechanical design, particularly in cam synthesis. Other software available pertains to *symbolic computations* and includes brand names such as Maple™, Mathematica™ and Macsyma, to cite the best known of this kind. We found symbolic calculations a great aid in building our theoretical framework, but soon discovered that none of these packages provides routines for algebraic manipulations with dual numbers, introduced in Chapter 2 as the main synthesis tool. We thus developed a few Mathematica routines for this purpose, which we include in Appendix C for the interested reader. Similar routines can be readily developed for Maple or Macsyma.

A feature of the book is its bibliography. Here, we compiled a list of 127 references that we include in Appendix A.

All in all, we have attempted to give new life to a rather languishing field of the theory of machines and mechanisms, that is currently undergoing a phase of stagnation with regard to theoretical and methodological developments. In doing this, we have also attempted to sensitize the reader to the new tools available that warrant a review of well-established but rather limited methods of mechanism synthesis. We illustrate this point in Chapter 8, where we revise the graphical methods of cam-profile synthesis. Here, we show that an improvement in accuracy and speed can be achieved virtually at no cost, by making proper use of current CAD tools. What is needed, however, is a change of attitude towards mechanical design that should bring about creative approaches. We have tested our novel synthesis procedures both in computer animations and in the machine-tool shop. Thus, we produced two prototypes of PRICAM, the pure-rolling indexing cam mechanisms that we synthesized as described in Chapter 5, detailed mechanical design drawings of which are included in Chapter 8. Our intention in including the latter was to highlight the realization of the abstract techniques that we introduce in Chapters 3 and 4, while establishing the link between the rather specific realm of mechanism synthesis and the much broader field of mechanical design.

For the accomplishment of this project we thank our main sources of funding, namely, *Consejo Nacional de Ciencia y Tecnología* (CONNACYT), of Mexico, and the Natural Sciences and Engineering Research Council (NSERC), of Canada. Our undertaking could have not been possible without the state-of-the-art facilities of McRCIM and the enthusiastic support of the machine-tool shop of the Department of Mechanical Engineering of McGill University. Here, we thank Mr. Arthur Clément very especially for his excellent work in producing most of the pieces required by our two prototypes of PRICAM. In the last stages of the manufacturing of the spherical mechanism, the *Centre de recherche industrielle du Québec* (CRIQ) came to our aid. Professor G. M. L. Gladwell, Editor of the Kluwer Series of Solid Mechanics and Its Applications, is duly acknowledged for his valuable suggestions. These comprised not only technical matters, but also English issues that would have otherwise obscured the ideas that we intended to communicate. The first author completed a substantial part of the work reported here as a doctoral candidate of the Department of Mechanical Engineering of McGill University. Last, but not least, the support and understanding of our families encouraged us to make it to our deadline with Kluwer Academic Publishers.

Montreal, July 1993

Max A. González-Palacios
Jorge Angeles

Contents

FOREWORD vii

NOMENCLATURE xv

1 PREAMBLE 1

1.1 Introductory Remarks 1

1.2 The Development of Cam Technology 3

1.2.1 Cam Synthesis 3

1.2.2 Screw Theory 11

1.3 Scope of the Book 13

1.4 Book Organization 13

2 SYNTHESIS TOOLS 15

2.1 Dual Numbers 15

2.2 The Aronhold-Kennedy Theorem 27

2.3 Ruled-Surface Geometry 30

3 KINEMATICS OF THREE-LINK MECHANISMS 37

3.1 Introduction 37

3.2 Surface Geometry 39

3.2.1 Revolute-Higher-Revolute (RHR) Mechanisms 42

3.2.2 Revolute-Higher-Prismatic (RHP) Mechanisms 43

3.2.3 Prismatic-Higher-Revolute (PHR) Mechanisms 45

3.2.4 Prismatic-Prismatic-Prismatic (PPP) Mechanisms 46

3.3 Pressure Angle 47

3.3.1 Pressure Angle of RHR Mechanisms 48

3.3.2 Pressure Angle of RHP Mechanisms 52

3.3.3 Pressure Angle of PHR Mechanisms 53
3.3.4 Pressure Angle of PPP Mechanisms 53

4 KINEMATICS OF FOUR-LINK MECHANISMS **55**
4.1 Introduction 55
4.2 Roller-Followers 56
4.2.1 Revolute-Higher-Cylindric-Revolute (RHCR) Mechanisms . . . 62
4.2.2 Revolute-Higher-Cylindric-Prismatic (RHCP) Mechanisms . . 76
4.2.3 Prismatic-Higher-Cylindric-Revolute (PHCR) Mechanisms . . 79
4.2.4 Prismatic-Higher-Cylindric-Prismatic (PHCP) Mechanisms . . 82
4.3 Pressure Angle 84
4.3.1 Pressure Angle of RHCR Mechanisms 85
4.3.2 Pressure Angle of RHCP Mechanisms 90
4.3.3 Pressure Angle of PHRR Mechanisms 91
4.3.4 Pressure Angle of PHRP Mechanisms 91
4.4 Three-Link Mechanisms with Constant Pressure Angle 93
4.4.1 R$\overline{\mathrm{HH}}$R Mechanisms 94
4.4.2 R$\overline{\mathrm{HH}}$P Mechanisms 96
4.4.3 P$\overline{\mathrm{HH}}$R Mechanisms 96
4.4.4 P$\overline{\mathrm{HH}}$P Mechanisms 98

5 INDEXING MECHANISMS **101**
5.1 Introduction 101
5.2 Input-Output Function of Indexing Cam Mechanisms (ICM) 102
5.3 ICM of RHR Type 103
5.4 ICM of RHCR Type 104
5.5 Pure Rolling with Indexing Cam Mechanisms 110
5.5.1 Primary Mechanism 113
5.5.2 Secondary Mechanism 114
5.6 Positive Action and Positive Motion 117

6 THE VOLUMETRIC PROPERTIES OF CAMS **125**
6.1 Introduction 125
6.2 General Formulation 127
6.2.1 Volume 129

6.2.2 First Moment 129
6.2.3 Second Moment 131
6.3 Planar Objects 132
6.4 Solids 135
6.5 An Outline of MOMENT 140
6.6 Volumetric Properties of Cams in Indexing Mechanisms 149
6.6.1 Planar PRICAM 149
6.6.2 Spherical PRICAM 155
6.6.3 Spatial Mechanisms 159

7 AN OUTLINE OF USYCAMS 163
7.1 Introduction 163
7.2 On-Line vs. Off-Line Synthesis 164
7.3 General Description of USYCAMS 164
7.4 Main Menu 166
7.4.1 Main Loop 167
7.5 Submenus 172
7.5.1 `sm_io_selection` 172
7.5.2 `di_3_osc` 174
7.5.3 `di_3_ind` 178
7.5.4 `di_4` 179
7.6 Synthesis Procedures 181
7.7 Application Example 194

8 THE CAD OF CAM MECHANISMS 199
8.1 Introduction 199
8.2 CAD-Based Methods. Planar Applications 200
8.2.1 RHRR Mechanisms 201
8.2.2 $R\overline{HH}R$ Mechanisms 205
8.2.3 RHRP Mechanisms 205
8.2.4 $R\overline{HH}P$ Mechanisms 208
8.3 Case Studies 211
8.3.1 The Manufacturing Drawings of the Planar PRICAM 213
8.3.2 The Manufacturing Drawings of the Spherical PRICAM 214

APPENDICES **220**

A BIBLIOGRAPHY **221**

B DISPLACEMENT PROGRAMS **231**

B.1 Generalized Input-Output Function 231

B.2 Cycloidal Function 232

B.3 Polynomial Functions 233

B.3.1 3-4-5 Polynomial 233

B.3.2 4-5-6-7 Polynomial 233

B.4 Combined Functions 233

B.4.1 Modified Trapezoidal Acceleration 234

C SYMBOLIC DUAL ALGEBRA **237**

C.1 CrossUnitDual 239

C.2 DecompDual 240

C.3 DivDual 241

C.4 SqrtDual 242

C.5 TanDual 243

C.6 UnitDual 244

INDEX **245**

NOMENCLATURE

a_1: distance between I_{21} and I_{31}
a_3: distance between I_{31} and I_{43}
a_4: distance between I_{43} and I_{42}
b_2: distance between I_{21} and I_{32}
b_3: distance between I_{32} and I_{43}
c: $\sqrt{\phi'^2 + 2\phi' \cos\alpha_1 + 1}$
d: $b_2 - a_1$
$\mathbf{e}_{ij}$: unit vector parallel to I_{ij}
$\hat{\mathbf{e}}_{ij}$: dual unit vector parallel to the polode $\hat{P}_{ij}$
h: amplitude of φ
$\mathbf{k}$: unit vector
$\mathbf{m}$: $\mathbf{p} \times \mathbf{e}$, the moment of a line with respect to a given point
$\mathbf{n}$: unit normal
$\mathbf{p}$: position vector of point P with respect to a given point O
$\mathbf{q}$: first moment of $\mathcal{R}$
$\mathbf{r}_i$: position vector of $\mathcal{R}_i$
$\mathbf{r}_p$: position vector of a point of the pitch surface
$\mathbf{r}_c$: position vector of a point of the cam surface
$\bar{\mathbf{r}}$: position vector of the centroid of $\mathcal{R}$
$\hat{\mathbf{s}}_i$: dual position vector of a point of the polode i
$\hat{\mathbf{s}}_p$: dual position vector of a point of the pitch surface
$\hat{\mathbf{s}}_c$: dual position vector of a point of the cam surface
v_2: dz_2/dt
v_3: dz_3/dt
v_{ij}: dz_{ij}/dt
$\mathbf{w}$: $\boldsymbol{\omega} \times \mathbf{r} / \| \boldsymbol{\omega} \times \mathbf{r} \|$
z_2: displacement of a point of body 2 lying on I_{21} with respect to body 1
z_3: displacement of a point of body 3 lying on I_{31} with respect to body 1
z_{ij}: displacement of a point of body i lying on I_{ij} with respect to body j

$\mathcal{C}$: closed contour
$\mathcal{C}_{ijk}$: great dual circle containing poles $\hat{P}_{ij}$, $\hat{P}_{ik}$ and $\hat{P}_{jk}$
I_{ij}: instantaneous screw axis of bodies i and j
$\mathcal{I}_k$: kth moment of $\mathcal{R}$
$\mathbf{I}$: second moment tensor of $\mathcal{R}$
$\mathcal{L}_{ijk}$: common perpendicular of I_{ij}, I_{ik} and I_{jk}
N: indexing number
$\hat{P}_{ij}$: pole of $\mathcal{S}_i$ with respect to $\mathcal{S}_j$
$\hat{\mathbf{Q}}(\hat{\nu})$: dual rotation through $\hat{\nu}$ about the X-axis
$\mathcal{R}$: bounded region
$\mathcal{R}_i$: ruled surface of body i
$\mathcal{R}_B$: bottom surface of the cam
$\mathcal{R}_C$: contacting ruled surface of the cam
$\mathcal{R}_T$: top surface of the cam
$\hat{\mathbf{R}}(\hat{\nu})$: dual rotation through $\hat{\nu}$ about the Y-axis
$\mathcal{S}$: boundary of $\mathcal{R}$
$\mathcal{S}_i$: dual sphere of body i
$\hat{\mathbf{S}}(\hat{\nu})$: dual rotation through $\hat{\nu}$ about the Z-axis
V: volume of $\mathcal{R}$
α_1: angle between I_{21} and I_{31}
$\hat{\alpha}_1$: $\alpha_1 + \epsilon a_1$
α_3: angle between I_{31} and I_{43}
$\hat{\alpha}_3$: $\alpha_3 + \epsilon a_3$
α_4: angle between I_{43} and I_{42}
$\hat{\alpha}_4$: $\alpha_4 + \epsilon a_4$
β: $\theta_2 - \alpha_1$
$\hat{\beta}$: $\beta + \epsilon d$
δ: angle between $\mathcal{L}_{123}$ and $\mathcal{L}_{341}$
$\hat{\delta}$: $\delta + \epsilon z_{32}$
ϵ: dual unit; $\epsilon^2 \equiv 0$
θ_2: angle between I_{21} and I_{32}
$\hat{\theta}_2$: $\theta_2 + \epsilon b_2$
θ_3: angle between I_{21} and I_{32}
$\hat{\theta}_3$: $\theta_3 + \epsilon b_3$
λ: real number
μ: pressure angle
ν: angle between $\mathcal{L}_{341}$ and $\mathcal{L}_{234}$
$\hat{\nu}$: $\nu + \epsilon z_{43}$

ξ^n: n-dimensional Euclidean space

$\boldsymbol{\xi}_i$: $\boldsymbol{\omega}_i \times \mathbf{r}_i$

ρ: radius of curvature

$\hat{\rho}$: dual radius of curvature

$\tau(x)$: normalized function of x so that x, $\tau \in [0, 1]$

ϕ: angle of rotation of the follower with respect to the frame

φ: generalized input-output function (pronounced *varphi*)

$\dot{\phi}$: $d\phi/dt$

ϕ': $d\phi/d\psi$

$\hat{\phi}$: $\phi + \epsilon z_{31}$

ψ: angle of rotation of the cam with respect to the frame

$\hat{\psi}$: $\psi + \epsilon z_{21}$

ω_{ij}: angular velocity between bodies i and j

$\hat{\omega}_{ij}$: $\omega_{ij} + \epsilon v_{ij}$

$\hat{\boldsymbol{\omega}}_{ij}$: $\hat{\omega}_{ij}\hat{\mathbf{e}}_{ij}$

$\|\cdot\|$: the Euclidean norm of its vector argument $(\cdot)$

Chapter 1

PREAMBLE

1.1 Introductory Remarks

A *mechanism*, according to the IFToMM Commission A for Standardization of Terminology (1991), is ***a system of bodies designed to convert motion of, and forces on, one or several bodies into constrained motions of, and forces on, other bodies.*** The way these bodies are connected is defined by their *kinematic pairs*. Basically, there are two kinds of kinematic pairs, namely, *higher* and *lower pairs*. The former refer to the coupling of two bodies in point or line contact, while the latter refer to the same in surface contact. Lower kinematic pairs can be of six types, namely, *revolute, prismatic, screw, cylindrical, spherical* and *planar* (Hartenberg and Denavit, 1967; Angeles, 1982).

A *cam mechanism* is defined as a mechanism whereby the motion is transmitted by a higher kinematic pair. The simplest cam mechanism is composed of three elements, *frame*, *cam* and *follower*, the cam being the driver, the follower the driven element. Another type of cam mechanism contains a fourth element, a *roller*, which is usually connected to the follower by a revolute pair, the higher pair coupling the cam with the roller. The coupling of a frame with a cam or a follower can be done by revolute, prismatic, cylindric or screw pairs.

Despite advances in robot technology, in which industrial manipulators have replaced mechanisms, cam mechanisms still find important industrial applications in the textile, food-processing and manufacturing industries. Moreover, many applications of cam mechanisms, specifically *indexing cam mechanisms*, arise in conjunction with industrial robots. For example, in Fig. 1.1*a*, a robot is shown installing fixtures on an automobile door; an indexing cam mechanism rotates the hexagonal table intermittently, thereby allowing the process to be automated. In Fig. 1.1*b* a piece is presented to a cluster of machine tools by means of a table driven with an intermittent motion supplied by an indexing cam mechanism. Notice that this layout permits various machining operations to take place simultaneously, with the consequent increase

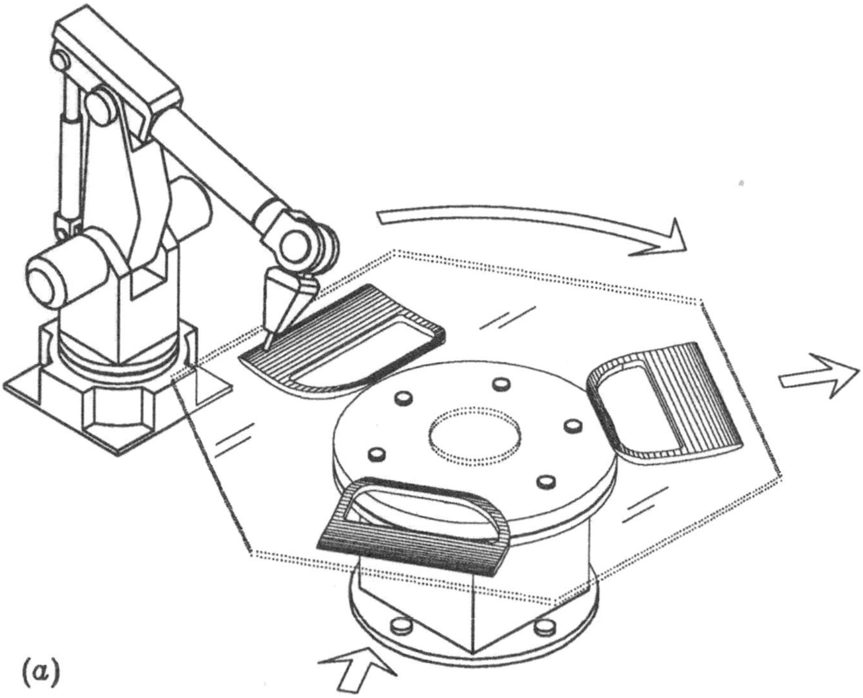

(a)

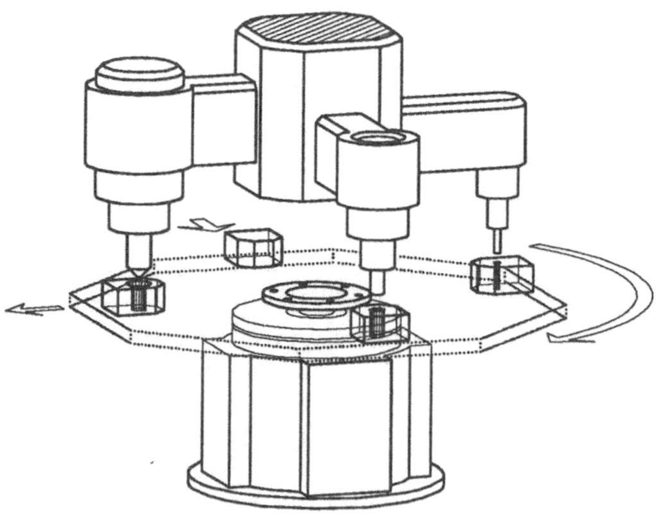

(b)

Figure 1.1 Industrial applications of cam mechanisms.

in productivity.

1.2 The Development of Cam Technology

This subsection is divided into two main parts. The first part presents an overview of previous work on cam mechanisms, while the second deals with the mathematical tools applied in this book.

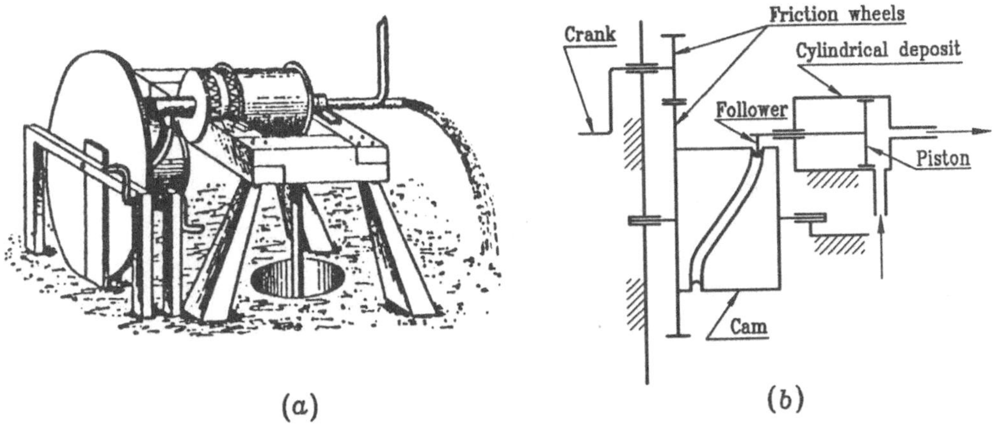

Figure 1.2 A pumping system.

1.2.1 Cam Synthesis

The origin of cam mechanisms can be traced back to the Paleolithic age, as claimed by Müller and Mauersberger (1988), who present an account of the evolution of cam mechanisms, their applications and their contributors. Cam mechanisms seem to have their origin in one of the simple mechanisms of the ancient times (Müller, 1987), namely, the wedge. Moreover, the modern design of cam mechanisms is considered to have been pioneered by Leonardo da Vinci, who designed one of the most significant applications of cam mechanisms, oriented to pumping systems, as shown in Fig. 1.2, where the rotary motion of the crank is transmitted via friction wheels to the cam; then, the rotary motion of the cam is transmitted to the lateral motion of the follower, which drives the piston. Furthermore, the motion of the piston in one direction pulls water from the well and fills the cylindrical deposit, while motion in the other direction pushes water out.

Half a century later, Agostino Rameli took Davinci's concept on pumps and designed ***multi-piston pumps***, arranging the pistons radially, as shown in Fig. 1.3. This type of mechanisms is now used in the well-known *axial piston pump*, which is shown in Fig. 1.4.

While there have been many significant contributions to the theory of cam mechanisms since the XV Century, only as recently as 1724 did J. Leupold report a method for the synthesis of the *profile* of a cam mechanism, whereas the concept of *pressure angle* was introduced only in 1826 by J. V. Poncelet, one of whose sketches is shown in Fig. 1.5. In the middle of the XIX Century, a complete survey on the theory of cam mechanisms was presented by R. Willis (1841).

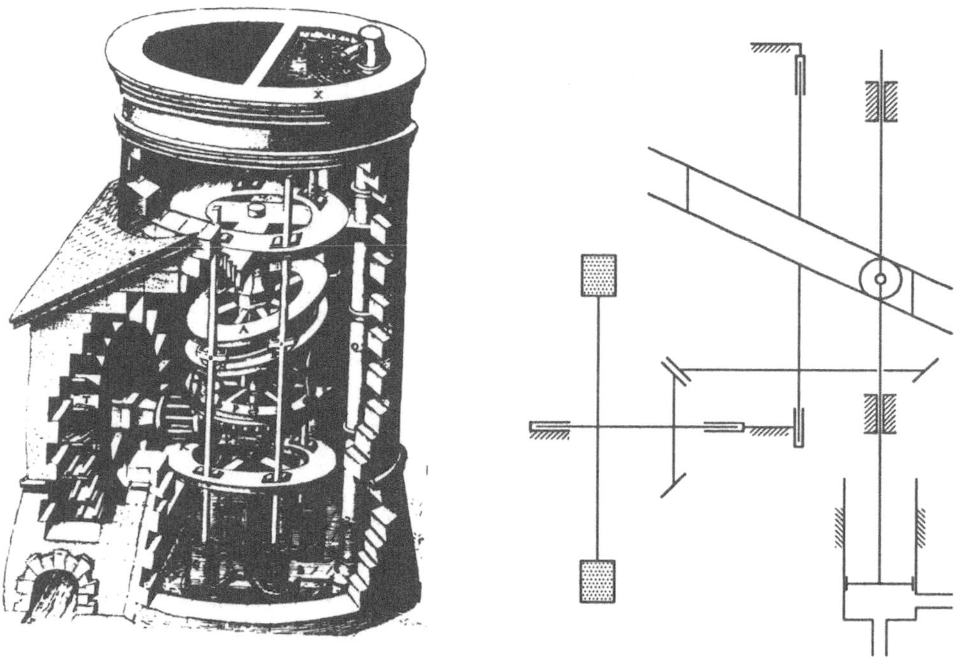

(a) (b)

Figure 1.3 A multi-piston pump mechanism.

The design of cam mechanisms in the first half of the XX Century was based on geometrical methods, while the cutting of the cam profile was done manually, as illustrated in Fig. 1.6. Below we present a survey of developments in the theory and practice of cam mechanisms, starting from the second half of the XX Century. This survey is divided into two main parts, namely, *kinematics* and *dynamics*.

Kinematics

In connection with ***kinematic synthesis***, a method based on *finite differences* was introduced by Johnson (1955) for the design of planar cam mechanisms. Later, this method was applied by the same author (Johnson, 1956*c*) to determine cam profiles with desired acceleration characteristics, while, with a different approach, Zigo (1967) devised an algorithm for the same purpose. Furthermore, some methods for the synthesis of the cam profile from prescribed *jerk* patterns were proposed by Di Benedetto (1975), while Fenton (1966*b*) proposed to avoid *noise in cams* with a proper selection of *cam-follower offset* for roller-follower *radial* cam mechanisms. A method of local analysis of the motion of mechanisms connected by a higher kinematic pair is discussed by Rosenberg and Křen (1987), while the theory of *conjugate surfaces* is applied to the synthesis of planar cam mechanisms by Zhong-Tang and Jing-Ping (1989).

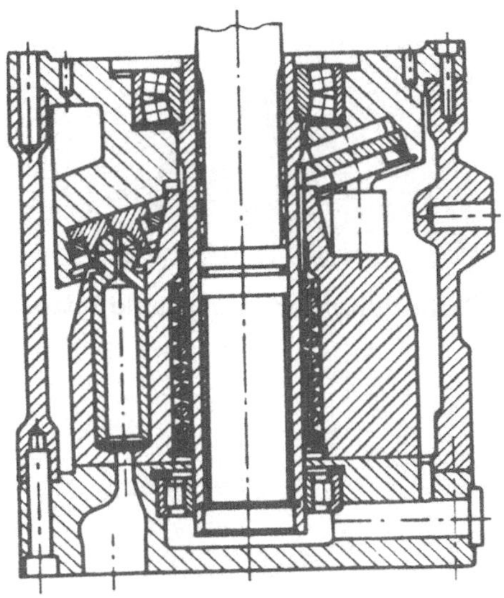

Figure 1.4 The axial piston pump.

The combination of linkages with cam mechanisms, in *five-link mechanisms*, was introduced by Hain (1970), who also proposed the optimization of these mechanisms (Hain, 1971). The same author reported later the optimization of a kinematic inversion of a five-link mechanism with a fixed cam (Hain, 1978). Furthermore, Amarnath and Gupta (1978) adopted Hain's work to design cam-linkage mechanisms for multiple-dwell generation, whereas Sadler and Yang (1990), using a different approach, reported the optimum design of five-link cam mechanisms. Shown in Fig. 1.7 is a typical arrangement for a mechanism of this kind.

Figure 1.5 Definition of pressure angle introduced by Poncelet.

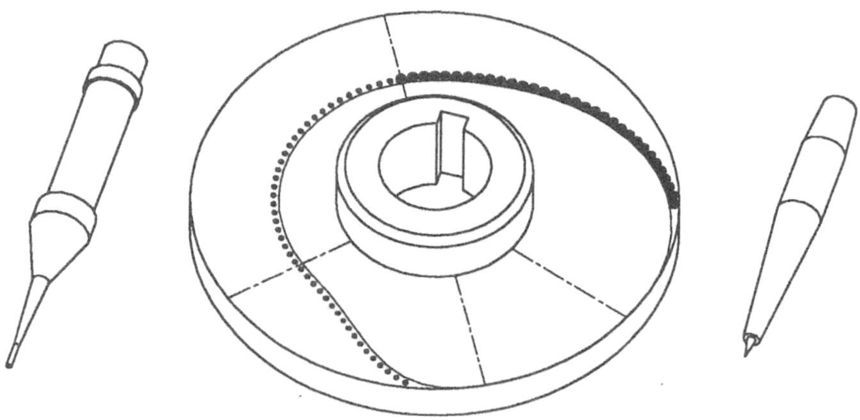

Figure 1.6 Primitive construction of a cam profile.

With the aim of improving the motion of cam mechanisms, Jackowski and Dubil (1967) and Wunderlich (1971) presented their design of *positive motion* with *single-disk planar cams* and *oscillating follower*. An example of their mechanism is shown in Fig. 1.8. Later, with a different approach, but the same concept, Hunt (1973) presented a study of *profiled-follower mechanisms*, a sample of which is shown in Fig. 1.9.

It is well known that *intermittent motion* can be achieved with a kind of kinematic inversion of the *slider-crank mechanism*, i.e., the *Geneva mechanism*, as shown in Fig. 1.10, which has been studied in the past to some extent (Bickford, 1965, 1972; Fenton, 1965, 1975*a*, 1975*b*; Olędzki and Szydlowski, 1975; Shadek et al., 1990).

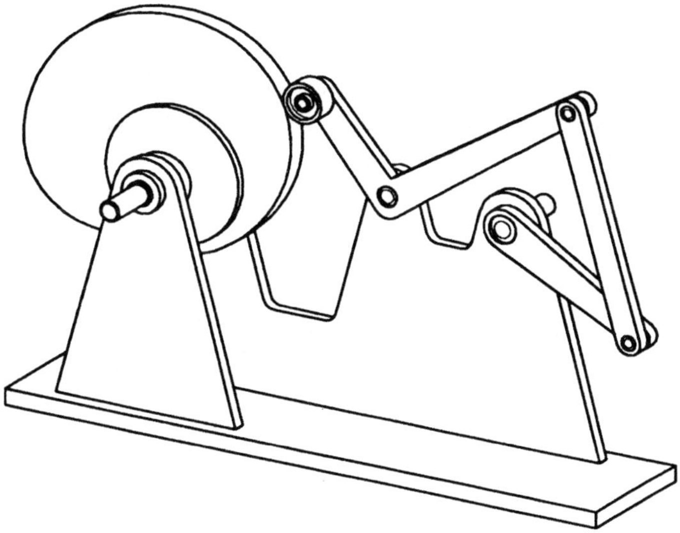

Figure 1.7 A five-link cam mechanism.

Intermittent motion has been produced with the aid of indexing cam mechanisms, like the one shown in Fig. 1.11, which exhibit a better dynamical behavior than the Geneva mechanism (Jacobs, 1949; Johnson, 1958; Makino, 1979). Moreover, the design of five-link cam mechanisms with multiple-dwell capability was reported by Amarnath and Gupta (1975).

The *optimization* of cam mechanisms can be considered from different points of view, depending on the practical problem at hand. Thus, the *minimization* of the cam size for planar cam mechanisms was reported by Fenton (1966*a*, 1975*c*) and Loeff and Soni (1975), while the minimization of cam forces is the subject of (Jones, 1978*b*), a comprehensive account on the optimization of planar cam mechanisms appearing in (Angeles and López-Cajún, 1991).

Experimental work to determine jump characteristics in planar cam-follower systems has been also reported (Rao and Raghavacharyulu, 1975), while experimental results on the changes of dynamic properties were presented by Bialkowicz et al. (1979). The problem of cutting the cam profile was studied by Jones (1978*a*), while Norton et al. (1988) analyzed the effect of *manufacturing* methods on cam performance. Furthermore, the *error* due to manufacturing and *assembly* was studied by Dhande and Chakraborty (1975) from a probabilistic viewpoint, as pertaining to some *planar* and *spatial* cam mechanisms. A method for the formula-based design of three-dimensional cams was reported by Raven (1959). Later, a unified approach to the design of this type of mechanisms was presented by Dhande et al. (1975). Moreover, Dittrich and Zakel (1979) reported a study of three-dimensional cam mechanisms based on the values of the pressure angle, while a study of *spherical cam mechanisms* was discussed earlier by Dittrich (1966). Recently, *envelope theory*, which had been limited to planar mechanisms, was applied to spatial cam surface geometry, as reported by Backhouse and Jones (1990).

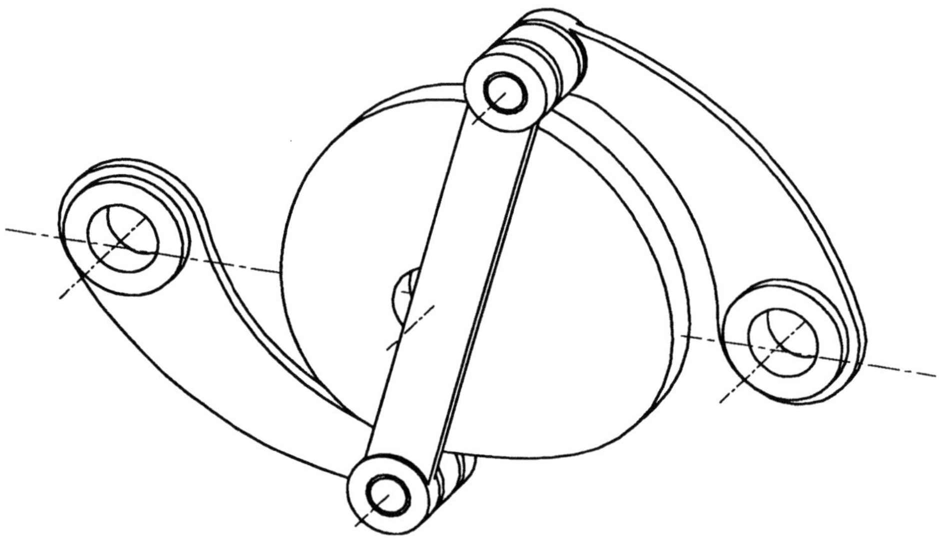

Figure 1.8 Positive motion with single disc cam mechanisms.

Dynamics

For cams rotating at high speed, impact loads become crucial in the design. Thus, if force characteristics are considered in the synthesis of cam mechanisms, the problem falls into the realm of ***dynamic synthesis***. A coordination of the ***polynomial equations of motion*** with the dynamic aspects of machine operation gives the ***polydyne*** cam design, which was discussed by Stoddart (1953*a*, 1953*b*). Another approach to dynamic synthesis was reported by Wiederrich and Roth (1975), who applied ***finite trigonometric series*** to their analysis. Furthermore, a procedure for the dynamic analysis of a cam mechanism with ***bearing clearances*** was proposed by Osman et al. (1987).

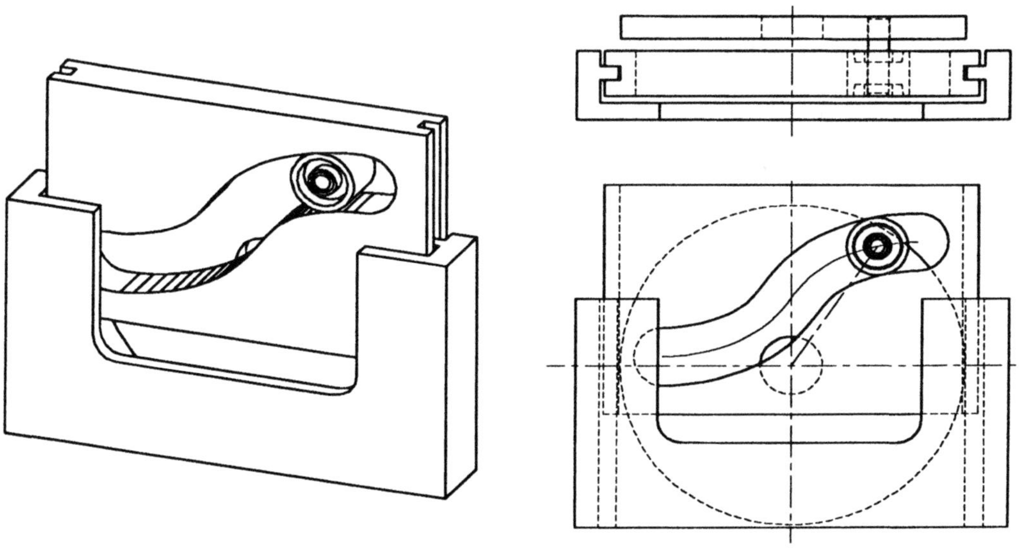

Figure 1.9 A profiled mechanism.

The optimization of cam and follower properties considering the dynamics of cam mechanisms has been reported by Johnson (1956*a*, 1956*b*) and Berzak and Freudenstein (1979), while the effects of cam-profile error on the dynamic behavior of follower cam systems was discussed by Grewal and Newcombe (1988).

In the realm of indexing cam mechanisms, *residual vibrations* were studied by Takano and Toyama (1979). The optimum configuration of planar external mechanisms of this kind is reported by Jones and Tsang (1987), whereas the optimum design of external and internal indexing cam mechanisms was addressed by Gouxun et al. (1988).

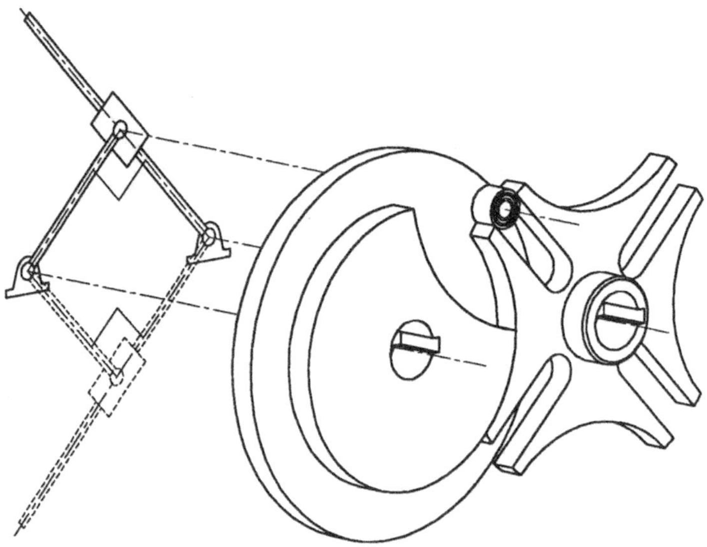

Figure 1.10 Geneva mechanism.

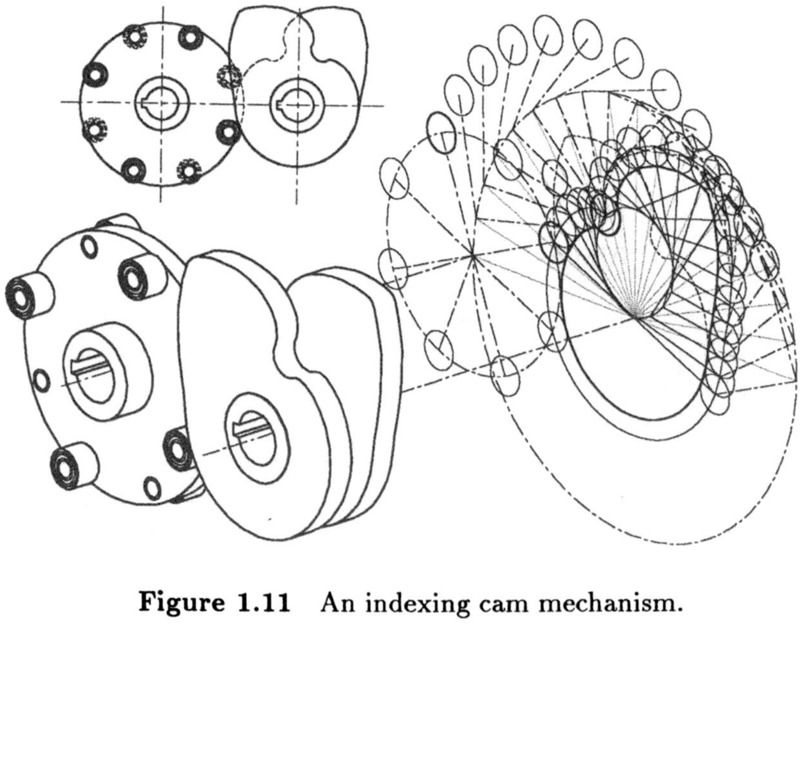

Figure 1.11 An indexing cam mechanism.

Backlash, squeeze and impact in planar cam mechanisms are simulated in the work reported by Koster (1975) and Kass and Chace (1975), while different methods of cutting a plate cam are compared by Norton (1988).

1.2.2 Screw Theory

The approach to the kinematic synthesis of cam mechanisms introduced here is based on *screw theory*. While it is difficult to state the date of the origin of this theory, Ball (1900) includes a list of contributors. A crucial development of screw theory is the concept of *cylindroid*, that is the locus of the instant screw axis of two bodies in relative motion that rotate about axes fixed to a common frame. The cylindroid is generated as the velocity ratio of the two bodies ranges from $-\infty$ to $+\infty$. Ball (1875) built a model of the cylindroid with steel wires mounted in a boxwood cylinder, as shown in Fig. 1.12*a*. The corresponding surface of the cylindroid is shown in Fig. 1.12*b*.

More recently, Yang (1974) presented a brief outline of the basic operations and geometric interpretations of the *calculus of screws*, while screw theory and *dual matrices* were applied by Pandrea and Voiculescu (1975) to determine the screw displacement of two arbitrary lines in space. The determination of *instantaneous screw* axes by relating screws and the algebraic concept of *motor* is introduced by Rooney (1975*a*). A theory that involves *repelling* and *contrary* screws is presented by Ohwovoriole and Roth (1981). Moreover, Sugimoto and Duffy (1982) applied elements of linear algebra to *screw systems*, and presented a general method for determining the system of instantaneous screws of any link in a closed-loop mechanism; later, Kerr and Sanger (1989) used this approach to evaluate the inner product of *reciprocal screws*. The correspondence of points, planes and lines of a set of four mutually reciprocal screws was outlined in turn by Hon-Cheung (1987), while Gibson and Hunt (1990*a* & *b*) presented all possible linear combinations of given instantaneous screws.

The *theorem of three axes* in spatial motion was introduced first by Beggs (1959) in German, while Phillips and Hunt (1964) published the same result for the first time in English. Presented in this book is a theoretical framework based on the application of this theorem. The analytical treatment of planar kinematics based on *canonical systems* and *instantaneous invariants* is extended to spatial kinematics in the work reported by Veldkamp (1967), who later presented an account of the algebra of dual quantities, the *differential geometry* of *dual curves* and their applications to theoretical spatial kinematics (Veldkamp, 1976). The theory pertaining to the kinematic synthesis of rigid-body motions for finitely- and infinitesimally-separated positions was reported by Chen and Roth (1969).

The theory of *ruled surfaces* has been considered an important tool for the study of screw theory. Along these lines, Yang et al. (1975) reported applications of screw theory to spatial mechanisms. Moreover, the differential properties of ruled surfaces

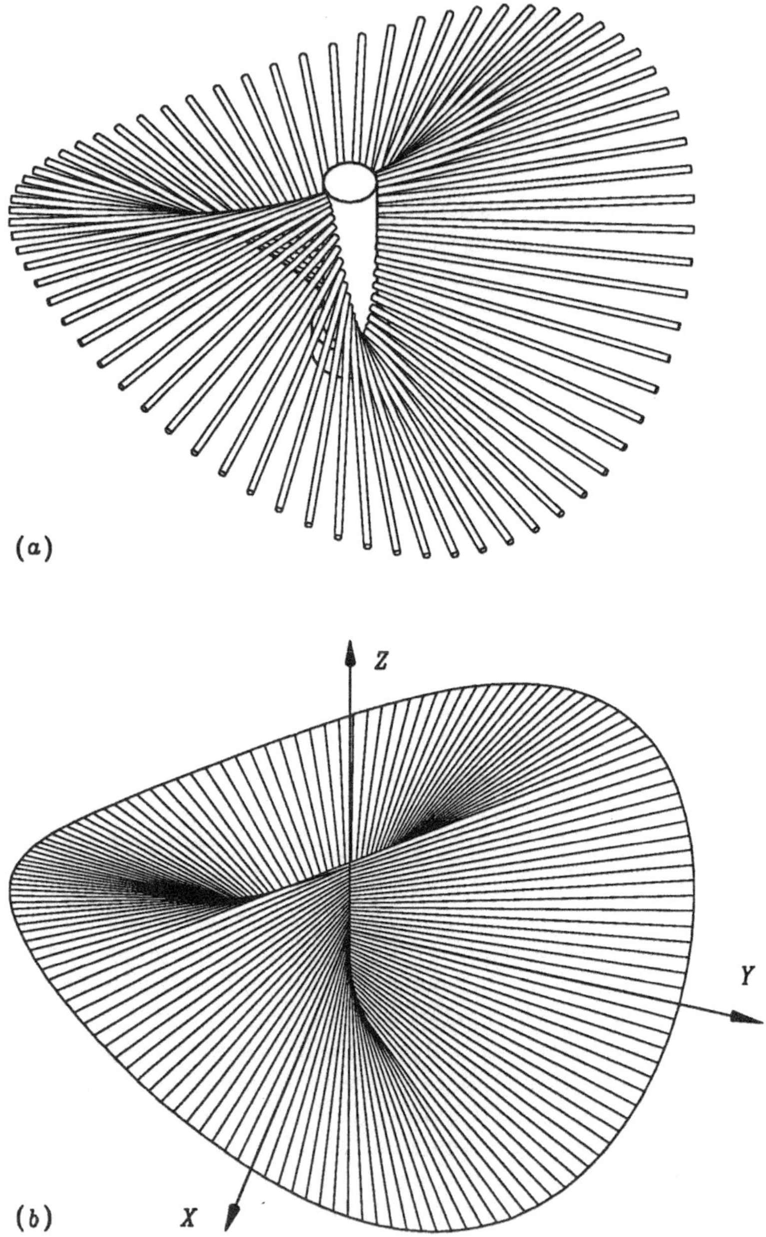

Figure 1.12 *a*) Ball's model of the cylindroid; *b*) the cylindroid as a ruled surface.

in a form applicable to spatial kinematics were reported by McCarthy and Roth (1981), while a study of ruled surfaces in dual space is given in Köse (1982*a*, 1982*b*). The geometry of axods is analyzed, in turn, with the aid of ruled-surface theory by Dizioğlu (1989), while a new dual integral invariant for a given closed ruled surface is introduced by Gürsoy (1990).

Contributions to the theory of mechanisms from the point of view of screw theory have been presented by different authors, e.g., Yang (1963), Yang and Freudenstein (1964), whose work is related to the application of *dual-quaternion algebra* to the study of spatial mechanisms; Hunt (1967*a* & *b*) presented a study of *spatial linkages* with prismatic pairs; Waldron (1972) reported a method for relating the geometry of contacting surfaces to motion capability in the joints of spatial linkages; Rooney (1975*b*) stated a principle enabling one to utilize laws derived for *spherical linkages* directly in the analysis of spatial linkages; Vadasz and Soni (1979) presented, in turn, a formula giving the ratios of two of the three direction cosines of the spatial Burmester lines for C-C cranks; Xiao and Yang (1989) reported more recently a derivation of the instantaneous kinematics of two arbitrarily chosen moving rigid bodies.

1.3 Scope of the Book

Cam mechanisms have been studied from different points of view, namely, kinematic synthesis, dynamic synthesis, analysis, design, optimization and manufacturing. This book is oriented to the kinematic synthesis of cam mechanisms in a unified framework. By this we mean that spatial, spherical and planar cam mechanisms are integrated in the same formulation. Traditionally, the synthesis of the three types of cam mechanisms has been approached using independent formulations. With a unified formulation, not only the well-known types of cam mechanisms, but also novel cam mechanisms can be synthesized, as shown in this book.

Moreover, since all the design parameters are considered in a unified framework, the optimization theory of cam mechanisms can be applied systematically, i.e., the criteria applied for the optimization of planar cam mechanisms, which can be found to some extent in the literature, can be complemented in order to provide general criteria for the optimization of spherical and spatial cam mechanisms.

1.4 Book Organization

The basic tools of cam synthesis, namely, dual-number algebra, the Aronhold-Kennedy Theorem and ruled-surface geometry, are included in Chapter 2. The unified synthesis of three-link mechanisms is presented in Chapter 3, where the geometry of contact surfaces and the significance of the pressure angle are discussed. Four types of three-link cam mechanisms are studied, which are classified according to their kinematic pairs, namely, RHR, PHR, RHP and PPP. The first and last letters stand for the type of pair of the input and output axes, respectively, while the middle letter, for the cam-follower pair. Moreover, R, P and H represent respectively, revolute, prismatic and higher pairs. The sliding velocity along the higher pair is minimized, and the contacting surfaces are obtained from this condition. In Chapter 4, the same philosophy is applied for the unified synthesis of four-link cam mechanisms. Now, the four types are defined as RHCR, RHCP, PHCR and PHCP. Similarly, the first and last letters refer, respectively, to the kinematic pair of the input and output axes, while the second letter refers to the cam-roller coupling, and the third letter, to the roller-follower pair, C standing for cylindric. A unified formulation for the pressure angle is introduced here. The concept of cam mechanism with constant pressure angle and an auxiliary roller is applied to the synthesis of cam mechanisms with flat-face followers. Here, four types are discussed, namely, R$\overline{\mathrm{HH}}$R, R$\overline{\mathrm{HH}}$P, P$\overline{\mathrm{HH}}$R and P$\overline{\mathrm{HH}}$P, where the symbol $\overline{\mathrm{HH}}$ is explained in Section 4.4

The theory discussed in Chapters 3 and 4 is now applied in Chapter 5 to the synthesis of indexing cam mechanisms with an extension to the synthesis of indexing cam mechanisms with pure rolling and positive motion. The volumetric properties of solids of arbitrary shapes are discussed in Chapter 6. The description of USY-CAMS, a software package developed for the interactive synthesis of cams, is outlined in Chapter 7. With the aid of a CAD system, graphical methods for planar cam mechanisms, based on the theory presented in Chapters 3 and 4, are discussed in Chapter 8, where the profile of the cam is obtained directly from the contact points between cam and follower. In this chapter, CAD drawings of the overall design of PRICAM, a novel *pure-rolling indexing cam mechanism*, are presented.

Three appendices are included for quick reference: a collection of 127 references is included in the Appendix A, while a few displacement functions applied for the synthesis of cam mechanisms are described in Appendix B, and the code of some functions to perform dual operations symbolically is included in Appendix C.

Chapter 2

SYNTHESIS TOOLS

2.1 Dual Numbers

A brief account of *dual numbers* is presented here as a quick reference for the reader. Those readers wanting to learn more on dual numbers are referred to (Yang, 1963, 1974; Yang and Freudenstein, 1964; Veldkamp, 1976; Bottema and Roth, 1979).

A *dual scalar*, *vector* or *matrix* quantity is represented as the sum of a *primal part* and a *dual part*, the latter beginning with the *dual unity* ϵ, which is nilpotent, i.e., $\epsilon^2 = 0$.

Let $\hat{a} = a + \epsilon a^*$ and $\hat{b} = b + \epsilon b^*$ be two dual scalars, with a, b, a^* and b^* being all real numbers. Equality, addition, multiplication and division are defined, respectively, as

$$\hat{a} = \hat{b} \Leftrightarrow a = b, \; a^* = b^* \tag{2.1a}$$

$$\hat{a} + \hat{b} = (a + b) + \epsilon(a^* + b^*) \tag{2.1b}$$

$$\hat{a}\hat{b} = ab + \epsilon(ab^* + a^*b) \tag{2.1c}$$

$$\frac{\hat{a}}{\hat{b}} = \frac{a}{b} - \epsilon\left(\frac{ab^* - a^*b}{b^2}\right), \quad (b \neq 0) \tag{2.1d}$$

Moreover, if a function $f(\hat{x})$, with $\hat{x} = x + \epsilon x^*$, has the derivative $df(\hat{x})/d\hat{x}$, we can define $f(\hat{x})$ in terms of its series expansion evaluated at $\hat{x} = x$, namely,

$$\begin{aligned} f(\hat{x}) &= f(x) + (\hat{x} - x)\frac{df(\hat{x})}{d\hat{x}}\Big|_{\hat{x}=x} + \dots + \frac{(\hat{x} - x)^i}{i!}\frac{d^i f(\hat{x})}{d\hat{x}^i}\Big|_{\hat{x}=x} + \dots \\ &= f(x) + \epsilon x^* \frac{df(\hat{x})}{d\hat{x}}\Big|_{\hat{x}=x} + \dots + \frac{\epsilon^i x^{*i}}{i!}\frac{d^i f(\hat{x})}{d\hat{x}^i}\Big|_{\hat{x}=x} + \dots \end{aligned}$$

but $\epsilon^i \equiv 0$, for $i = 2, 3, 4, \ldots$, and the series expansion reduces to

$$f(\hat{x}) = f(x) + \epsilon x^* f'(x), \qquad f'(x) \equiv \frac{df(\hat{x})}{d\hat{x}}\Big|_{\hat{x}=x} \tag{2.2}$$

For example, with application of eq.(2.2), one can readily obtain an expression for the square root of a dual number, namely,

$$\sqrt{\hat{x}} = \sqrt{x} + \epsilon\frac{x^*}{2\sqrt{x}}, \qquad (x > 0) \tag{2.3}$$

Example 2.1.1: *Given* $\hat{x} = x + \epsilon x^*$, *prove that*

$$e^{\epsilon x^*} = 1 + \epsilon x^* \tag{2.4}$$

Solution:

First, we consider the function $f(\hat{x}) = e^{\hat{x}}$: from eq.(2.2) we have

$$e^{\hat{x}} = e^x + \epsilon x^* e^x = e^x(1 + \epsilon x^*) \tag{a}$$

On the other hand,

$$e^{\hat{x}} = e^{x+\epsilon x^*} = e^x e^{\epsilon x^*} \tag{b}$$

Thus, by equating the rightmost-hand sides of eqs.(a & b), the intended proof is done. Alternatively, relation (2.4) can also be derived by a straightforward development of the series expansion of the exponential. ■

With the aid of eq.(2.4), a dual number $\hat{a} = a + \epsilon a^*$, can be expressed in exponential form, namely,

$$\hat{a} = a(1 + \epsilon q) = ae^{\epsilon q}, \qquad q = \frac{a^*}{a}, \qquad a \neq 0 \tag{2.5}$$

where q is called the *pitch*, an attribute that characterizes the dual number $\hat{a}$. Thus, when $q \neq 0$, $\hat{a}$ is a *proper dual number*; when $q \to \infty$, $\hat{a}$ is a *pure dual number* and when $q = 0$, $\hat{a}$ reduces to the real number a.

Henceforth, vectors are regarded as column arrays. When vectors are represented as row arrays to save space, a superscript T on this array is used to remind us of our convention.

Now, an ordered triple of dual numbers $[\hat{a}_1,\ \hat{a}_2,\ \hat{a}_3]^T$, is called a dual vector $\hat{\mathbf{a}}$, its components being $\hat{a}_i$, for $i = 1, 2, 3$. If the components of $\hat{\mathbf{a}}$ are real, then the vector is real. Moreover, let $\hat{\mathbf{a}}$ and $\hat{\mathbf{b}}$ be two dual vectors; their equality, multiplication by a dual scalar, inner product and cross-product are defined, respectively, as

$$\hat{\mathbf{a}} = \hat{\mathbf{b}} \Leftrightarrow \hat{a}_i = \hat{b}_i, \quad i = 1,2,3 \tag{2.6a}$$

$$\hat{c}\hat{\mathbf{a}} = [\hat{c}\hat{a}_1,\ \hat{c}\hat{a}_2,\ \hat{c}\hat{a}_3]^T \tag{2.6b}$$

$$\hat{\mathbf{a}} \cdot \hat{\mathbf{b}} = \hat{a}_1\hat{b}_1 + \hat{a}_2\hat{b}_2 + \hat{a}_3\hat{b}_3 \tag{2.6c}$$

$$\hat{\mathbf{a}} \times \hat{\mathbf{b}} = [\hat{a}_2\hat{b}_3 - \hat{a}_3\hat{b}_2,\ \hat{a}_3\hat{b}_1 - \hat{a}_1\hat{b}_3,\ \hat{a}_1\hat{b}_2 - \hat{a}_2\hat{b}_1]^T \tag{2.6d}$$

In most cases, the definitions of the rules of vector algebra apply to dual vectors. Thus, a dual vector $\hat{\mathbf{a}} \equiv [\hat{a}_1, \hat{a}_2, \hat{a}_3]^T$ can be written as a linear combination of basis vectors, namely,

$$\hat{\mathbf{a}} = \hat{a}_1\mathbf{i} + \hat{a}_2\mathbf{j} + \hat{a}_3\mathbf{k} \tag{2.7}$$

Consequently,

$$\begin{aligned}\hat{\mathbf{a}} &= (a_1 + \epsilon a_1^*)\mathbf{i} + (a_2 + \epsilon a_2^*)\mathbf{j} + (a_3 + \epsilon a_3^*)\mathbf{k} \\ &= a_1\mathbf{i} + a_2\mathbf{j} + a_3\mathbf{k} + \epsilon(a_1^*\mathbf{i} + a_2^*\mathbf{j} + a_3^*\mathbf{k}) \\ &= \mathbf{a} + \epsilon\mathbf{a}^*\end{aligned} \tag{2.8}$$

From the foregoing discussion one might be induced to regard dual vectors as vector spaces over a corresponding field. However, dual numbers, contrary to real or complex numbers, *do not form a field.* Indeed, to form a field, a set of numbers must contain the inverse of every non-zero number under multiplication. However, from the definition of division of dual numbers, eq.(2.1d), it is apparent that pure dual numbers, although non-zero, do not have such an inverse. Nevertheless, dual vectors share many properties of vectors defined over the real or complex field, as we will show presently. Moreover, from eq.(2.8) one can readily derive frame-invariant representations of relations (2.6a–d), namely, if

$$\hat{\mathbf{a}} \equiv \mathbf{a} + \epsilon\mathbf{a}^*, \quad \hat{\mathbf{b}} \equiv \mathbf{b} + \epsilon\mathbf{b}^* \quad \text{and} \quad \hat{c} \equiv c + \epsilon c^*,$$

then

$$\hat{\mathbf{a}} = \hat{\mathbf{b}} \Leftrightarrow \mathbf{a} = \mathbf{b} \quad \text{and} \quad \mathbf{a}^* = \mathbf{b}^* \tag{2.9a}$$

$$\hat{c}\hat{\mathbf{a}} = c\mathbf{a} + \epsilon(c^*\mathbf{a} + c\mathbf{a}^*) \tag{2.9b}$$

$$\hat{\mathbf{a}} \cdot \hat{\mathbf{b}} = \mathbf{a} \cdot \mathbf{b} + \epsilon(\mathbf{a} \cdot \mathbf{b}^* + \mathbf{a}^* \cdot \mathbf{b}) \tag{2.9c}$$

$$\hat{\mathbf{a}} \times \hat{\mathbf{b}} = \mathbf{a} \times \mathbf{b} + \epsilon(\mathbf{a} \times \mathbf{b}^* + \mathbf{a}^* \times \mathbf{b}) \tag{2.9d}$$

In particular, if $\hat{\mathbf{b}} = \hat{\mathbf{a}}$, eq.(2.9c) leads to

$$\hat{\mathbf{a}} \cdot \hat{\mathbf{a}} = \| \mathbf{a} \|^2 + \epsilon 2\mathbf{a} \cdot \mathbf{a}^* \tag{2.9e}$$

Now, the *Euclidean norm* $\| \hat{\mathbf{a}} \|$ of a dual vector $\hat{\mathbf{a}} = \mathbf{a} + \epsilon\mathbf{a}^*$ is defined as

$$\| \hat{\mathbf{a}} \| \equiv \sqrt{\hat{\mathbf{a}} \cdot \hat{\mathbf{a}}} = \sqrt{\mathbf{a} \cdot \mathbf{a} + \epsilon 2(\mathbf{a} \cdot \mathbf{a}^*)}$$

Moreover, with the aid of eq.(2.3), one can readily obtain

$$\| \hat{\mathbf{a}} \| = \| \mathbf{a} \| + \epsilon \frac{\mathbf{a} \cdot \mathbf{a}^*}{\| \mathbf{a} \|}, \qquad (\mathbf{a} \neq \mathbf{0}) \tag{2.10}$$

where $\| \mathbf{a} \|$ denotes the Euclidean norm of $\mathbf{a}$, a notation that will be used throughout. Furthermore, $\hat{\mathbf{a}}$ is called a unit dual vector if $\| \hat{\mathbf{a}} \| = 1$. Any dual vector can thus be expressed as

$$\hat{\mathbf{a}} = \| \hat{\mathbf{a}} \| \, \mathrm{duv}(\hat{\mathbf{a}}) \tag{2.11}$$

where $\mathrm{duv}(\hat{\mathbf{a}})$ is the *dual unit vector* of $\hat{\mathbf{a}}$, and is computed as

$$\mathrm{duv}(\hat{\mathbf{a}}) \equiv \frac{\hat{\mathbf{a}}}{\| \hat{\mathbf{a}} \|} = \frac{\mathbf{a}}{\| \mathbf{a} \|} + \epsilon \frac{(\mathbf{a} \times \mathbf{a}^*) \times \mathbf{a}}{\| \mathbf{a} \|^3} \tag{2.12}$$

The proof that the last expression of eq.(2.12) is indeed a dual unit vector is left as an exercise.

Furthermore, a line $\mathcal{L}$ can be defined via the dual unit vector $\hat{\mathbf{e}}$, defined as

$$\hat{\mathbf{e}} = \mathbf{e} + \epsilon\mathbf{m}, \qquad \mathbf{m} \equiv \mathbf{e}^* \tag{2.13}$$

where $\mathbf{e}^T\mathbf{e} = 1$ and $\mathbf{e}^T\mathbf{m} = 0$. Here, $\mathbf{e}$ defines the *direction* of $\mathcal{L}$, while $\mathbf{m}$ the *moment* of $\mathcal{L}$ with respect to a self-understood point O, namely,

$$\mathbf{m} = \mathbf{p} \times \mathbf{e} \tag{2.14}$$

$\mathbf{p}$ being the vector directed from O to an arbitrary point P of $\mathcal{L}$. Moreover, $\mathbf{e}$ and $\mathbf{m}$ are called the *primal* and *dual* parts of $\hat{\mathbf{e}}$, respectively. Thus, the components of $\hat{\mathbf{e}}$ are defined as the *line coordinates* of $\mathcal{L}$ (Yang et al., 1975), while $\hat{\mathbf{e}}$ defines the dual point $\hat{P}$ on the *dual unit sphere* $\mathcal{S}$ as shown in Fig. 2.1*a*. Furthermore, the six components of the two vectors $\mathbf{e}$ and $\mathbf{m}$ in eq. (2.13) constitute the *Plücker coordinates* of $\mathcal{L}$. An alternative representation of $\mathcal{L}$ is given as shown in Fig. 2.1*b*, where

$$\mathbf{r} = \mathbf{p} + \lambda\mathbf{e} \tag{2.15}$$

and λ is a real number. The components of $\mathbf{r}$ are defined as the *point coordinates* of $\mathcal{L}$. The relation between line coordinates and point coordinates is derived from eq. (2.14), upon cross-multiplying its two sides by vector $\mathbf{e}$, namely,

$$\mathbf{e} \times \mathbf{m} = \mathbf{e} \times (\mathbf{p} \times \mathbf{e}) \tag{2.16}$$

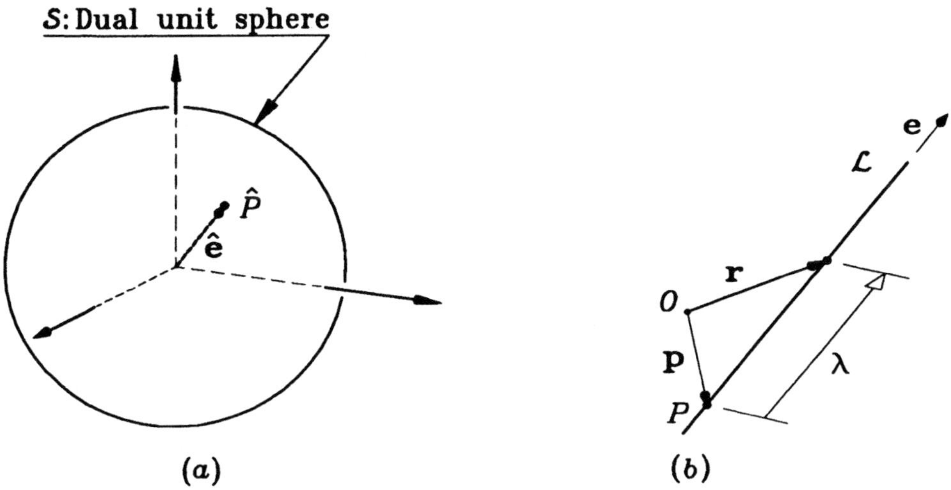

Figure 2.1 A line represented *a*) in *line coordinates* and *b*) in *point coordinates.*

Expanding the right–hand side of eq.(2.16), one obtains

$$\mathbf{e} \times \mathbf{m} = \mathbf{p} - \mathbf{e}^T\mathbf{p}\mathbf{e} \tag{2.17}$$

If P is chosen as that point of $\mathcal{L}$ lying closest to the origin, henceforth denoted by P_0 of position vector $\mathbf{p}_0$, then, from eq.(2.17),

$$\mathbf{p}_0 = \mathbf{e} \times \mathbf{m} \tag{2.18}$$

and vector $\mathbf{r}$ of eq.(2.15) is given by

$$\mathbf{r} = \mathbf{e} \times \mathbf{m} + \lambda\mathbf{e} \tag{2.19}$$

Now, let $\mathcal{L}_1$ and $\mathcal{L}_2$ be, in general, two skew lines as shown in Fig. 2.2, the *dual angle* between them being defined as

$$\hat{\nu} \equiv \nu + \epsilon h \tag{2.20}$$

where ν is the angle between $\mathbf{e}_1$ and $\mathbf{e}_2$ and h is the distance between $\mathcal{L}_1$ and $\mathcal{L}_2$. As such, h is nonnegative. Now, with the aid of eq.(2.2), the trigonometric functions of $\hat{\nu}$ can be expressed as

$$\cos\hat{\nu} = \cos\nu - \epsilon h \sin\nu \tag{2.21a}$$

$$\sin\hat{\nu} = \sin\nu + \epsilon h\cos\nu \tag{2.21b}$$

$$\tan\hat{\nu} = \tan\nu + \epsilon h(1+\tan^2\nu) \tag{2.21c}$$

$$\cot\hat{\nu} = \cot\nu - \epsilon\frac{h}{\sin^2\nu}, (\sin\nu \neq 0) \tag{2.21d}$$

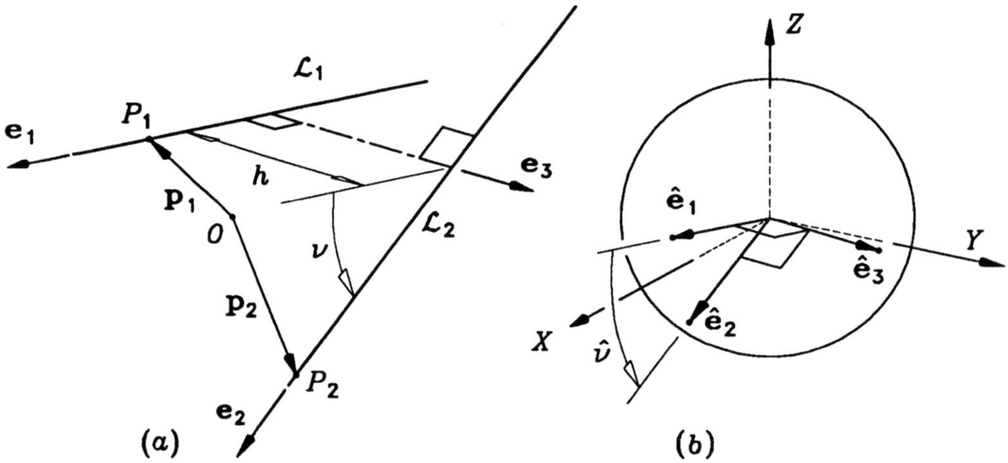

Figure 2.2 Definition of the dual angle $\hat{\nu} \equiv \nu + \epsilon h$.

Moreover, let $\hat{\mathbf{e}}_1$ and $\hat{\mathbf{e}}_2$ be two dual unit vectors, and $\hat{\mathbf{e}}_3$ be the dual unit vector of the cross product $\hat{\mathbf{e}}_1 \times \hat{\mathbf{e}}_2$, as depicted in Fig. 2.2, i.e.,

$$\hat{\mathbf{e}}_3 \equiv \frac{\hat{\mathbf{e}}_1 \times \hat{\mathbf{e}}_2}{\| \hat{\mathbf{e}}_1 \times \hat{\mathbf{e}}_2 \|} \tag{2.22}$$

Thus, the dual angle between $\hat{\mathbf{e}}_1$ and $\hat{\mathbf{e}}_2$ is defined as

$$\cos\hat{\nu} = \hat{\mathbf{e}}_1 \cdot \hat{\mathbf{e}}_2 \tag{2.23a}$$

$$\sin\hat{\nu} = \hat{\mathbf{e}}_1 \times \hat{\mathbf{e}}_2 \cdot \hat{\mathbf{e}}_3 \tag{2.23b}$$

Below we illustrate the geometric interpretation of eqs.(2.23). First, by application of relation (2.9c),

$$\hat{\mathbf{e}}_1 \cdot \hat{\mathbf{e}}_2 = \mathbf{e}_1 \cdot \mathbf{e}_2 + \epsilon(\mathbf{e}_1 \cdot \mathbf{m}_2 + \mathbf{m}_1 \cdot \mathbf{e}_2) \tag{2.24}$$

i.e.,

$$\cos\hat{\nu} = \mathbf{e}_1 \cdot \mathbf{e}_2 + \epsilon(\mathbf{e}_1 \cdot \mathbf{m}_2 + \mathbf{m}_1 \cdot \mathbf{e}_2) \tag{2.25}$$

Upon equating the real and dual parts of the right-hand sides of eqs.(2.21a) and (2.25), we obtain

$$\mathbf{e}_1 \cdot \mathbf{e}_2 = \cos\nu \tag{2.26a}$$

$$\mathbf{e}_1 \cdot \mathbf{m}_2 + \mathbf{m}_1 \cdot \mathbf{e}_2 = -h\sin\nu \tag{2.26b}$$

Moreover, from Fig. 2.2*a*,

$$\mathbf{m}_i = \mathbf{p}_i \times \mathbf{e}_i, \quad i = 1, 2 \tag{2.26c}$$

If now eqs.(2.26c) are substituted into eq.(2.26b), it is apparent that

$$[(\mathbf{p}_2 - \mathbf{p}_1) \times \mathbf{e}_2] \cdot \mathbf{e}_1 = -h\sin\nu \tag{2.27}$$

The left-hand side of eq.(2.27) represents the projection onto line $\mathcal{L}_1$ of the moment of $\mathcal{L}_2$ with respect to P_1. Thus, eq.(2.27) states that the foregoing projection equals the negative of the product of the distance between the two lines times the sine of the angle between the two lines. The static interpretation of the same equation is straightforward: the right-hand side of eq.(2.27) represents the moment of a unit force, whose line of action is $\mathcal{L}_2$, with respect to $\mathcal{L}_1$. Obviously, the component of this force parallel to $\mathcal{L}_1$ exerts zero moment about this line, while the component perpendicular to $\mathcal{L}_1$, of magnitude $\sin\nu$, exerts a moment about $\mathcal{L}_1$, that equals $-h\sin\nu$, the negative sign arising from the definitions of $\mathbf{e}_1$ and ν in Fig. 2.2. We then have the relation

$$\hat{\mathbf{e}}_1 \cdot \hat{\mathbf{e}}_2 = \cos\nu - \epsilon h \sin\nu \tag{2.28}$$

Now we turn to relation (2.23a). First, notice that

$$\hat{\mathbf{e}}_1 \times \hat{\mathbf{e}}_2 \cdot \hat{\mathbf{e}}_3 \equiv \hat{\mathbf{e}}_1 \times \hat{\mathbf{e}}_2 \cdot \frac{\hat{\mathbf{e}}_1 \times \hat{\mathbf{e}}_2}{\| \hat{\mathbf{e}}_1 \times \hat{\mathbf{e}}_2 \|} \tag{2.29}$$

i.e.,

$$\hat{\mathbf{e}}_1 \times \hat{\mathbf{e}}_2 \cdot \hat{\mathbf{e}}_3 = \| \hat{\mathbf{e}}_1 \times \hat{\mathbf{e}}_2 \| \tag{2.30}$$

where, upon expansion,

$$\hat{\mathbf{e}}_1 \times \hat{\mathbf{e}}_2 = \mathbf{e}_1 \times \mathbf{e}_2 + \epsilon(\mathbf{m}_1 \times \mathbf{e}_2 + \mathbf{e}_1 \times \mathbf{m}_2) \tag{2.31}$$

If we now apply relation (2.10) to calculate the Euclidean norm of $\mathbf{e}_1 \times \mathbf{e}_2$, as given by eq.(2.10), the associated dual expression is derived, namely,

$$\| \hat{\mathbf{e}}_1 \times \hat{\mathbf{e}}_2 \| = \| \mathbf{e}_1 \times \mathbf{e}_2 \| + \epsilon \frac{(\mathbf{e}_1 \times \mathbf{e}_2) \cdot (\mathbf{m}_1 \times \mathbf{e}_2 + \mathbf{e}_1 \times \mathbf{m}_2)}{\| \mathbf{e}_1 \times \mathbf{e}_2 \|} \tag{2.32}$$

Upon substitution of eqs.(2.31) and (2.32) into eq.(2.23b), we obtain

$$\| \mathbf{e}_1 \times \mathbf{e}_2 \| + \epsilon \frac{(\mathbf{e}_1 \times \mathbf{e}_2) \cdot (\mathbf{m}_1 \times \mathbf{e}_2 + \mathbf{e}_1 \times \mathbf{m}_2)}{\| \mathbf{e}_1 \times \mathbf{e}_2 \|} = \sin\hat{\nu} \tag{2.33}$$

Moreover, from eq.(2.21b), eq.(2.33) leads to the relations

$$\| \mathbf{e}_1 \times \mathbf{e}_2 \| = \sin \nu \geq 0 \tag{2.34a}$$

and

$$\frac{\mathbf{e}_1 \times \mathbf{e}_2}{\| \mathbf{e}_1 \times \mathbf{e}_2 \|} \cdot (\mathbf{m}_1 \times \mathbf{e}_2 + \mathbf{e}_1 \times \mathbf{m}_2) = h \cos \nu \tag{2.34b}$$

Equation (2.34a) thus states that ν should be defined so that $0 \leq \nu \leq \pi$; otherwise, this equation is just a consequence of eq.(2.26a). Now, upon expansion of the left-hand side of eq.(2.34b), we readily obtain

$$\left[\frac{\mathbf{e}_1 \times \mathbf{e}_2}{\| \mathbf{e}_1 \times \mathbf{e}_2 \|} \cdot (\mathbf{p}_2 - \mathbf{p}_1)\right] (\mathbf{e}_1 \cdot \mathbf{e}_2) = h \cos \nu \tag{2.35}$$

and, by virtue of eq.(2.26a) and the definition of $\mathbf{e}_3$, the latter relation reduces to

$$\mathbf{e}_3 \cdot (\mathbf{p}_2 - \mathbf{p}_1) = h \tag{2.36}$$

which simply states that the projection of segment P_1P_2 onto the common normal to $\mathcal{L}_1$ and $\mathcal{L}_2$ equals the distance between these two lines. We can thus summarize the foregoing results in the equation below:

$$\hat{\mathbf{e}}_1 \times \hat{\mathbf{e}}_2 \cdot \hat{\mathbf{e}}_3 = \sin \nu + \epsilon h \cos \nu \tag{2.37}$$

In the discussion above we assumed implicitly that lines $\mathcal{L}_1$ and $\mathcal{L}_2$ are not parallel; if they are, then $\nu = 0$ and eq.(2.37) leads to $\hat{\mathbf{e}}_1 \times \hat{\mathbf{e}}_2 \cdot \hat{\mathbf{e}}_3 = h$, i.e., the box product of the dual unit vectors of the two lines and their common perpendicular is well defined, *even though the common perpendicular to the two parallel lines is undefined.* Indeed, $\hat{\mathbf{e}}_3$ itself, as given by eq.(2.22), is not defined, for, in this case, the product $\hat{\mathbf{e}}_1 \times \hat{\mathbf{e}}_2$ is a dual vector with zero real part, thus violating the requirement that the real part of a dual vector be not zero in defining the associated Euclidean norm, as in eq.(2.10). What we need, then, is an alternative means of defining $\hat{\mathbf{e}}_3$ in the particular case of two parallel lines, an issue that we address in the paragraphs below. To this end, we resort to a concept introduced in (González-Palacios, Angeles and Ranjbaran, 1993).

Let the point coordinates of two parallel lines be defined as

$$\mathcal{L}_i: \ \mathbf{r}_i = \mathbf{q}_i + \lambda \mathbf{e}, \qquad i = 1, 2 \tag{2.38}$$

Now, their line-coordinate representations are given as

$$\hat{\mathbf{e}}_i = \mathbf{e} + \epsilon \mathbf{m}_i \tag{2.39}$$

with $\mathbf{m}_i = \mathbf{q}_i \times \mathbf{e}$. We now have

Theorem 2.1: *Let two lines $\mathcal{L}_1$ and $\mathcal{L}_2$ passing through points Q_1 and Q_2, be parallel to the unit vector* $\mathbf{e}$. *Moreover, let P_1 and P_2 be the points of $\mathcal{L}_1$ and $\mathcal{L}_2$ closest to a point O. Then, the line passing through P_1 and P_2 is perpendicular to* $\mathbf{e}$, *and hence, to $\mathcal{L}_1$ and $\mathcal{L}_2$.*

Proof: Let $\mathbf{p}_i$ be the position vector of P_i. Then, from eq.(2.18),

$$\mathbf{p}_i = \mathbf{e} \times \mathbf{m}_i \tag{2.40}$$

Hence

$$\mathbf{p}_2 - \mathbf{p}_1 = \mathbf{e} \times (\mathbf{m}_2 - \mathbf{m}_1)$$

which is obviously perpendicular to $\mathbf{e}$ and the theorem follows.

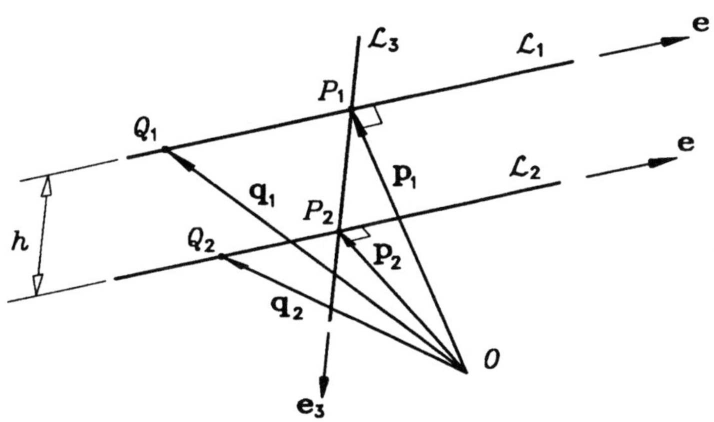

Figure 2.3 The common perpendicular to two parallel lines.

Now, considering Theorem 2.1, the line $\mathcal{L}_3$ perpendicular to both $\mathcal{L}_1$ and $\mathcal{L}_2$ passing through points P_1 and P_2 of $\mathcal{L}_1$ and $\mathcal{L}_2$, respectively, as shown in Fig. 2.3, is defined by its vector of point coordinates as

$$\mathbf{r}_3 = \mathbf{p}_1 + \lambda \mathbf{e}_3 \tag{2.41}$$

where

$$\mathbf{e}_3 = \frac{\mathbf{p}_2 - \mathbf{p}_1}{h} \tag{2.42}$$

and $h \equiv \| \mathbf{p}_2 - \mathbf{p}_1 \|$.

Moreover, the line-coordinate representation of $\mathcal{L}_3$ is given as

$$\hat{\mathbf{e}}_3 = \mathbf{e}_3 + \epsilon \mathbf{p}_1 \times \mathbf{e}_3 \tag{2.43}$$

Upon substitution of eq.(2.40) into eq.(2.43), one can readily obtain the dual representation of $\mathcal{L}_3$ as

$$\hat{\mathbf{e}}_3 = \frac{\mathbf{p}_2 - \mathbf{p}_1}{h} + \epsilon \frac{\mathbf{p}_1 \times \mathbf{p}_2}{h} \tag{2.44}$$

Example 2.1.2: *Given the unit vectors* $\mathbf{e}_i$*, for* $i = 1, 2$*, representing the directions of two lines* $\mathcal{L}_i$*, and the position vectors* $\mathbf{q}_i$ *of arbitrary points on each line, find the position vectors of the points* P_i *that are the intersection of the two lines with* $\mathcal{L}_3$*, their common perpendicular.*

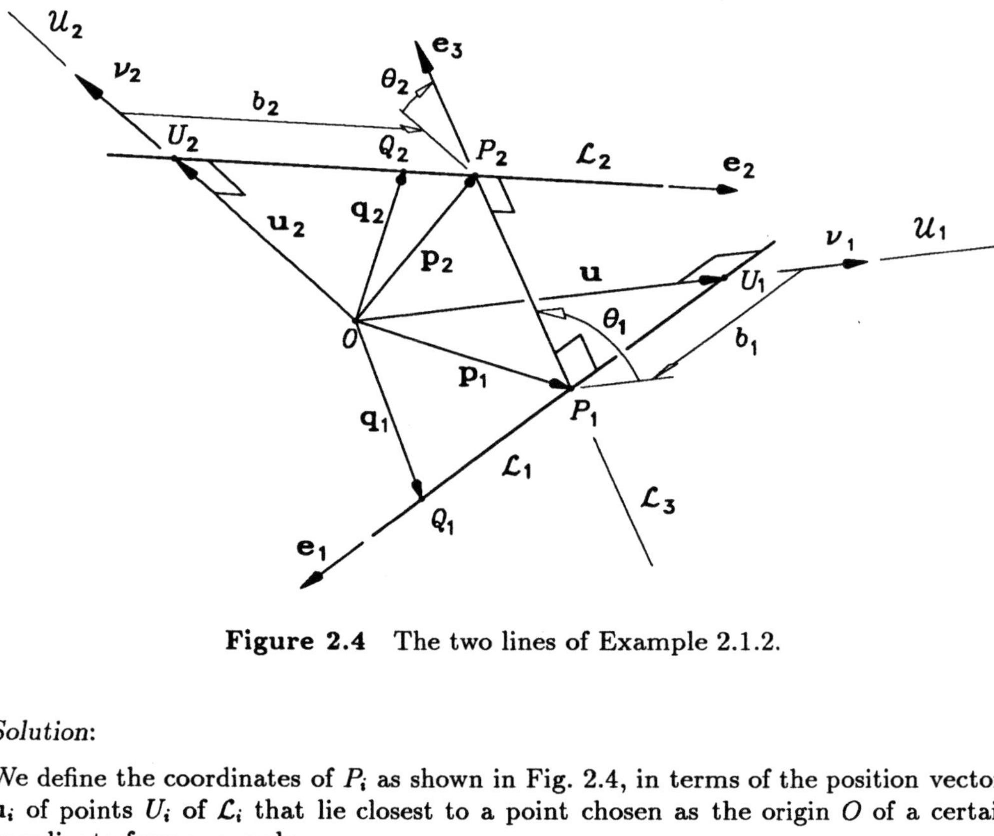

Figure 2.4 The two lines of Example 2.1.2.

Solution:

We define the coordinates of P_i as shown in Fig. 2.4, in terms of the position vectors $\mathbf{u}_i$ of points U_i of $\mathcal{L}_i$ that lie closest to a point chosen as the origin O of a certain coordinate frame, namely,

$$\mathbf{p}_i = \mathbf{u}_i + b_i \mathbf{e}_i \tag{a}$$

Note that, in Fig. 2.4, $\boldsymbol{\nu}_1$ and $\boldsymbol{\nu}_2$ are unit vectors parallel to $\mathbf{u}_1$ and $\mathbf{u}_2$, respectively. Now, the equation of $\mathcal{L}_i$ in terms of $\mathbf{q}_i$ and $\mathbf{e}_i$ is given in point-coordinate form as

$$\mathbf{r}_i = \mathbf{q}_i + \lambda_i \mathbf{e}_i \tag{b}$$

or, in line-coordinate form as

$$\hat{\mathbf{e}}_i = \mathbf{e}_i + \epsilon \mathbf{q}_i \times \mathbf{e}_i \tag{c}$$

But, from eq.(2.19),

$$\mathbf{r}_i = \mathbf{e}_i \times (\mathbf{q}_i \times \mathbf{e}_i) + \lambda_i \mathbf{e}_i \tag{d}$$

The first term of the right hand-side of eq.(d) represents the position vector of point U_i, i.e.,

$$\mathbf{u}_i = \mathbf{q}_i - (\mathbf{q}_i \cdot \mathbf{e}_i)\mathbf{e}_i \tag{e}$$

Substitution of eq.(e) into eq.(a) leads to

$$\mathbf{p}_i = \mathbf{q}_i + (b_i - \mathbf{q}_i \cdot \mathbf{e}_i)\mathbf{e}_i \tag{f}$$

Now, in order to obtain b_i, we notice that $\mathcal{L}_i$ is perpendicular to $\boldsymbol{\nu}_i$ and $\mathbf{e}_3$, where $\boldsymbol{\nu}_i$ and $\mathbf{e}_3$ are the unit vectors indicated in Fig. 2.4. Now, we define the dual unit vectors $\hat{\boldsymbol{\nu}}_i$ and $\hat{\mathbf{e}}_3$ as

$$\hat{\boldsymbol{\nu}}_i \equiv \frac{\mathbf{u}_i}{\| \mathbf{u}_i \|} + \epsilon \mathbf{0}$$

and

$$\hat{\mathbf{e}}_3 \equiv \mathrm{duv}(\hat{\mathbf{e}}_1 \times \hat{\mathbf{e}}_2)$$

All we need now is the dual angle $\hat{\theta}_i = \theta_i + \epsilon b_i$, for $i = 1, 2$, between the common perpendicular to $\mathcal{L}_1$ and $\mathcal{L}_2$, denoted by $\mathcal{L}_3$, and lines $\mathcal{U}_1$ and $\mathcal{U}_2$ passing through O and U_1 and U_2, respectively. This dual angle can be readily evaluated from the expression below:

$$\tan \hat{\theta}_i = \frac{\hat{\nu}_i \times \hat{\mathbf{e}}_3 \cdot \hat{\mathbf{e}}_i}{\hat{\nu}_i \cdot \hat{\mathbf{e}}_3} \tag{g}$$

■

Now, let $\mathcal{F}_1$ and $\mathcal{F}_2$ be two initially coincident coordinate frames. Assume that $\mathcal{F}_2$ is rotated through an angle ν about its X-axis and translated a distance h along the same axis, as shown in Fig. 2.5. Thus, the transformation from $\mathcal{F}_2$- to $\mathcal{F}_1$-coordinates is given by the ***dual screw matrix*** shown below:

$$\hat{\mathbf{Q}}(\hat{\nu}) = \begin{bmatrix} 1 & 0 & 0 \\ 0 & \cos\hat{\nu} & -\sin\hat{\nu} \\ 0 & \sin\hat{\nu} & \cos\hat{\nu} \end{bmatrix} \tag{2.45a}$$

In other words, $\hat{\mathbf{Q}}(\hat{\nu})$ represents a *dual rotation* through $\hat{\nu}$ about the X-axis. Similarly, the dual rotations through $\hat{\nu}$ about the Y- and Z-axes are given by

$$\hat{\mathbf{R}}(\hat{\nu}) = \begin{bmatrix} \cos\hat{\nu} & 0 & \sin\hat{\nu} \\ 0 & 1 & 0 \\ -\sin\hat{\nu} & 0 & \cos\hat{\nu} \end{bmatrix} \tag{2.45b}$$

and

$$\hat{\mathbf{S}}(\hat{\nu}) = \begin{bmatrix} \cos\hat{\nu} & -\sin\hat{\nu} & 0 \\ \sin\hat{\nu} & \cos\hat{\nu} & 0 \\ 0 & 0 & 1 \end{bmatrix} \tag{2.45c}$$

respectively.

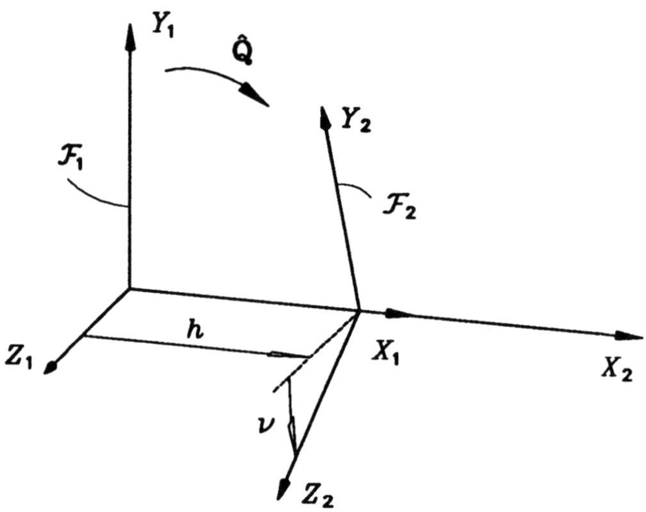

Figure 2.5 Dual rotation of frame $\mathcal{F}$ from $\mathcal{F}_1$ to $\mathcal{F}_2$.

Example 2.1.3: *The transformation from one frame to another defined by a set of dual rotations is a common exercise in the application of dual numbers. In this example we present a transformation matrix that occurs very frequently in robotics. Recalling Example 2.1.2, assume that we have two frames $\mathcal{F}_i$, for $i = 1, 2$, such that their Z_i-axes lie along the line $\mathcal{L}_i$ with direction $\mathbf{e}_i$, while their X_i-axes lie along the lines $\mathcal{U}_1$ and $\mathcal{L}_3$ with directions $\boldsymbol{\nu}_1$ and $\mathbf{e}_3$, respectively, as shown in Fig. 2.6. Find the transformation matrix $\hat{\mathbf{T}}$ to transfer components of vectors from frame $\mathcal{F}_2$ to components in $\mathcal{F}_1$.*

Solution:

Let us consider an intermediate frame $\mathcal{F}'$ resulting from the dual rotation of $\mathcal{F}_1$ by the dual angle $\hat{\theta}_1 \equiv \theta_1 + \epsilon b_1$ about the Z_1-axis. Then, the dual rotation of $\mathcal{F}'$ through a dual angle $\hat{\alpha} = \alpha + \epsilon a$ about the X'-axis takes $\mathcal{F}'$ to $\mathcal{F}_2$. Thus, $\hat{\mathbf{T}}$ is defined as

$$\hat{\mathbf{T}} = \hat{\mathbf{S}}(\hat{\theta}_1)\hat{\mathbf{Q}}(\hat{\alpha}) \tag{a}$$

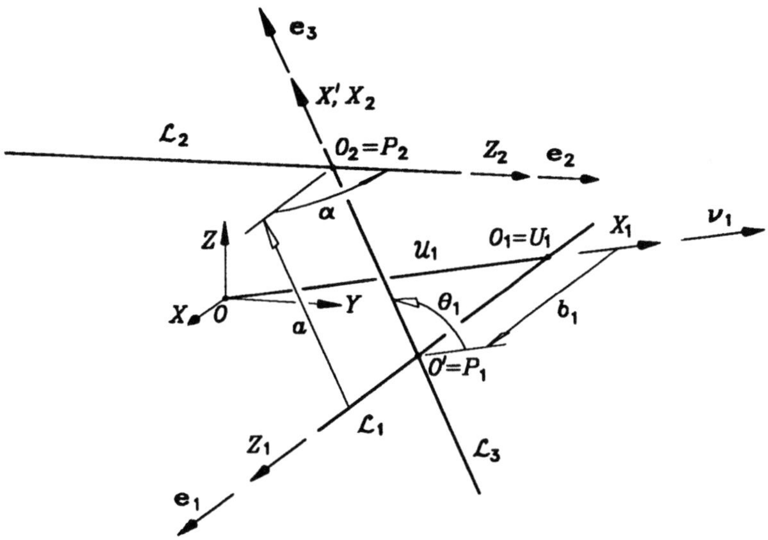

Figure 2.6 The two dual rotations of Example 2.1.3.

Now, by means of eqs.(2.45), $\hat{\mathbf{T}}$ is written as

$$\hat{\mathbf{T}} = \begin{bmatrix} \cos\hat{\theta}_1 & -\sin\hat{\theta}_1\cos\hat{\alpha} & -\sin\hat{\theta}_1\sin\hat{\alpha} \\ \sin\hat{\theta}_1 & \cos\hat{\theta}_1\cos\hat{\alpha} & -\cos\hat{\theta}_1\sin\hat{\alpha} \\ 0 & \sin\hat{\alpha} & \cos\hat{\alpha} \end{bmatrix} \qquad (b)$$

Notice that the dual matrix $\hat{\mathbf{T}}$ contains both translation and rotation information, and hence, plays the role of the rather popular 4×4 transformation matrices of robotics (Craig, 1986). ■

2.2 The Aronhold-Kennedy Theorem

Below we include the discussion of the *Aronhold-Kennedy Theorem* in three dimensions, first stated by Beggs (1959), as proposed by Veldkamp (1976). At the outset, we recall the concept of *instantaneous screw axis*, or *ISA* for brevity. When two bodies $\mathcal{A}$ and $\mathcal{B}$ are in relative motion, a set of points of these bodies exists along which the relative velocity has a minimum magnitude. Moreover, this set is a line, which is termed the ISA of $\mathcal{B}$ with respect to $\mathcal{A}$, or of $\mathcal{A}$ with respect to $\mathcal{B}$ (Angeles, 1982). This result finds applications in the design of *hypoid gears* (Dudley, 1962), whose pitch surfaces are hyperboloids generated by the motion of the ISA about the screw axes of the pinion and the gear, thereby minimizing power losses, for contact takes place along the line of minimum sliding.

When three bodies $\mathcal{A}$, $\mathcal{B}$ and $\mathcal{C}$ are in relative motion, then three different ISAs can be defined as shown in Fig. 2.7. What the Aronhold-Kennedy Theorem states is that the foregoing ISAs are not arbitrarily located in space, but all three share one common perpendicular $\mathcal{N}$, as illustrated in Fig. 2.8.

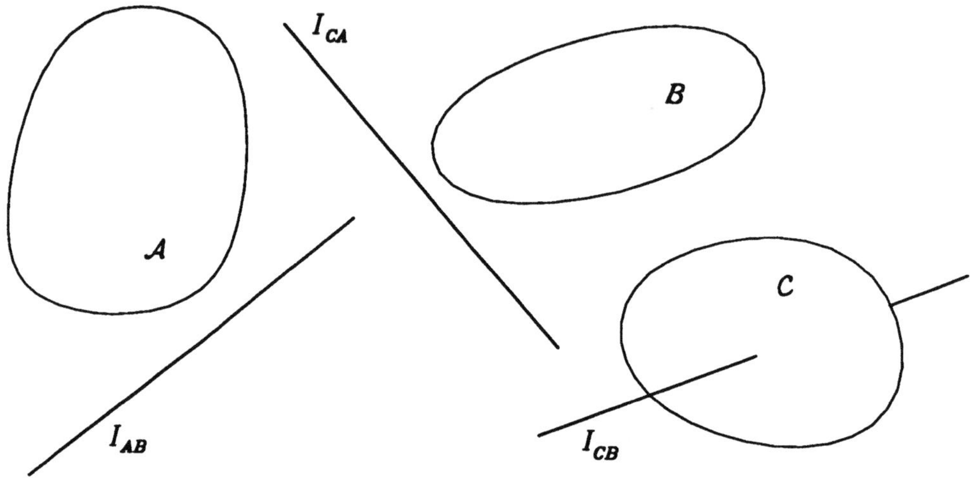

Figure 2.7 The three ISAs of three rigid bodies in relative motion.

In Veldkamp's discussion we will need the concept of *dual unit sphere.* The dual unit sphere, or *dual sphere* for short, is the locus of points of the *dual space* lying a unit distance away from the origin of this space. Furthermore, the dual space is the space of points whose position vectors are dual vectors. Thus, all position vectors of points on the dual sphere are dual unit vectors, and hence, represent lines in the Euclidean space.

By virtue of the definition of the dual sphere, then, the screw axis of a rigid-body motion becomes a point on the dual sphere. As the body moves arbitrarily in the Euclidean space, then, its motion is mapped into a curve on the dual sphere, which is the dual image of the *polode* of the body. the polode being the ruled surface traced by the ISA of the body with respect to a fixed body.

The relative screw motion of two rigid bodies, then, becomes relative spherical motion in the dual space. Thus, one can imagine these bodies in the dual space as two concentric dual unit spheres $\mathcal{S}_1$ and $\mathcal{S}_2$. The *dual angular velocity* of $\mathcal{S}_2$ with respect to $\mathcal{S}_1$ is given as

$$\hat{\omega}_{21} = \hat{\omega}_{21}\hat{\mathbf{e}}_{21} \equiv (\omega_{21} + \epsilon v_{21})\hat{\mathbf{e}}_{21} \tag{2.46}$$

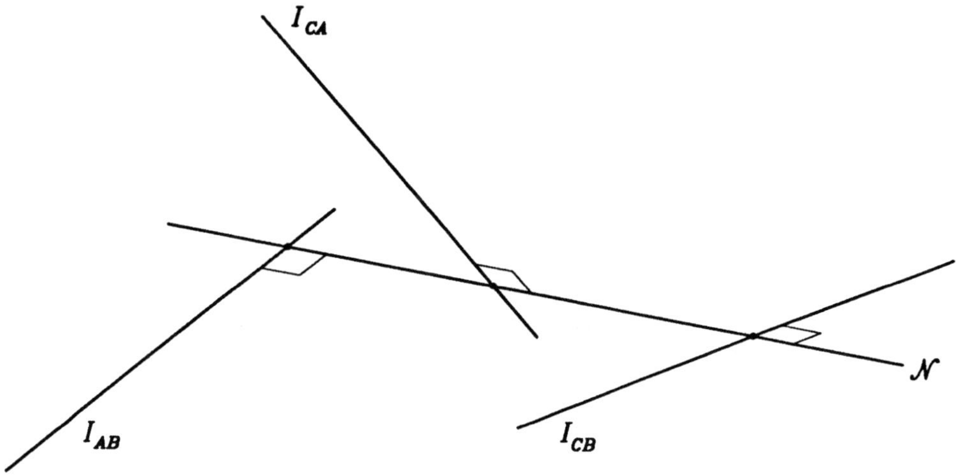

Figure 2.8 A geometric representation of the Aronhold-Kennedy Theorem.

where $\hat{\mathbf{e}}_{21}$ is a unit dual vector defining point $\hat{P}_{21}$, the two components of the dual scalar $\hat{\omega}_{21}$, ω_{21} and v_{21}, representing, in real space, the signed magnitudes of the angular velocity and the velocity of the points lying on the screw axis. The velocity of any point $\hat{P}$ of $\mathcal{S}_2$, defined by the unit dual vector $\hat{\mathbf{p}}$, is given by

$$\hat{\mathbf{v}}_{21} = \hat{\boldsymbol{\omega}}_{21} \times \hat{\mathbf{p}} \tag{2.47}$$

It is clear that $\hat{\mathbf{v}}_{21} = \mathbf{0}$ if $\hat{P}$ coincides with $\hat{P}_{21}$, which is known as the pole of the motion.

Now, the three dual angular velocities of the unit dual spheres $\mathcal{S}_1$, $\mathcal{S}_2$ and $\mathcal{S}_3$ in relative motion are related as

$$\hat{\boldsymbol{\omega}}_{31} = \hat{\boldsymbol{\omega}}_{32} + \hat{\boldsymbol{\omega}}_{21} \tag{2.48}$$

Similarly, the dual velocities of a point are expressed as

$$\hat{\mathbf{v}}_{31} = \hat{\mathbf{v}}_{32} + \hat{\mathbf{v}}_{21} \tag{2.49}$$

where

$$\hat{\mathbf{v}}_{21} = \hat{\boldsymbol{\omega}}_{21} \times \hat{\mathbf{p}}_{21} \tag{2.50}$$

with similar definitions for $\hat{\mathbf{v}}_{32}$ and $\hat{\mathbf{v}}_{31}$.

Equation (2.48) can also be expressed in the form

$$\hat{\omega}_{31}\hat{\mathbf{e}}_{31} = \hat{\omega}_{32}\hat{\mathbf{e}}_{32} + \hat{\omega}_{21}\hat{\mathbf{e}}_{21} \tag{2.51}$$

where $\hat{\mathbf{e}}_{ij}$ indicates the three poles $\hat{P}_{ij}$, and $\hat{\omega}_{ij}$ are the signed norms of the dual vectors $\hat{\boldsymbol{\omega}}_{ij}$, defined as

$$\hat{\omega}_{ij} = \omega_{ij} + \epsilon v_{ij} \tag{2.52}$$

As a result of eq.(2.51), the Aronhold-Kennedy Theorem is expressed as (Veldcamp, 1976)

Theorem 2.2 (Aronhold-Kennedy) *The poles $\hat{P}_{21}$, $\hat{P}_{31}$ and $\hat{P}_{32}$ of three unit dual spheres in relative motion lie on the same great circle.*

Thus, the Aronhold-Kennedy Theorem defined in the dual space involves pure rotations, pure translations and general screw motions. The dual representation of Theorem 2.2 is included in Fig. 2.9.

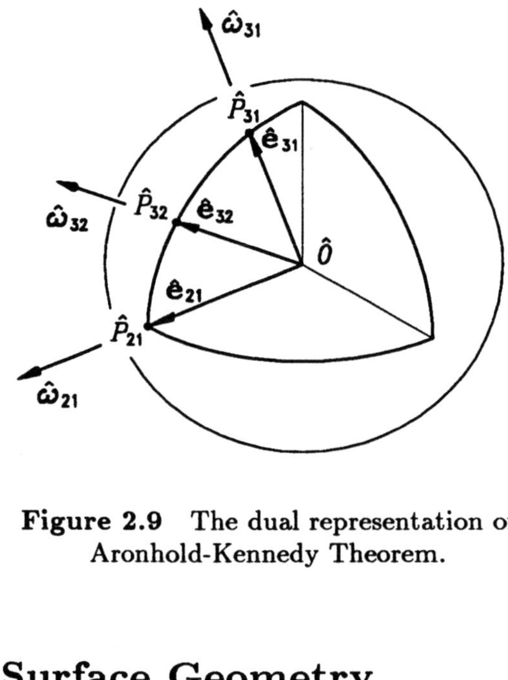

Figure 2.9 The dual representation of Aronhold-Kennedy Theorem.

2.3 Ruled-Surface Geometry

A surface $\mathcal{R}$ generated by the motion of a line $\mathcal{L}$, defined as in eq.(2.15), is called a *ruled surface*, $\mathcal{L}$ being its *generatrix* (Struik, 1961). Thus, the position vector of $\mathcal{R}$ is given by

$$\mathbf{r}(\psi, \lambda) = \mathbf{p}(\psi) + \lambda \mathbf{e}(\psi) \tag{2.53}$$

where ψ and λ are the parameters of $\mathcal{R}$. The curve defined by the points $\mathbf{p}(\psi)$ is called the *directrix* of $\mathcal{R}$, and the vectors $\mathbf{e}(\psi)$ with origin at the center of a unit sphere describe the *spherical indicatrix* of $\mathcal{L}$. Moreover, if $\mathbf{p}(\psi)$ is constant, $\mathcal{R}$ is a cone, whereas $\mathcal{R}$ is a cylinder if $\mathbf{e}$ is constant.

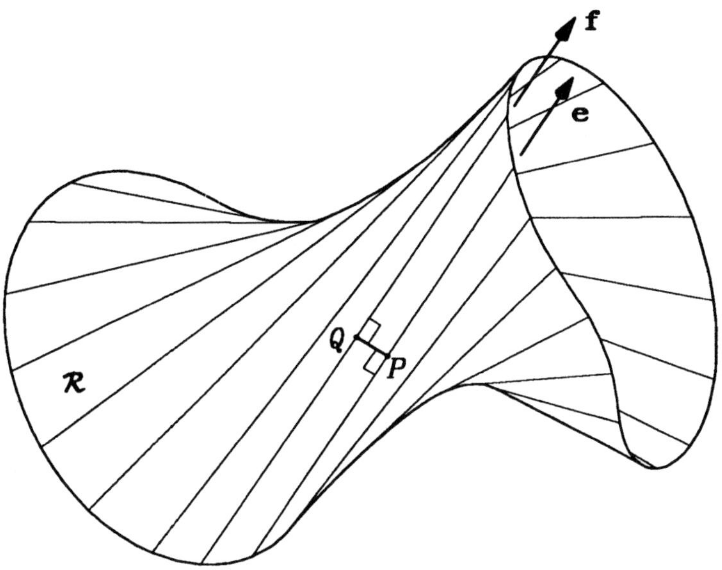

Figure 2.10 Graphical interpretation of the striction curve.

We now introduce the concept of ***striction curve***. Given a ruled surface $\mathcal{R}$, consider two infinitesimally separated generators $\mathbf{e}$ and $\mathbf{f} \equiv \mathbf{e} + d\mathbf{e}$ and their common perpendicular PQ, as shown in Fig. 2.10. Then, point P is the *central point* of the generator $\mathbf{e}$, the locus of the central points defining the striction curve.

The striction curve is traced by the position vector $\mathbf{g}(\psi)$ defined as (DoCarmo 1976)

$$\mathbf{g}(\psi) = \mathbf{p} - \frac{\mathbf{e}' \cdot \mathbf{p}'}{\mathbf{e}' \cdot \mathbf{e}'}\mathbf{e} \tag{2.54}$$

the prime denoting differentiation with respect to ψ.

Thus, the central circle of an axially-symmetric hyperboloid $\mathcal{H}$ is the striction curve of $\mathcal{H}$, whereas the striction curve is undetermined in cylindrical surfaces and reduces to a point in conical surfaces.

Now, the *unit normal* of $\mathcal{R}$ can be derived from eq. (2.53) as

$$\mathbf{n}(\psi, \lambda) = \frac{\mathbf{u}}{\| \mathbf{u} \|} \tag{2.55}$$

where

$$\mathbf{u} = \frac{\partial \mathbf{r}}{\partial \psi} \times \frac{\partial \mathbf{r}}{\partial \lambda} \tag{2.56}$$

In terms of eq. (2.53), $\mathbf{n}(\psi, \lambda)$ becomes

$$\mathbf{n} = \frac{(\mathbf{p}' + \lambda \mathbf{e}') \times \mathbf{e}}{\| \mathbf{u} \|} \tag{2.57}$$

The unit normal takes a limiting value as λ decreases to $-\infty$, this value being called the *asymptotic normal*, which is defined as (McCarthy, 1987)

$$\mathbf{a} \equiv \lim_{\lambda \to -\infty} \mathbf{n}(\psi, \lambda) = \lim_{\lambda \to -\infty} \frac{(\mathbf{p}' + \lambda \mathbf{e}') \times \mathbf{e}}{\sqrt{(\mathbf{p}' \cdot \lambda \mathbf{e}')^2 - (\mathbf{e} \cdot \mathbf{p}')^2}} = \frac{-\mathbf{e}' \times \mathbf{e}}{\| \mathbf{e}' \|} \tag{2.58}$$

A third vector, called the *central normal*, is defined from $\mathbf{a}$ and $\mathbf{e}$ as follows:

$$\mathbf{c} \equiv \mathbf{a} \times \mathbf{e} = \frac{\mathbf{e}'}{\| \mathbf{e}' \|} \tag{2.59}$$

Since $\mathbf{c}$ is a unit vector, we conclude that $\mathbf{a}$ and $\mathbf{e}$ are perpendicular. Thus, the triad $\{\mathbf{a}, \mathbf{e}, \mathbf{c}\}$ is called the *natural trihedron* of $\mathcal{R}$ and is defined on the striction curve of $\mathcal{R}$.

Example 2.3.1: *The expression below represents a family of hyperboloids of one sheet, where fixed values of p and α define one particular surface:*

$$\mathbf{r}(\psi, \lambda) = p \begin{bmatrix} \cos\psi \\ \sin\psi \\ 0 \end{bmatrix} + \lambda \begin{bmatrix} -\sin\psi \sin\alpha \\ \cos\psi \sin\alpha \\ \cos\alpha \end{bmatrix}, \quad 0 \le \psi \le 2\pi, \quad -\infty \le \lambda \le +\infty \tag{2.60}$$

Find:

i) the equation of the striction curve;

ii) the expression of the unit normal and the natural trihedron;

What happens if $p \equiv 0$ or $\alpha \equiv 0$?

Solution:

i) We notice that, if eq.(2.60) is written in the form of eq.(2.53), $\mathbf{p}'$ and $\mathbf{e}'$ are readily obtained as

$$\mathbf{p}' = p \begin{bmatrix} -\sin\psi \\ \cos\psi \\ 0 \end{bmatrix}, \quad \mathbf{e}' = \begin{bmatrix} -\cos\psi \sin\alpha \\ -\sin\psi \sin\alpha \\ 0 \end{bmatrix}$$

Substituting these values into eq.(2.54), the points of the striction curve are defined as

$$\mathbf{g}(\psi) = p\begin{bmatrix}\cos\psi\\ \sin\psi\\ 0\end{bmatrix}$$

and hence, the striction curve is the circle lying in the XY plane, centered at the origin of this plane, and of radius p.

ii) The unit normal is obtained from eq.(2.57), namely,

$$\mathbf{n} = \frac{1}{\nu}\begin{bmatrix}p\cos\psi\cos\alpha - \lambda\sin\psi\sin\alpha\cos\alpha\\ p\sin\psi\cos\alpha + \lambda\cos\psi\sin\alpha\cos\alpha\\ -\lambda\sin^2\alpha\end{bmatrix}, \qquad \nu \equiv \sqrt{p^2\cos^2\alpha + \lambda^2\sin^2\alpha}$$

Now, from eqs.(2.58, 2.60 & 2.59), we obtain, respectively,

$$\mathbf{a} = \begin{bmatrix}\sin\psi\cos\alpha\\ -\cos\psi\cos\alpha\\ \sin\alpha\end{bmatrix}, \quad \mathbf{e} = \begin{bmatrix}-\sin\psi\sin\alpha\\ \cos\psi\sin\alpha\\ \cos\alpha\end{bmatrix}, \quad \mathbf{c} = \begin{bmatrix}\cos\psi\\ \sin\psi\\ 0\end{bmatrix}$$

thereby defining the natural trihedron. For purposes of illustration, we assume now the numerical values $p = 1$, $\alpha = 35°$ and $-5 \le \lambda \le 5$, the corresponding hyperboloid, the striction curve $\mathcal{C}$, the natural trihedron and the direction of $\mathbf{n}$ along the directrix in which $\psi = 0$, being shown in Fig. 2.11.

It is clear now that, if $p \equiv 0$, we have a family of regular cones and, if $\alpha \equiv 0$, a family of regular cylinders. ■

Now, if $\mathbf{g}(\psi)$ and $\mathbf{e}(\psi)$ are reparameterized by the arc length s of the spherical indicatrix, we obtain the equation of the position vector of $\mathcal{R}$ in the form

$$\mathbf{r}(s, \lambda) = \mathbf{p}(s) + \lambda\mathbf{e}(s) \tag{2.61}$$

where s is defined as

$$s = \int_0^{\psi} \| \mathbf{e}' \| \, d\psi \tag{2.62}$$

Moreover, the *geodesic Frenet* equations of the $\{\mathbf{a},\ \mathbf{c},\ \mathbf{e}\}$ triad are given by

$$d\mathbf{e}/ds = \mathbf{c} \tag{2.63}$$

$$d\mathbf{c}/ds = \gamma\mathbf{a} - \mathbf{e} \tag{2.64}$$

$$d\mathbf{a}/ds = -\gamma\mathbf{c} \tag{2.65}$$

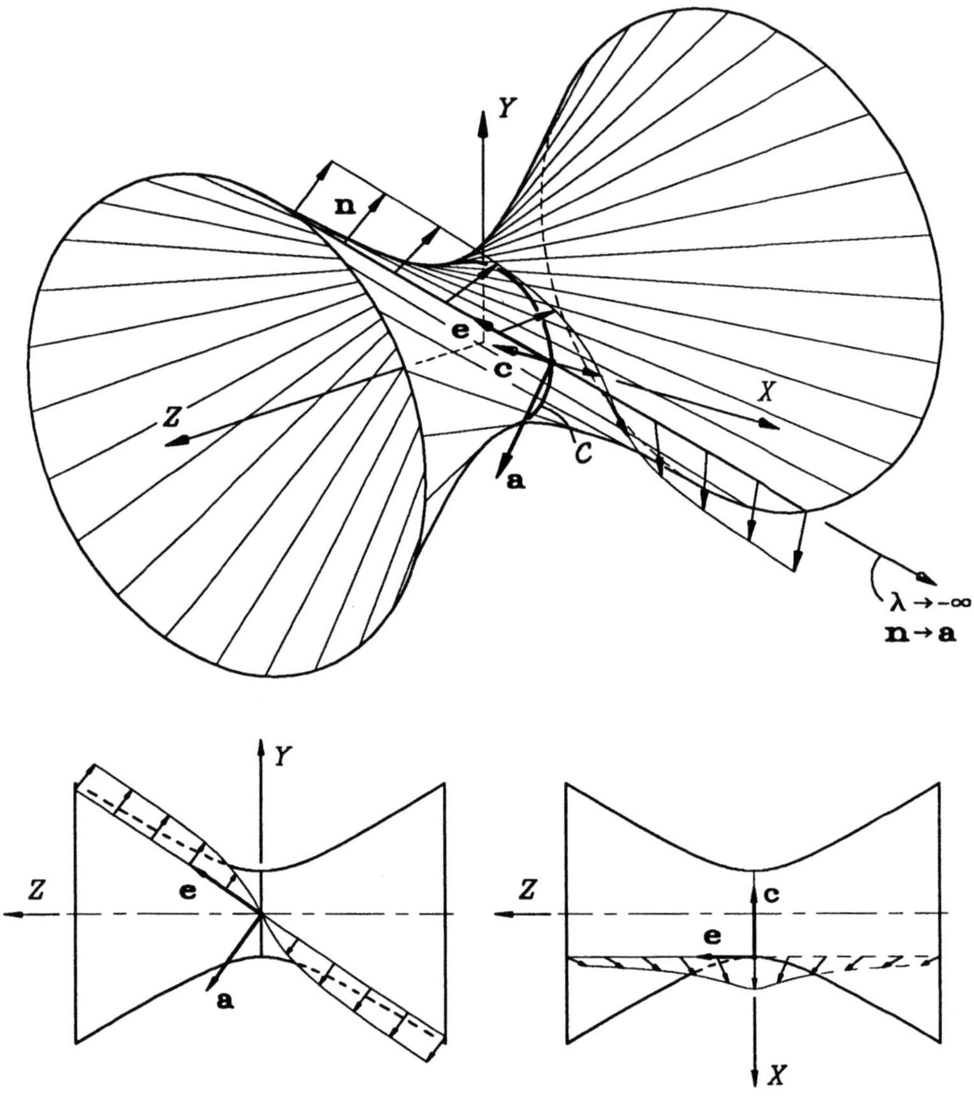

Figure 2.11 A hyperboloid of one sheet and its natural trihedron at $\psi = 0$.

where s represents the arc length of the spherical indicatrix of $\mathbf{e}$, while γ is the *geodesic curvature*. The latter is expressed in terms of $\mathbf{e}$ and its derivatives with respect to ψ, namely,

$$\gamma = \frac{\mathbf{e} \times \mathbf{e}' \cdot \mathbf{e}''}{\| \mathbf{e}' \|^3} \tag{2.66}$$

The positional variation of the trihedron defined by $\mathbf{a}$, $\mathbf{c}$ and $\mathbf{e}$ is given by

$$\frac{d\mathbf{g}}{ds} = \Delta \mathbf{a} + \Gamma \mathbf{e} \tag{2.67}$$

where

$$\Delta \equiv \frac{\mathbf{p}' \cdot \mathbf{e} \times \mathbf{e}'}{\| \mathbf{e}' \|^2} \tag{2.68}$$

and

$$\Gamma \equiv \frac{\mathbf{p}' \cdot \mathbf{e}}{\| \mathbf{e}' \|} - \frac{1}{\| \mathbf{e}' \|} \frac{d}{dt} \left(\frac{\mathbf{p}' \cdot \mathbf{e}}{\| \mathbf{e}' \|^2} \right) \tag{2.69}$$

Now, the dual spherical *radius of curvature* of $\mathcal{R}$, $\hat{\rho} = \rho + \epsilon r$, is given by (McCarthy, 1987)

$$\tan \hat{\rho} = \frac{1/\hat{\kappa}}{\hat{\gamma}/\hat{\kappa}} \tag{2.70}$$

where

$$\hat{\kappa} = \kappa + \epsilon(K - \kappa\Delta) \tag{2.71}$$

$$\hat{\gamma} = \gamma + \epsilon(\Gamma - \gamma\Delta \tag{2.72}$$

and the definitions below:

$$\kappa \equiv \sqrt{1 + \gamma^2} \tag{2.73}$$

$$K \equiv \frac{\gamma\Gamma + \Delta}{\kappa} \tag{2.74}$$

Chapter 3

KINEMATICS OF THREE-LINK MECHANISMS

3.1 Introduction

This chapter is confined to the study of cam mechanisms composed of three rigid links, namely, the fixed frame, the *driving element* or cam and the *driven element* or follower. While the study is general, special attention is given to indexing cam mechanisms (ICM).

At the outset, we describe the objective of this Chapter. As a matter of fact, Chapters 4, 5 and 8 share the same objective. Our objective is, then, to *synthesize* the contact surfaces of two rigid bodies, the cam and the follower, so as to produce a *prescribed* motion of the latter as a function of the motion of the former. Both the prescribed follower motion and the cam motion are described by one variable, and hence, the prescribed motion is fully specified by a scalar function of a scalar variable. This function is termed henceforth the *input-output function.*

The synthesis of the contact surfaces of both cam and follower as well as the quality of the transmission, quantified via the pressure angle, are discussed in this chapter in a unified fashion. The contact surfaces are designed as ruled surfaces, and motion is transmitted along a common line, which gives rise to a higher kinematic pair. Two more kinematic pairs arise, namely, the cam-frame and the follower-frame pairs, which belong to the class of lower kinematic pairs, and can be of any of three types, namely, revolute, prismatic or screw (Angeles, 1982). This generalization allows the analysis of those mechanisms not only with *rotating* but also with *translating followers*, i.e., if the output motion is attained via a prismatic pair.

By virtue of the theorem on the existence of an instantaneous screw axis, already recalled in Section 2.2, three ISAs (Is) arise within the kind of mechanisms under study. Moreover, the input and output pairs are represented by their instant screw axes I_{21} and I_{31}, where 1, 2 and 3 stand for frame, cam and follower, respectively. For given input and output pairs and a given *input-output function* there is a unique pair of cam and follower surfaces that produce the given motion with a *minimum sliding* on the higher pair. This solution is achieved when the contact line coincides with the instant screw axis I_{32}. In some degenerate cases the surfaces may collapse into a line. Other solutions can be obtained if the contact line and I_{32} are not coincident, but parallel, e.g., in cam mechanisms with *flat-face followers*.

In this book, planar, spherical and spatial cam mechanisms are classified according to the relative location of their ISAs. Thus, the mechanism is planar if every ISA is either parallel to one direction or perpendicular to it; the mechanism is spherical if all ISAs are concurrent; spatial, if none of the two foregoing conditions is met.

The contact line of the higher pair is an element of both surfaces, and is derived by application of the Aronhold-Kennedy Theorem, first proposed by Aronhold (1872) and Kennedy (1886) independently for planar motion and later extended to the three dimensional case by Beggs (1959) and Phillips and Hunt (1964). The formulation is based on a mapping of the motion onto the dual space, where the three bodies are considered as dual unit spheres $\mathcal{S}_1$, $\mathcal{S}_2$ and $\mathcal{S}_3$ in relative motion, and the *poles* $\hat{P}_{ij}$ are the images of the axes I_{ij} under the aforementioned mapping.

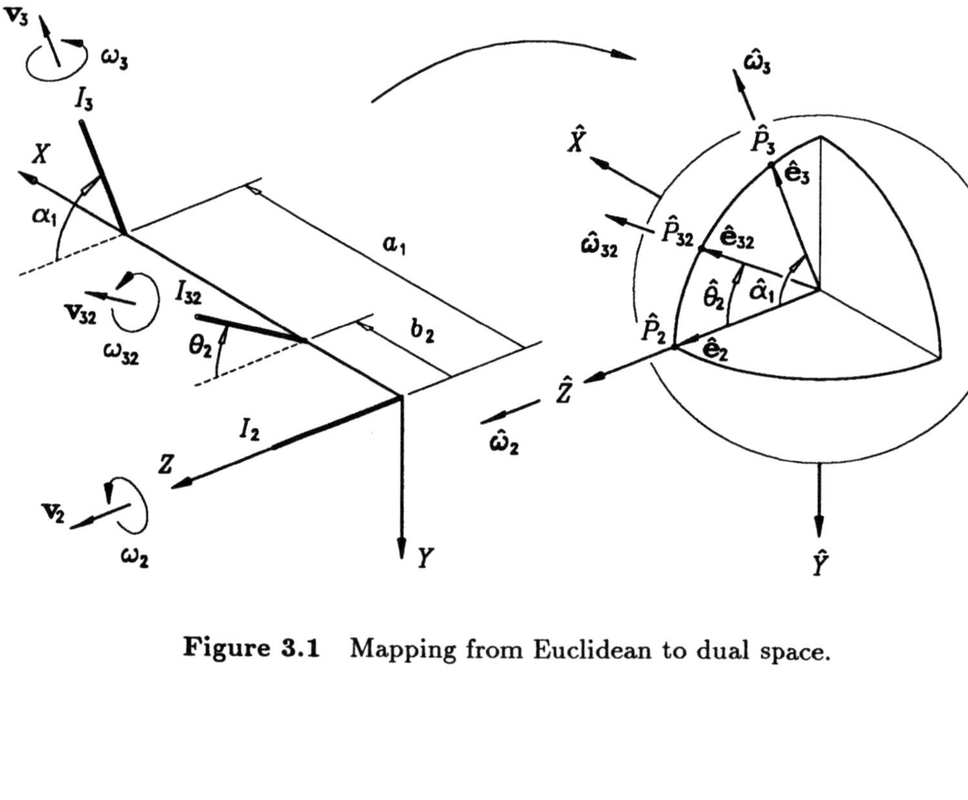

Figure 3.1 Mapping from Euclidean to dual space.

3.2 Surface Geometry

As mentioned in Section 3.1, the surfaces of the cam and the follower are generated by the sweeping action of I_{32} onto each of the two bodies. The first part of this section is devoted to the derivation of the parameters defining I_{32}, the second to the definition of the surfaces.

Let $\mathcal{S}_1$, $\mathcal{S}_2$ and $\mathcal{S}_3$ be the dual unit spheres representing, in the dual space, the frame, cam and follower, respectively. We recall briefly here that *dual quantities* are denoted with a hat ($\hat{}$) and are composed of a primal part, usually a real scalar, vector or matrix, and a dual part, correspondingly a real scalar, vector or matrix, preceded by the *dual unity* ϵ with the property $\epsilon^2 \equiv 0$. The relative dual angular velocities of the foregoing dual spheres are related as in eq.(2.51), i.e.,

$$\hat{\omega}_{31}\hat{\mathbf{e}}_{31} = \hat{\omega}_{32}\hat{\mathbf{e}}_{32} + \hat{\omega}_{21}\hat{\mathbf{e}}_{21}$$

In the above equation, the subscripts (ij) denote the dual angular velocity of the ith dual sphere with respect to the jth dual sphere, but, in order to simplify the nomenclature, all dual angular velocities with respect to 1, the frame, will be indicated only with the label of the other sphere, eq.(2.51) thus taking on the form

$$\hat{\omega}_3\hat{\mathbf{e}}_3 = \hat{\omega}_{32}\hat{\mathbf{e}}_{32} + \hat{\omega}_2\hat{\mathbf{e}}_2 \tag{3.1}$$

Now, three coordinate frames are defined so that their $\hat{X}$-axes are collinear and each of the $\hat{Z}$-axes is oriented towards one of the poles. Thus, the poles are locally represented by the dual unit vector

$$\hat{\mathbf{k}} = [0, \quad 0, \quad 1]^T \tag{3.2}$$

with zero dual part.

Furthermore, $\hat{\alpha}_1$ is the dual angle from $\hat{\mathbf{e}}_2$ to $\hat{\mathbf{e}}_3$, whereas $\hat{\theta}_2$ is the dual angle from $\hat{\mathbf{e}}_2$ to $\hat{\mathbf{e}}_{32}$, both of which are defined as

$$\hat{\alpha}_1 = \alpha_1 + \epsilon a_1 \tag{3.3a}$$

$$\hat{\theta}_2 = \theta_2 + \epsilon b_2 \tag{3.3b}$$

The geometric representation of the foregoing mapping is shown in Fig. 3.1, where $\mathbf{v}_{ij} = v_{ij}\mathbf{e}_{ij}$, while v_{ij} and ω_{ij} are the components of $\hat{\omega}_{ij}$, as defined in eq.(2.52). Thus, the unit dual vectors of the poles are given as

$$\hat{\mathbf{e}}_3 = \hat{\mathbf{Q}}(\hat{\alpha}_1)\hat{\mathbf{k}} \tag{3.4a}$$

$$\hat{\mathbf{e}}_{32} = \hat{\mathbf{Q}}(\hat{\theta}_2)\hat{\mathbf{k}} \tag{3.4b}$$

$$\hat{\mathbf{e}}_2 = \hat{\mathbf{k}} \tag{3.4c}$$

where $\hat{\mathbf{Q}}$ is defined in eq.(2.45a).

Next, substituting the values of eqs.(3.4) into eq.(3.1), one obtains

$$\hat{\omega}_3 \begin{bmatrix} 0 \\ -\sin\hat{\alpha}_1 \\ \cos\hat{\alpha}_1 \end{bmatrix} = \hat{\omega}_{32} \begin{bmatrix} 0 \\ -\sin\hat{\theta}_2 \\ \cos\hat{\theta}_2 \end{bmatrix} + \hat{\omega}_2 \begin{bmatrix} 0 \\ 0 \\ 1 \end{bmatrix} \tag{3.5}$$

Two dual equations are obtained from eq.(3.5), namely,

$$\hat{\omega}_{32} \sin\hat{\theta}_2 = \hat{\omega}_3 \sin\hat{\alpha}_1 \tag{3.6a}$$

$$\hat{\omega}_{32} \cos\hat{\theta}_2 = \hat{\omega}_3 \cos\hat{\alpha}_1 - \hat{\omega}_2 \tag{3.6b}$$

Thus, the parameters of the pole $\hat{P}_{32}$ can be written in terms of those of the poles $\hat{P}_2$ and $\hat{P}_3$. Consequently, $\hat{\omega}_{32}$ is obtained by the addition of the squared terms of eqs.(3.6), while $\hat{\theta}_2$ is obtained upon dividing both sides of eq.(3.6a) by the corresponding sides of eq.(3.6b), namely,

$$\hat{\omega}_{32}^2 = \hat{\omega}_3^2 - 2\hat{\omega}_3\hat{\omega}_2 \cos\hat{\alpha}_1 + \hat{\omega}_2^2 \tag{3.7}$$

$$\tan\hat{\theta}_2 = \frac{\hat{\omega}_3 \sin\hat{\alpha}_1}{\hat{\omega}_3 \cos\hat{\alpha}_1 - \hat{\omega}_2} \tag{3.8}$$

Moreover, we are interested in calculating the numerical values of the real angular velocity and the real *sliding velocity* of the follower with respect to the cam, ω_{32} and v_{32}, respectively. The former is the signed magnitude of the difference $\boldsymbol{\omega}_3 - \boldsymbol{\omega}_2$, while the latter is the signed minimum magnitude of the difference $\mathbf{v}_{P3} - \mathbf{v}_{P2}$, where $\mathbf{v}_{P3}$ is the velocity of point P of the follower and $\mathbf{v}_{P2}$ is the velocity of the same point of the cam. Now, in order to obtain the real angular velocity ω_{32} and the real sliding velocity v_{32}, one can expand eq.(3.7) considering eqs.(2.52 & 3.3) and the definitions given in eqs.(2.1 & 2.21), which readily lead to

$$\omega_{32} = \pm\sqrt{{\omega_3}^2 - 2\omega_3\omega_2 \cos\alpha_1 + {\omega_2}^2} \tag{3.9a}$$

$$v_{32} = \frac{\omega_3 v_3 + \omega_3\omega_2 a_1 \sin\alpha_1 - (\omega_3 v_2 + \omega_2 v_3)\cos\alpha_1 + \omega_2 v_2}{\omega_{32}} \tag{3.9b}$$

A similar procedure is followed to obtain θ_2 and b_2 from eq.(3.8), namely,

$$\tan\theta_2 = \frac{\omega_3 \sin\alpha_1}{\omega_3 \cos\alpha_1 - \omega_2} \tag{3.10a}$$

$$b_2 = \frac{({\omega_3}^2 - \omega_3\omega_2\cos\alpha_1)a_1 + (\omega_3 v_2 - \omega_2 v_3)\sin\alpha_1}{\omega_{32}^2} \tag{3.10b}$$

In general, eqs.(3.9 & 3.10) provide all the necessary information to define I_{32}. Moreover, we claim that these are the general equations from which one can derive any type of the known cam mechanisms satisfying the condition of *minimum sliding* velocity at the contact line and new types depending on the combination used on the input and output screw parameters, namely, ω_2, v_2, ω_3, v_3, α_1 and a_1. However, special treatment may be needed if both the input- and output-pairs are prismatic, as discussed in Subsection 3.2.4.

Now, if $\mathcal{S}_2$ rotates with respect to $\mathcal{S}_1$ through the dual angle $\hat{\psi} = \psi + \epsilon z_2$, where both ψ and z_2 are functions of time, $\hat{P}_{32}$ will trace a polode on $\mathcal{S}_2$, which is given by the dual unit vector $\hat{\mathbf{s}}_2(\hat{\psi})$ defined below:

$$\hat{\mathbf{s}}_2(\hat{\psi}) = \hat{\mathbf{S}}^T(\hat{\psi})\hat{\mathbf{Q}}(\hat{\theta}_2)\hat{\mathbf{k}} \tag{3.11a}$$

where the dual rotation $\hat{\mathbf{S}}$ is defined in eq.(2.45c).

Similarly, if $\mathcal{S}_3$ rotates with respect to $\mathcal{S}_1$ through the dual angle $\hat{\phi} = \phi + \epsilon z_3$, where both ϕ and z_3 are functions of time t, $\hat{P}_{32}$ will trace another polode on $\mathcal{S}_3$, which is given by the dual unit vector $\hat{\mathbf{s}}_3(\hat{\phi})$, defined as

$$\hat{\mathbf{s}}_3(\hat{\phi}) = \hat{\mathbf{S}}^T(\hat{\phi})\hat{\mathbf{Q}}(\hat{\theta}_2 - \hat{\alpha}_1)\hat{\mathbf{k}} \tag{3.11b}$$

Furthermore, the polodes defined by the dual unit vectors $\hat{\mathbf{s}}_2$ and $\hat{\mathbf{s}}_3$ are the images of the ruled surfaces $\mathcal{R}_2$ and $\mathcal{R}_3$, the contact surfaces of cam and follower, respectively. Now, upon expansion, eqs.(3.11) take on the form

$$\hat{\mathbf{s}}_2(t) = \begin{bmatrix} -s\psi(t)s\theta_2 \\ -c\psi(t)s\theta_2 \\ c\theta_2 \end{bmatrix} - \epsilon \begin{bmatrix} b_2 s\psi(t)c\theta_2 + z_2(t)c\psi(t)s\theta_2 \\ b_2 c\psi(t)c\theta_2 - z_2(t)s\psi(t)s\theta_2 \\ b_2 s\theta_2 \end{bmatrix} \tag{3.12a}$$

$$\hat{\mathbf{s}}_3(t) = \begin{bmatrix} -s\phi(t)s\beta \\ -c\phi(t)s\beta \\ c\beta \end{bmatrix} - \epsilon \begin{bmatrix} ds\phi(t)c\beta + z_3(t)c\phi(t)s\beta \\ dc\phi(t)c\beta - z_3(t)s\phi(t)s\beta \\ ds\beta \end{bmatrix} \tag{3.12b}$$

where $d \equiv b_2 - a_1$ and $\beta \equiv \theta_2 - \alpha_1$. Furthermore, $c \leftarrow \cos$ and $s \leftarrow \sin$.

By means of eq.(2.19), the point coordinates of $\mathcal{R}_2$ and $\mathcal{R}_3$ can be readily obtained from eqs.(3.12), namely,

$$\mathbf{r}_2(t,\lambda) = \begin{bmatrix} b_2 c\psi(t) - z_2(t)s\psi(t)s\theta_2 c\theta_2 \\ -b_2 s\psi(t) - z_2(t)c\psi(t)s\theta_2 c\theta_2 \\ -z_2(t)s^2\theta_2 \end{bmatrix} + \lambda \begin{bmatrix} -s\psi(t)s\theta_2 \\ -c\psi(t)s\theta_2 \\ c\theta_2 \end{bmatrix} \tag{3.13a}$$

and

$$\mathbf{r}_3(t,\lambda) = \begin{bmatrix} dc\phi(t) - z_3(t)s\phi(t)s\beta c\beta \\ -ds\phi(t) - z_3(t)c\phi(t)s\beta c\beta \\ -z_3(t)s^2\beta \end{bmatrix} + \lambda \begin{bmatrix} -s\phi(t)s\beta \\ -c\phi(t)s\beta \\ c\beta \end{bmatrix} \tag{3.13b}$$

3.2.1 Revolute-Higher-Revolute (RHR) Mechanisms

A cam mechanism with both input and output revolute pairs and with the contact line coinciding with I_{32} is called a *revolute-higher-revolute* (RHR) mechanism. Its input-output function can be expressed as

$$\phi = \phi(\psi) \tag{3.14}$$

and, from the type of input and output pairs,

$$z_2 = v_2 = 0 \tag{3.15a}$$

$$z_3 = v_3 = 0 \tag{3.15b}$$

Next, from eq.(3.14), $\dot{\phi}$ can be expressed as

$$\dot{\phi} \equiv \frac{d\phi}{dt} = \frac{d\phi}{d\psi}\frac{d\psi}{dt} = \phi'\omega_2 \tag{3.15c}$$

Now, substituting the values of eqs.(3.15) into eqs.(3.9 & 3.10), the screw parameters of I_{32} are readily derived, namely,

$$\omega_{32} = \pm\omega_2 c \tag{3.16a}$$

$$v_{32} = \pm\frac{\phi' \sin\alpha_1}{c} a_1\omega_2 \tag{3.16b}$$

$$\tan\theta_2 = \frac{\phi' \sin\alpha_1}{\phi' \cos\alpha_1 - 1} \tag{3.16c}$$

$$b_2 = \frac{\phi'^2 - \phi'\cos\alpha_1}{c^2} a_1 \tag{3.16d}$$

where

$$c \equiv \sqrt{\phi'^2 - 2\phi'\cos\alpha_1 + 1} \tag{3.17}$$

Furthermore, combining eqs.(3.16) with eqs.(3.13), the cam and follower surfaces are derived from

$$\mathbf{r}_2(\psi,\lambda) = b_2\begin{bmatrix}\cos\psi \\ -\sin\psi \\ 0\end{bmatrix} + \lambda\begin{bmatrix}-\sin\psi\sin\theta_2 \\ -\cos\psi\sin\theta_2 \\ \cos\theta_2\end{bmatrix} \tag{3.18a}$$

and

$$\mathbf{r}_3(\phi,\lambda) = (b_2 - a_1)\begin{bmatrix}\cos\phi \\ -\sin\phi \\ 0\end{bmatrix} + \lambda\begin{bmatrix}-\sin\phi\sin(\theta_2-\alpha_1) \\ -\cos\phi\sin(\theta_2-\alpha_1) \\ \cos(\theta_2-\alpha_1)\end{bmatrix} \tag{3.18b}$$

in agreement with the results presented in (González-Palacios and Angeles, 1990), using an alternative approach, based on the application of vector algebra.

Now we can draw some conclusions, namely,

i) The surfaces generated by $\mathbf{r}_i$ correspond to any of the three classes of cam mechanisms. Thus, the classification below is given in terms of the values of a_1 and α_1:

- Planar: $a_1 \neq 0$, $\alpha_1 = 0$.
- Spherical: $a_1 = 0$, $\alpha_1 \neq 0$.
- Spatial: $a_1 \neq 0$, $\alpha_1 \neq 0$.

ii) As expected, *pure rolling* is achieved in the first two cases mentioned above, which is apparent from eq.(3.16b), since v_{32} vanishes with the corresponding values of a_1 and α_1.

3.2.2 Revolute-Higher-Prismatic (RHP) Mechanisms

A cam mechanism with a revolute pair as input and a prismatic pair as output, with line of contact I_{32}, is called a *revolute-higher-prismatic* (RHP) mechanism. Moreover, the output motion can be expressed as a function of the input motion, namely,

$$z_3 = z_3(\psi) \tag{3.19}$$

From the types of input and output pairs, one concludes that

$$z_2 = v_2 = 0 \tag{3.20a}$$

$$\phi = \omega_3 = 0 \tag{3.20b}$$

Now, upon considering eq.(3.19), we can express v_3 as

$$v_3 \equiv \frac{dz_3}{dt} = \frac{dz_3}{d\psi}\frac{d\psi}{dt} = z_3'\omega_2 \tag{3.20c}$$

Moreover, in light of eqs.(3.20), eqs.(3.9 & 3.10) take the form

$$\omega_{32} = -\omega_2 \tag{3.21a}$$

$$v_{32} = -z_3'\omega_2 \cos\alpha_1 \tag{3.21b}$$

$$\tan\theta_2 = 0 \tag{3.21c}$$

$$b_2 = -z_3' \sin\alpha_1 \tag{3.21d}$$

Furthermore, by substituting eqs.(3.21) into eqs.(3.13), the cam and follower surfaces are readily derived in terms of the position vectors $\mathbf{r}_2(\psi, \lambda)$ and $\mathbf{r}_3(z_3, \lambda)$, namely,

$$\mathbf{r}_2(\psi, \lambda) = z_3' \sin\alpha_1 \begin{bmatrix} -\cos\psi \\ \sin\psi \\ 0 \end{bmatrix} + \lambda \begin{bmatrix} 0 \\ 0 \\ 1 \end{bmatrix} \tag{3.22a}$$

and

$$\mathbf{r}_3(z_3, \lambda) = \begin{bmatrix} -(z_3' \sin\alpha_1 + a_1) \\ z_3 \sin\alpha_1 \cos\alpha_1 \\ -z_3 \sin\alpha_1 \end{bmatrix} + \lambda \begin{bmatrix} 0 \\ \sin\alpha_1 \\ \cos\alpha_1 \end{bmatrix} \tag{3.22b}$$

Equations (3.22) represent a family of cam mechanisms, characterized by specific values of α_1. Because the output is a prismatic pair, a_1 can be arbitrarily defined without changing the geometry of the follower surface. In other words, a_1 only gives the position of the follower frame. Now, by looking at eq.(3.21b), v_{32} vanishes if $\alpha_1 = \pi/2$ or $3\pi/2$. Indeed, substitution of these values into eqs.(3.22) leads to the surfaces given below:

$$\mathbf{r}_2(\psi, \lambda) = \pm z_3' \begin{bmatrix} -\cos\psi \\ \sin\psi \\ 0 \end{bmatrix} + \lambda \begin{bmatrix} 0 \\ 0 \\ 1 \end{bmatrix} \tag{3.23a}$$

and

$$\mathbf{r}_3(z_3, \lambda) = \mp \begin{bmatrix} (z_3' + a_1) \\ 0 \\ z_3 \end{bmatrix} \pm \lambda \begin{bmatrix} 0 \\ 1 \\ 0 \end{bmatrix} \tag{3.23b}$$

the upper sign prevailing when $\alpha_1 = \pi/2$.

From these results we can draw a few conclusions, namely,

i) The surfaces are *cylindrical.*

ii) The cam mechanism has a translating follower.

iii) Because $\alpha_1 \neq 0$ and $b_2 \neq 0$, the ISAs are non-parallel and non-intersecting. Therefore, according to the given definition, these two mechanisms are spatial cams.

On the other hand, v_{32} of eq.(3.21b) is a maximum if $\alpha_1 = 0$ or π, the contacting surfaces being obtained by substituting either of the two values of α_1 into eqs.(3.22), namely,

$$\mathbf{r}(\lambda) = \pm\lambda \begin{bmatrix} 0 \\ 0 \\ 1 \end{bmatrix} \tag{3.24a}$$

and

$$\mathbf{f}(\lambda) = \begin{bmatrix} -a_1 \\ 0 \\ 0 \end{bmatrix} \pm \lambda \begin{bmatrix} 0 \\ 1 \\ 0 \end{bmatrix} \tag{3.24b}$$

from which we can draw a few additional conclusions, i.e.,

i) The surfaces degenerate into a common line, which is coincident with I_{32}.

ii) The solution is ideally a *cylindrical cam mechanism.* In order to have a feasible mechanism, either the locus of the higher pair should be different from the locus of I_{32} or an intermediate rigid body should be included, e.g., a roller.

iii) I_{32} is parallel to I_2 and I_3, which leads to a planar mechanism.

3.2.3 Prismatic-Higher-Revolute (PHR) Mechanisms

A *prismatic-higher-revolute* (PHR) mechanism is similar to an RHP mechanism, the only difference being that now the input has a prismatic pair and the output a revolute pair. The output motion is expressed as

$$\phi = \phi(z_2) \tag{3.25}$$

Now, considering the input and output pairs, one obtains

$$\psi = \omega_2 = 0 \tag{3.26a}$$

$$z_3 = v_3 = 0 \tag{3.26b}$$

Furthermore, ω_3 is expressed as

$$\omega_3 \equiv \frac{d\phi}{dt} = \frac{d\phi}{dz_2}\frac{dz_2}{dt} = \phi' v_2 \tag{3.27}$$

Substitution of eqs.(3.26) into eqs.(3.9 & 3.10) readily leads to

$$\omega_{32} = \pm\phi' v_2 \tag{3.28a}$$

$$v_{32} = \mp v_2 \cos\alpha_1 \tag{3.28b}$$

$$\tan\theta_2 = \tan\alpha_1 \tag{3.28c}$$

$$b_2 = \frac{\phi' a_1 + \sin\alpha_1}{\phi'} \tag{3.28d}$$

Thus, the general surfaces for a PHR mechanism are obtained by substitution of eqs.(3.28) into eqs.(3.13), namely,

$$\mathbf{r}_2(z_2, \lambda) = \begin{bmatrix} a_1 + \sin\alpha_1/\phi' \\ -z_2 s a_1 \cos\alpha_1 \\ -z_2 \sin^2\alpha_1 \end{bmatrix} + \lambda \begin{bmatrix} 0 \\ -\sin\alpha_1 \\ \cos\alpha_1 \end{bmatrix} \tag{3.29a}$$

$$\mathbf{r}_3(\phi, \lambda) = \frac{\sin\alpha_1}{\phi'} \begin{bmatrix} \cos\phi \\ -\sin\phi \\ 0 \end{bmatrix} + \lambda \begin{bmatrix} 0 \\ 0 \\ 1 \end{bmatrix} \tag{3.29b}$$

Moreover, in order to have pure rolling, i.e., $v_{32} = 0$, α_1 must be $\pi/2$ or $3\pi/2$. Since the results are analogous to those of section 3.2.2, one obtains the same conclusions given there. According to the results presented in Sections 3.2.2 and 3.2.3, the theorems below are readily formulated:

Theorem 3.1: *For RHP or PHR mechanisms, the contact line of minimum sliding velocity is parallel to the axis of the revolute pair.*

Theorem 3.2: *Pure rolling of RHP or PHR mechanisms is achieved if the axis of the prismatic pair is perpendicular to both the contact line and the axis of the revolute pair.*

3.2.4 Prismatic-Prismatic-Prismatic (PPP) Mechanisms

If both the input and output pairs are prismatic in a cam mechanism, the third pair is also prismatic and so, the mechanism is termed *prismatic-prismatic-prismatic* (PPP). Properly speaking, then, PPP mechanisms are lower-pair mechanisms, and hence, deserve a special treatment. The underlying input-output function can be expressed as

$$z_3 = z_3(z_2) \tag{3.30}$$

This type of mechanism is the exception to the general formulation given in eqs.(3.9 & 3.10), where the screw parameters, except for ω_{32}, are undefined. Consequently, a particular derivation should be considered, i.e., instead of relative angular velocities, relative translational velocities of the contact point will be analyzed. These are related as

$$\hat{\mathbf{v}}_{31} = \hat{\mathbf{v}}_{32} + \hat{\mathbf{v}}_{21} \tag{2.49}$$

which can be rewritten as

$$v_3\hat{\mathbf{e}}_3 = v_{32}\hat{\mathbf{e}}_{32} + v_2\hat{\mathbf{e}}_2 \tag{3.31}$$

By comparing eq.(3.31) with eq.(3.1), one can apply a procedure similar to the one used to derive eqs.(3.7 & 3.8), to obtain

$$v_{32}^2 = v_3^2 - 2v_3v_2\cos\hat{\alpha}_1 + v_2^2 \tag{3.32a}$$

and

$$\tan\hat{\theta}_2 = \frac{v_3\sin\hat{\alpha}_1}{v_3\cos\hat{\alpha}_1 - v_2} \tag{3.32b}$$

Upon separation of eqs.(3.32) into primal and dual parts, we can readily obtain

$$v_{32} = \pm\sqrt{v_3^2 - 2v_3v_2\cos\alpha_1 + v_2^2} + \epsilon v_3v_2a_1\sin\alpha_1 \tag{3.33a}$$

and

$$\tan\theta_2 + \epsilon b_2(1+\tan^2\theta_2) = \frac{v_3\sin\alpha_1}{v_3\cos\alpha_1 - v_2} + \epsilon\left[\frac{v_3^2 - v_3 v_2\cos\alpha_1}{(v_3\cos\alpha_1 - v_2)^2}\right] a_1 \qquad (3.33b)$$

The dual part of the left-hand side of eq.(3.33a) is zero, and hence,

$$v_3 v_2 a_1 \sin\alpha_1 = 0$$

In general, $v_i \neq 0$, for $i = 2, 3$, and $\sin\alpha_1 \neq 0$. Thus, for the above equation to hold, a_1 must vanish. Consequently, from the dual part of eq.(3.33b), $b_2 = 0$.

Moreover, from eq.(3.30), one can define the relationship

$$v_3 \equiv \frac{dz_3}{dt} = \frac{dz_3}{dz_2}\frac{dz_2}{dt} = z_3' v_2 \qquad (3.34)$$

Thus, with the aid of eq.(3.34), v_{32} and θ_2 can be written as

$$v_{32} = \pm\left(\sqrt{z_3'^2 - 2z_3'\cos\alpha_1 + 1}\right) v_2 \qquad (3.35a)$$

$$\tan\theta_2 = \frac{z_3'\sin\alpha_1}{z_3'\cos\alpha_1 - 1} \qquad (3.35b)$$

from which we can draw the conclusions below:

i) The coupling between cam and follower is a lower pair, namely, prismatic.

ii) Since $a_1 = b_2 = 0$, the three ISAs are coplanar.

3.3 Pressure Angle

In some cam mechanisms, as we will see in Chapter 5, the cam may play temporarily the role of the driven element; correspondingly, the follower may play temporarily the role of the driving element. For this reason, we distinguish here between direct and inverse operations of the mechanism. The *direct operation* is defined as that in which the cam is the driving element, the *inverse operation* being that in which the cam is the driven element.

Henceforth we assume that the cam and follower are bounded by the ruled surfaces $\mathcal{R}_2$ and $\mathcal{R}_3$, respectively. The pressure angle μ_i is then defined as that subtended between the direction of the unit normal to $\mathcal{R}_i$ and the direction of the velocity of the contact point as pertaining to the driven element, which is bounded by $\mathcal{R}_i$, for

$i = 2, 3$. Since contact takes place along a line, μ_i is derived as a function of λ, which is a parameter defined along that line.

Let $\mathbf{r}_i(\psi, \lambda)$ be the position vector of a point of the ruled surface $\mathcal{R}_i$, defined as

$$\mathbf{r}_i(\psi, \lambda) = \mathbf{p}_i(\psi) + \lambda \mathbf{e}_i(\psi) \tag{3.36}$$

The normal vector $\mathbf{n}_i$ of $\mathcal{R}_i$ is determined as in eq.(2.55), namely,

$$\mathbf{n}_i(\psi, \lambda) = \frac{\mathbf{u}_i}{\| \mathbf{u}_i \|}; \quad \mathbf{u}_i = \frac{\partial \mathbf{r}_i}{\partial \psi} \times \frac{\partial \mathbf{r}_i}{\partial \lambda}$$

In terms of eq. (3.36), $\mathbf{n}_i(\psi, \lambda)$ becomes

$$\mathbf{n}_i = \frac{\mathbf{m}_i}{\| \mathbf{m}_i \|}; \qquad \mathbf{m}_i = (\mathbf{p}_i' + \lambda \mathbf{e}_i') \times \mathbf{e}_i \tag{3.37}$$

the prime denoting differentiation with respect to ψ.

Moreover, the unit vector $\mathbf{w}_i$, parallel to the follower velocity at the contact point, is obtained as

$$\mathbf{w}_i = \frac{\boldsymbol{\xi}_i}{\| \boldsymbol{\xi}_i \|}; \qquad \boldsymbol{\xi}_i = \boldsymbol{\omega}_i \times \mathbf{r}_i \tag{3.38}$$

Once $\mathbf{n}_i$ and $\mathbf{w}_i$ are known, the pressure angle is derived from the relation

$$\tan \mu_i = \frac{\| \mathbf{n}_i \times \mathbf{w}_i \|}{\mathbf{n}_i \cdot \mathbf{w}_i} \tag{3.39}$$

The value of μ_i for the spatial mechanism changes at every point of the contact line. In order to derive a significant value of μ_i for this case, we define it along the *striction curve* of $\mathcal{R}_i$, as defined in Section 2.3. At the striction curve, λ attains the value λ_S given below:

$$\lambda_S = -\frac{\mathbf{p}_i' \cdot \mathbf{e}_i'}{\mathbf{e}_i' \cdot \mathbf{e}_i'} \tag{3.40}$$

and hence, μ_i is defined as

$$\mu_i \equiv \mu_i(\lambda_S) \tag{3.41}$$

3.3.1 Pressure Angle of RHR Mechanisms

General expressions of the pressure angle of RHR cam mechanisms for both direct and inverse operations are now derived. The pressure angle of other types can be obtained following the same procedure presented here.

Direct Operation

The pressure angle is obtained in this case for $i = 3$. By following the procedure mentioned above, the unit normal is written as

$$\mathbf{n}_3 = \frac{(\mathbf{p}_3' + \lambda \mathbf{e}_3') \times \mathbf{e}_3}{\| \mathbf{m}_3 \|} \tag{3.42}$$

where the prime denotes differentiation with respect to ψ, and m_3 is defined in eq.(3.37).

From eq. (3.18*b*), we obtain

$$\mathbf{p}_3' = \begin{bmatrix} -d\phi' \sin\phi + b_2' \cos\phi \\ -d\phi' \cos\phi - b_2' \sin\phi \\ 0 \end{bmatrix} \tag{3.43}$$

and

$$\mathbf{e}_3' = \begin{bmatrix} -\phi' \cos\phi \sin\beta - \theta_2' \sin\phi \cos\beta \\ \phi' \sin\phi \sin\beta - \theta_2' \cos\phi \cos\beta \\ -\theta_2' \sin\beta \end{bmatrix} \tag{3.44}$$

θ_2' and b_2' being derived from eqs. (3.16c & 3.16d), namely,

$$\theta_2' = -\frac{\phi'' \sin\alpha_1}{c} \tag{3.45a}$$

and

$$b_2' = \frac{2\phi' - (\phi'^2 + 1)\cos\alpha_1}{c^2} \phi'' a_1 \tag{3.45b}$$

with c defined as in eq. (3.17). Combining eqs. (3.42–3.44), we obtain the desired expression for $\mathbf{n}_3$, i.e.,

$$\mathbf{n}_3 = \begin{bmatrix} -(d\phi' c\phi + b_2' s\phi)c\beta + \lambda(\phi' s\phi s\beta c\beta - \theta_2' c\phi) \\ (d\phi' s\phi - b_2' c\phi)c\beta + \lambda(\phi' c\phi s\beta c\beta + \theta_2' s\phi) \\ -b_2' s\beta + \lambda\phi' s^2\beta \end{bmatrix} \frac{1}{\| \mathbf{m}_3 \|} \tag{3.46}$$

Now $\mathbf{w}_3$ can be written as

$$\mathbf{w}_3 = \begin{bmatrix} d\sin\phi + \lambda\cos\phi\sin\beta \\ d\cos\phi - \lambda\sin\phi\sin\beta \\ 0 \end{bmatrix} \frac{1}{\| \boldsymbol{\xi}_3 \|} \tag{3.47}$$

the cross product of the right-hand side of eq. (3.39) being computed as

$$\mathbf{n}_3 \times \mathbf{w}_3 = \begin{bmatrix} db_2' c\phi s\beta + \lambda s^2\beta(b_2' s\phi + \phi' d c\phi) - \lambda^2 \phi' s\phi s^3\beta \\ -db_2' s\phi s\beta - \lambda s^2\beta(b_2' c\phi - \phi' d s\phi) + \lambda^2\phi' c\phi s^3\beta \\ -d^2\phi' c\beta - \lambda(d\theta_2' - b_2' s\beta c\beta) - \lambda^2\phi' s^2\beta c\beta \end{bmatrix} \frac{1}{\| \boldsymbol{\xi}_3 \| \| \mathbf{m}_3 \|} \tag{3.48}$$

whose magnitude is given by

$$\| \mathbf{n}_3 \times \mathbf{w}_3 \| = \frac{\sqrt{A\lambda^4 + B\lambda^3 + C\lambda^2 + D\lambda + E}}{\| \boldsymbol{\xi}_3 \| \, \| \mathbf{m}_3 \|} \tag{3.49}$$

with the definitions shown below:

$$A \equiv \phi''^2 s^4\beta \tag{3.50a}$$

$$B \equiv F s^2\beta \tag{3.50b}$$

$$C \equiv (b_2'^2 + \phi''^2 d^2)s^2\beta + (\theta_2'^2 + \phi'^2 s^2\beta c^2\beta)d^2 - 2db_2'\theta_2' s\beta c\beta \tag{3.50c}$$

$$D \equiv F d^2 \tag{3.50d}$$

$$E \equiv d^2(b_2'^2 s^2\beta + d^2\phi'^2 c^2\beta) \tag{3.50e}$$

$$F \equiv 2\phi'(d\theta_2' c\beta - b_2' s\beta) \tag{3.50f}$$

Furthermore, the dot product of the right–hand side of eq. (3.39) is readily obtained as

$$\mathbf{n}_3 \cdot \mathbf{w}_3 = -\frac{db_2' \cos\beta + \lambda^2\theta_2' \sin\beta}{\| \boldsymbol{\xi}_3 \| \, \| \mathbf{m}_3 \|} \tag{3.51}$$

Finally, combining eqs. (3.39, 3.48 & 3.51), the tangent of the pressure angle takes on the form

$$\tan\mu_3 = -\frac{\sqrt{A\lambda^4 + B\lambda^3 + C\lambda^2 + D\lambda + E}}{db_2' \cos\beta + \lambda^2\theta_2' \sin\beta} \tag{3.52}$$

For the planar case, $\alpha_1 = 0$, and hence, $\theta_2 = \beta = \theta_2' = 0$, $A = B = C = D = 0$, and $E = d^4\phi'^2$. Substituting all these values into eq. (3.52), $\tan\mu_3$ reduces to

$$\tan\mu_3 = -\frac{d}{b_2'}\phi' = \frac{\phi'(\phi' - 1)}{\phi''} \tag{3.53}$$

For the spherical case, the pressure angle is obtained by taking the limit of $\tan\mu_3$ as λ tends to infinity (DoCarmo, 1976) in eq. (3.52), namely,

$$\tan\mu_3 = \lim_{\lambda\to\infty} -\frac{\sqrt{A\lambda^4 + B\lambda^3 + C\lambda^2 + D\lambda + E}}{db_2' \cos\beta + \lambda^2\theta_2' \sin\beta} = -\frac{\phi' \sin\beta}{\theta_2'} \tag{3.54}$$

Manipulating the expression for $\sin\beta$, eq. (3.54) can be readily rewritten as

$$\tan\mu_3 = \frac{\phi'}{\phi''}\sqrt{c} \tag{3.55}$$

with c defined as in eq. (3.17).

If c is evaluated for $\alpha_1 = 0$, $\sqrt{c} = \phi' - 1$. Therefore, eq. (3.53) is a particular case of eq. (3.55).

For the spatial case, eq. (3.52) is evaluated at the striction point, i.e., at λ_S as given by eq. (3.40), which, in terms of eqs. (3.43 & 3.44), becomes

$$\lambda_S = -\phi' \frac{\theta_2' d \cos\beta - b_2' \sin\beta}{\theta_2'^2 + \phi'^2 \sin^2\beta} \tag{3.56}$$

Moreover,

$$\sin\beta \equiv \sin(\theta_2 - \alpha_1) = \frac{\sin\alpha_1}{\sqrt{c}} \tag{3.57a}$$

$$\cos\beta \equiv \cos(\theta_2 - \alpha_1) = \frac{\phi' - \cos\alpha_1}{\sqrt{c}} \tag{3.57b}$$

$$d \equiv b_2 - a_1 = \frac{\phi' \cos\alpha_1 - 1}{c} a_1 \tag{3.57c}$$

and now, substituting eqs.(3.45 and 3.57) into eq.(3.56), one obtains,

$$\lambda_S = \frac{\phi'^2 \phi'' \sin\alpha_1 \sqrt{c}}{(\phi'^2 c + \phi''^2) c} a_1 \tag{3.58}$$

which is the expression determining the point at which the pressure angle is evaluated.

Inverse operation

Here, the cam is considered as the driven element; therefore, we derive the expression for the pressure angle using $i = 2$. If we follow the same steps as those for the direct operation, we can readily obtain the expression for the pressure angle as

$$\tan\mu_2 = -\frac{\sqrt{\tilde{A}\lambda^4 + \tilde{B}\lambda^3 + \tilde{C}\lambda^2 + \tilde{D}\lambda + \tilde{E}}}{b_2 b_2' \cos\theta + \lambda^2 \theta_2' \sin\theta} \tag{3.59}$$

where

$$\tilde{A} \equiv s^4\theta_2 \tag{3.60a}$$

$$\tilde{B} \equiv \tilde{F} s^2\theta_2 \tag{3.60b}$$

$$\tilde{C} \equiv (b_2'^2 + b_2^2) s^2\theta_2 + (\theta_2'^2 + s^2\theta_2 c^2\theta_2) b_2^2 - 2 b_2 b_2' \theta_2' s\theta_2 c\theta_2 \tag{3.60c}$$

$$\tilde{D} \equiv \tilde{F} b_2^2 \tag{3.60d}$$

$$\tilde{E} \equiv b_2^2 (b_2'^2 s^2\theta_2 + b_2^2 \cos^2\theta_2) \tag{3.60e}$$

$$\tilde{F} \equiv 2(b_2 \theta_2' \cos\theta_2 - b_2' \sin\theta_2) \tag{3.60f}$$

For the planar case, $\alpha_1 = 0$, and hence, $\theta_2 = \theta_2' = 0$, $\tilde{A} = \tilde{B} = \tilde{C} = \tilde{D} = 0$ and $\tilde{E} = b_2^4$.

Upon substitution of all these values into eq. (3.59), we can reduce $\tan\mu_2$ to

$$\tan\mu_2 = -\frac{b_2}{b_2'} = \frac{\phi'(\phi'-1)}{\phi''} \tag{3.61}$$

For the spherical case, the pressure angle is obtained by taking the limit of $\tan\mu_2$ as λ tends to infinity in eq. (3.59), namely,

$$\tan\mu_2 = \lim_{\lambda\to\infty} -\frac{\sqrt{\tilde{A}\lambda^4 + \tilde{B}\lambda^3 + \tilde{C}\lambda^2 + \tilde{D}\lambda + \tilde{E}}}{b_2 b_2' \cos\theta + \lambda^2\theta_2'\sin\theta} = -\frac{\sin\theta}{\theta_2'} \equiv \frac{\phi'}{\phi''}\sqrt{c} \tag{3.62}$$

where c is defined as in eq.(3.17).

For the spatial case, eq. (3.52) is evaluated at the value of λ given by eq. (3.40), namely, at

$$\lambda_S = -\frac{b_2\theta_2'\cos\theta_2 - b_2'\sin\theta_2}{\theta_2'^2 + \sin^2\theta_2} \tag{3.63}$$

But

$$\sin\theta_2 = \frac{\phi'\sin\alpha_1}{\sqrt{c}} \tag{3.64a}$$

$$\cos\theta_2 = \frac{\phi'\cos\alpha_1 - 1}{\sqrt{c}} \tag{3.64b}$$

and hence, eq.(3.63) leads to

$$\lambda_S = \frac{\phi'^2\phi''\sin\alpha\sqrt{c}}{(\phi'^2 c + \phi''^2)c}a_1 \tag{3.65}$$

3.3.2 Pressure Angle of RHP Mechanisms

The pressure angle of RHP mechanisms is also derived with the procedure introduced in Section 3.3.1. Thus, from eqs.(3.22), the pressure angle for both direct and inverse operations can be readily obtained as

$$\tan\mu_3 = \frac{\sqrt{(z_3''s\alpha_1 c\alpha_1)^2 + z_3'^2 s^2\alpha_1(c^2\alpha_1 + s\alpha_1)^2}}{-z_3''s\alpha_1} \tag{3.66a}$$

$$\tan\mu_2 = \frac{\sqrt{(z_3''s\alpha_1 c\alpha_1)^2 + z_3'^2 s^2\alpha_1(c^2\alpha_1 + 1)}}{-z_3''s\alpha_1} \tag{3.66b}$$

Notice that, for $\alpha_1 = \pm\pi/2$,

$$\tan\mu_3 = \tan\mu_2 = \frac{z_3'}{-z_3''} \tag{3.67}$$

3.3.3 Pressure Angle of PHR Mechanisms

The pressure angle of PHR mechanisms for direct and inverse operations is obtained from eqs.(3.29), namely,

$$\tan\mu_3 = \tan\mu_2 = \frac{\sqrt{\phi''^2 c^2\alpha_1 + \phi'^4}}{s\alpha_1} \tag{3.68}$$

3.3.4 Pressure Angle of PPP Mechanisms

From the results derived in Section 3.2.4, one can obtain the expressions for the contact surfaces of PPP mechanisms, and then, the pressure angle can be computed. By following this procedure, the pressure angle is found to be $\pi/2$ and no transmission is possible with this type of mechanisms. Nevertheless, with the introduction of a fourth body, i.e., a roller, a feasible mechanism is possible, as discussed in Chapter 4.

Chapter 4

KINEMATICS OF FOUR-LINK MECHANISMS

4.1 Introduction

This chapter is devoted to the study of cam mechanisms with four links. The study of three-link mechanisms with constant pressure angle is included because these mechanisms are synthesized within our unified framework with the aid of an *auxiliary roller*, which thus leads to a four-link mechanism.

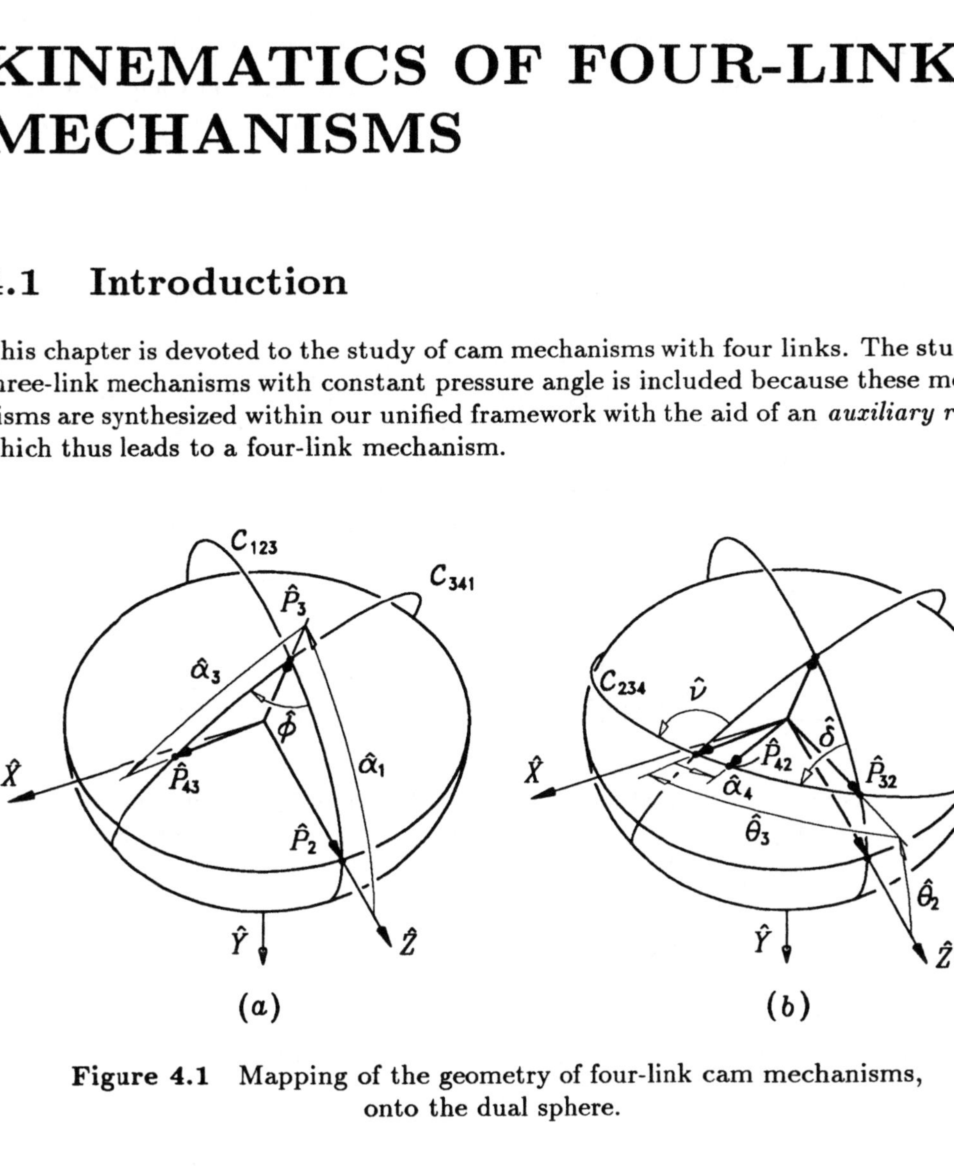

Figure 4.1 Mapping of the geometry of four-link cam mechanisms, onto the dual sphere.

One fourth dual sphere $\mathcal{S}_4$, which represents the roller, is added to the three already considered in Chapter 3. Thus, a total of six ISAs arise now and, according to the Aronhold-Kennedy Theorem in the dual sphere, we have four great circles, each with three poles, namely, $\mathcal{C}_{123}$: $\{\hat{P}_{21}, \hat{P}_{32}, \hat{P}_{31}\}$, $\mathcal{C}_{234}$: $\{\hat{P}_{32}, \hat{P}_{42}, \hat{P}_{43}\}$, $\mathcal{C}_{341}$: $\{\hat{P}_{43}, \hat{P}_{31}, \hat{P}_{41}\}$ and $\mathcal{C}_{412}$: $\{\hat{P}_{41}, \hat{P}_{42}, \hat{P}_{21}\}$. The polodes, as defined in Section 2.2, generated by $\hat{P}_{43}$ and $\hat{P}_{42}$ over $\mathcal{S}_2$, are the dual curves of the pitch and contact surfaces of the cam, the former being used to derive the pressure angle.

Now, a great circle on a dual unit sphere can be defined with at least two points. Thus, $\mathcal{C}_{123}$ and $\mathcal{C}_{341}$, shown in Fig. 4.1*a*, are defined by the given poles $\hat{P}_{21}$, $\hat{P}_{31}$ and $\hat{P}_{43}$.

Moreover, $\hat{P}_{32}$ is computed according to the results obtained in Chapter 3. Once $\hat{P}_{32}$ is obtained, both $\hat{P}_{43}$ and $\hat{P}_{32}$ define $\mathcal{C}_{234}$, as shown in Fig. 4.1*b*. Finally, $\hat{P}_{42}$ is located by the given dual radius of the roller, represented by $\hat{\alpha}_4$. The given parameters as well as the unknown variables for the synthesis of four-link cam mechanisms are shown in Table 4.1.

Table 4.1 Notation used for the synthesis of four-link cam mechanisms

Given parameters:	
$\hat{\alpha}_1 = \alpha_1 + \epsilon a_1$	dual angle between input and output axes
$\hat{\alpha}_3 = \alpha_3 + \epsilon a_3$	dual angle between output and roller axes
$\hat{\alpha}_4 = \alpha_4 + \epsilon a_4$	dual angle of the roller
$\hat{\psi} = \psi + \epsilon z_2$	dual angle of rotation of the cam
$\hat{\phi} = \phi + \epsilon z_3$	dual angle of rotation of the follower
Unknown variables:	
$\hat{\theta}_2 = \theta_2 + \epsilon b_2$	dual arc $\hat{P}_2\hat{P}_{32}$
$\hat{\theta}_3 = \theta_3 + \epsilon b_3$	dual arc $\hat{P}_{32}\hat{P}_{43}$
$\hat{\delta} = \delta + \epsilon z_{32}$	dual angle between arc $\hat{P}_{32}\hat{P}_3$ and arc $\hat{P}_{32}\hat{P}_{43}$
$\hat{\nu} = \nu + \epsilon z_{43}$	dual angle between arc $\hat{P}_3\hat{P}_{34}$ and arc $\hat{P}_{32}\hat{P}_{43}$

The components of the dual unit vector $\hat{\mathbf{e}}$ in the frame $\mathcal{F}_i$ are represented by $[\hat{\mathbf{e}}]_i$. In order to simplify the notation, the symbol $[\cdot]_i$ will be omitted for those vectors given in the frame fixed to $\mathcal{S}_2$.

4.2 Roller-Followers

As mentioned above, the dual curve generated by the motion of $\hat{P}_{43}$ on $\mathcal{S}_2$ represents the pitch surface of the cam, which is given by the family of lines

$$\hat{\mathbf{s}}_p(t) \equiv \mathbf{e}_{43} + \epsilon\mathbf{m}_{43} = \mathbf{S}^T(\hat{\psi})\mathbf{Q}(\hat{\alpha}_1)\mathbf{S}(\hat{\phi})\mathbf{Q}(\hat{\alpha}_3)\mathbf{k} \tag{4.1}$$

with $\hat{\mathbf{Q}}$, $\hat{\mathbf{S}}$ and $\mathbf{k}$ defined, respectively, in eqs.(2.45a, 2.45c & 3.2).

From eq. (2.19), the point coordinates of the *pitch ruled surface* $\mathcal{R}_p$ are defined as

$$\mathbf{r}_p(t,\lambda) = \mathbf{p}_{43}(t) + \lambda\mathbf{e}_{43}(t) \tag{4.2}$$

where $\mathbf{p}_{43} = \mathbf{e}_{43} \times \mathbf{m}_{43}$.

Thus, $\mathbf{e}_{43}$, $\mathbf{m}_{43}$ and $\mathbf{p}_{43}$ can be readily computed as

$$\mathbf{e}_{43} = [-h_1, -h_2, k_1]^T \tag{4.3a}$$

$$\mathbf{m}_{43} = \begin{bmatrix} -k_1 s\psi a_1 - h_3 a_3 - h_2 z_2 + k_5 s\alpha_3 z_3 \\ -k_1 c\psi a_1 - h_4 a_3 + h_1 z_2 - k_6 s\alpha_3 z_3 \\ -k_3 a_1 - k_4 a_3 + k_7 z_3 \end{bmatrix} \tag{4.3b}$$

$$\mathbf{p}_{43} = \begin{bmatrix} (h_2 k_3 + k_1^2 c\psi)a_1 + (h_2 k_4 + h_4 k_1)a_3 \\ -k_1 h_1 z_2 + (k_1 k_6 s\alpha_3 - k_7 h_2)z_3 \\ \\ -(h_1 k_3 + k_1^2 s\psi)a_1 - (h_1 k_4 + h_3 k_1)a_3 \\ -k_1 h_2 z_2 + (k_1 k_5 s\alpha_3 + k_7 h_1)z_3 \\ \\ (h_1 c\psi - h_2 s\psi)k_1 a_1 + (h_1 h_4 - h_2 h_3)a_3 \\ -(h_1^2 + h_2^2)z_2 + (h_2 k_5 + h_1 k_6)s\alpha_3 z_3 \end{bmatrix} \tag{4.3c}$$

where

$$k_1 = c\alpha_1 c\alpha_3 - s\alpha_1 s\alpha_3 c\phi \tag{4.4a}$$

$$k_2 = c\alpha_1 c\alpha_3 c\phi - s\alpha_1 s\alpha_3 \tag{4.4b}$$

$$k_3 = s\alpha_1 c\alpha_3 + c\alpha_1 s\alpha_3 c\phi \tag{4.4c}$$

$$k_4 = s\alpha_3 c\alpha_1 + c\alpha_3 s\alpha_1 c\phi \tag{4.4d}$$

$$k_5 = c\psi c\phi + s\psi s\phi c\alpha_1 \tag{4.4e}$$

$$k_6 = s\psi c\phi - c\psi s\phi c\alpha_1 \tag{4.4f}$$

$$k_7 = s\alpha_1 s\alpha_3 s\phi \tag{4.4g}$$

$$h_1 = k_3 s\psi - s\alpha_3 s\phi c\psi \tag{4.4h}$$

$$h_2 = k_3 c\psi + s\alpha_3 s\phi s\psi \tag{4.4i}$$

$$h_3 = k_2 s\psi - c\alpha_3 s\phi c\psi \tag{4.4j}$$

$$h_4 = k_2 c\psi + c\alpha_3 s\phi s\psi \tag{4.4k}$$

Now, the dual curve representing the surface of the cam is defined by the trajectory that $\hat{P}_{42}$ traces on $\mathcal{S}_2$, namely,

$$\hat{\mathbf{s}}_C(t) \equiv \mathbf{e}_{42} + \epsilon\mathbf{m}_{42} = \hat{\mathbf{S}}^T(\hat{\psi})\hat{\mathbf{Q}}(\hat{\theta}_2)\hat{\mathbf{S}}(\hat{\delta})\hat{\mathbf{Q}}(\hat{\eta})\mathbf{k} \tag{4.5}$$

Furthermore, in order to completely define $\hat{\mathbf{s}}_C$, $\hat{\theta}_2$ is computed from eqs. (3.10). The dual angles $\hat{\eta}$ and $\hat{\delta}_2$ are derived below. The former is defined as

$$\hat{\eta} \equiv \eta + \epsilon(b_3 - a_4) = \hat{\theta}_3 - \hat{\alpha}_4 \tag{4.6}$$

$\hat{\alpha}_4$ representing the dual radius of the roller.

Moreover, $\hat{\theta}_3$ is obtained from the relation

$$\tan\hat{\theta}_3 = \frac{\hat{\mathbf{e}}_{32} \times \hat{\mathbf{e}}_{43} \cdot \hat{\mathbf{e}}_{234}}{\hat{\mathbf{e}}_{32} \cdot \hat{\mathbf{e}}_{43}} \tag{4.7}$$

where the vectors appearing in the right-hand side can be expressed in any frame, as long as all are in the same frame, for the cross and dot products are frame-invariant. If these vectors are expressed in frame $\mathcal{F}_1$, fixed to $\mathcal{S}_1$, then,

$$[\hat{\mathbf{e}}_{32}]_1 = \hat{\mathbf{Q}}(\hat{\theta}_2)\mathbf{k} \tag{4.8a}$$

$$[\hat{\mathbf{e}}_{43}]_1 = \hat{\mathbf{Q}}(\hat{\alpha}_1)\hat{\mathbf{S}}(\hat{\phi})\hat{\mathbf{Q}}(\hat{\alpha}_3)\mathbf{k} \tag{4.8b}$$

and $[\hat{\mathbf{e}}_{234}]_1$ is defined as

$$[\hat{\mathbf{e}}_{234}]_1 \equiv \frac{[\hat{\mathbf{e}}_{32}]_1 \times [\hat{\mathbf{e}}_{43}]_1}{\| [\hat{\mathbf{e}}_{32}]_1 \times [\hat{\mathbf{e}}_{43}]_1 \|} \tag{4.9}$$

Furthermore, the dual angle $\hat{\delta}$, shown in Figure 4.1, is obtained as

$$\tan\hat{\delta} = \frac{[\hat{\mathbf{e}}_{123}]_1 \times [\hat{\mathbf{e}}_{234}]_1 \cdot [\hat{\mathbf{e}}_{32}]_1}{[\hat{\mathbf{e}}_{123}]_1 \cdot [\hat{\mathbf{e}}_{234}]_1} \tag{4.10}$$

where

$$[\hat{\mathbf{e}}_{123}]_1 = [\,1, \quad 0, \quad 0\,]^T \tag{4.11}$$

From eqs. (2.19 & 4.5), the point coordinates of the *contact surface* of the cam are given as

$$\mathbf{r}_C(t,\lambda) = \mathbf{p}_{42}(t) + \lambda\mathbf{e}_{42}(t) \tag{4.12}$$

where $\mathbf{p}_{42}(t) = \mathbf{e}_{42} \times \mathbf{m}_{42}$.

The transformation dual matrices in eq.(4.5) are the same as those in eq.(4.1), but with different arguments; therefore, the results given for $\mathbf{e}_{42}$, $\mathbf{m}_{42}$ and $\mathbf{p}_{42}$ are analogous to those of eqs.(4.3), namely,

$$\mathbf{e}_{42} = [-\tilde{h}_1, -\tilde{h}_2, \tilde{k}_1]^T \tag{4.13a}$$

$$\mathbf{m}_{42} = \begin{bmatrix} -\tilde{k}_1 s\psi b_2 - \tilde{h}_3(b_3 - a_4) - \tilde{h}_2 z_2 + \tilde{k}_5 s\eta z_{32} \\ -\tilde{k}_1 c\psi b_2 - \tilde{h}_4(b_3 - a_4) + \tilde{h}_1 z_2 - \tilde{k}_6 s\eta z_{32} \\ -\tilde{k}_3 b_2 - \tilde{k}_4(b_3 - a_4) + \tilde{k}_7 z_{32} \end{bmatrix} \tag{4.13b}$$

$$\mathbf{p}_{42} = \begin{bmatrix} (\tilde{h}_2\tilde{k}_3 + \tilde{k}_1^2 c\psi)b_2 + (\tilde{h}_2\tilde{k}_4 + \tilde{h}_4\tilde{k}_1)(b_3 - a_4) \\ -\tilde{k}_1\tilde{h}_1 z_2 + (\tilde{k}_1\tilde{k}_6 s\eta - \tilde{k}_7\tilde{h}_2)z_{32} \\ \\ -(\tilde{h}_1\tilde{k}_3 + \tilde{k}_1^2 s\psi)b_2 - (\tilde{h}_1\tilde{k}_4 + \tilde{h}_3\tilde{k}_1)(b_3 - a_4) \\ -\tilde{k}_1\tilde{h}_2 z_2 + (\tilde{k}_1\tilde{k}_5 s\eta + \tilde{k}_7\tilde{h}_1)z_{32} \\ \\ (\tilde{h}_1 c\psi - \tilde{h}_2 s\psi)\tilde{k}_1 b_2 + (\tilde{h}_1\tilde{h}_4 - \tilde{h}_2\tilde{h}_3)(b_3 - a_4) \\ -(\tilde{h}_1^2 + \tilde{h}_2^2)z_2 + (\tilde{h}_2\tilde{k}_5 + \tilde{h}_1\tilde{k}_6)s\eta z_{32} \end{bmatrix} \tag{4.13c}$$

where

$$\tilde{k}_1 = c\theta_2 c\eta - s\theta_2 s\eta c\delta \tag{4.14a}$$

$$\tilde{k}_2 = c\theta_2 c\eta c\delta - s\theta_2 s\eta \tag{4.14b}$$

$$\tilde{k}_3 = s\theta_2 c\eta + c\theta_2 s\eta c\delta \tag{4.14c}$$

$$\tilde{k}_4 = s\eta c\theta_2 + c\eta s\theta_2 c\delta \tag{4.14d}$$

$$\tilde{k}_5 = c\psi c\delta + s\psi s\delta c\theta_2 \tag{4.14e}$$

$$\tilde{k}_6 = s\psi c\delta - c\psi s\delta c\theta_2 \tag{4.14f}$$

$$\tilde{k}_7 = s\theta_2 s\eta s\delta \tag{4.14g}$$

$$\tilde{h}_1 = \tilde{k}_3 s\psi - s\eta s\delta c\psi \tag{4.14h}$$

$$\tilde{h}_2 = \tilde{k}_3 c\psi + s\eta s\delta s\psi \tag{4.14i}$$

$$\tilde{h}_3 = \tilde{k}_2 s\psi - c\eta s\delta c\psi \tag{4.14j}$$

$$\tilde{h}_4 = \tilde{k}_2 c\psi + c\eta s\delta s\psi \tag{4.14k}$$

In order to avoid *undercutting*, which is a phenomenon that occurs whenever the follower or the cam cutter cannot follow the desired cam profile (Chen, 1982), the dual angle $\hat{\alpha}_4$ must be constrained by

$$\hat{\alpha}_4 < \hat{\rho}_{min} \tag{4.15}$$

where $\hat{\rho}_{min}$ is the minimum value of the dual radius of curvature of the *dual pitch curve* of the cam, and is evaluated according to eq.(2.70).

Now, the dual angle $\hat{\nu} = \nu + \epsilon z_{43}$ can be computed as

$$\tan \hat{\nu} = \frac{\hat{\mathbf{e}}_{341} \times \hat{\mathbf{e}}_{234} \cdot \hat{\mathbf{e}}_{43}}{\hat{\mathbf{e}}_{341} \cdot \hat{\mathbf{e}}_{234}} \tag{4.16}$$

where

$$\hat{\mathbf{e}}_{341} = \frac{\hat{\mathbf{e}}_{31} \times \hat{\mathbf{e}}_{43}}{\sin \hat{\alpha}_3} \tag{4.17}$$

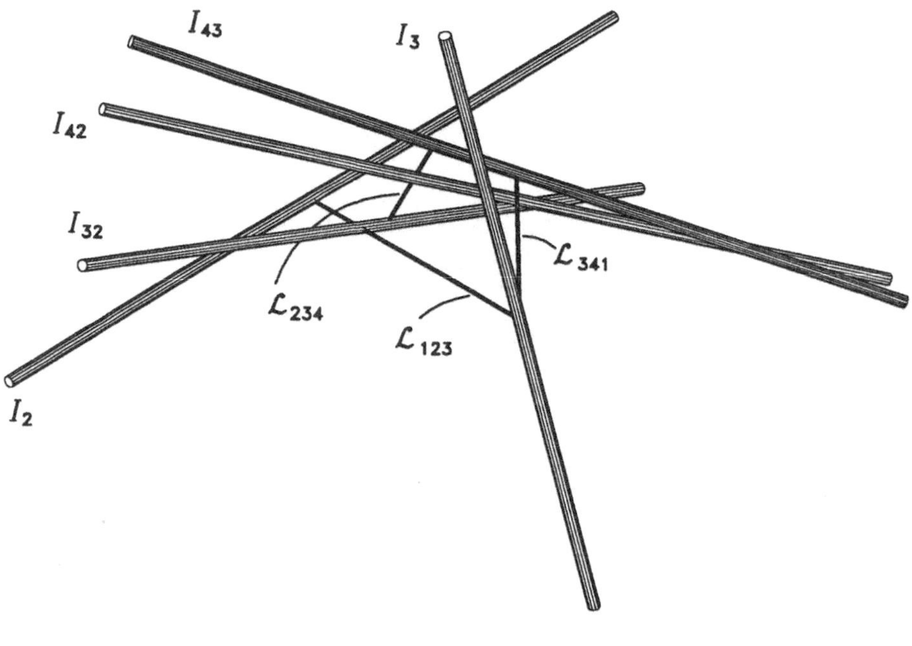

Figure 4.2 ISAs of an RHCR cam mechanism.

The axial motion of the roller with respect to the follower is defined by z_{43}. In general, $z_{43} = z_{43}(t)$, the roller-follower pair being cylindrical; however, a revolute pair is present when z_{43} is constant.

Now, in order to derive the expressions for $\hat{\omega}_{42}$ and $\hat{\omega}_{43}$, the Aronhold-Kennedy Theorem, as stated in eq.(2.51), is applied to the poles of the great circle $\mathcal{C}_{234}$, namely,

$$\hat{\omega}_{32}[\hat{\mathbf{e}}_{32}]_1 = \hat{\omega}_{42}[\hat{\mathbf{e}}_{42}]_1 + \hat{\omega}_{43}[\hat{\mathbf{e}}_{43}]_1 \tag{4.18}$$

with

$$[\hat{\mathbf{e}}_{32}]_1 = \hat{\mathbf{Q}}(\hat{\theta}_2)\mathbf{k} \tag{4.19a}$$

$$[\hat{\mathbf{e}}_{42}]_1 = \hat{\mathbf{Q}}(\hat{\alpha}_1)\hat{\mathbf{S}}(\hat{\phi})\hat{\mathbf{Q}}(\hat{\alpha}_3)\hat{\mathbf{S}}(\hat{\nu})\hat{\mathbf{Q}}(\hat{\alpha}_4)\mathbf{k} \tag{4.19b}$$

$$[\hat{\mathbf{e}}_{43}]_1 = \hat{\mathbf{Q}}(\hat{\alpha}_1)\hat{\mathbf{S}}(\hat{\phi})\hat{\mathbf{Q}}(\hat{\alpha}_3)\mathbf{k} \tag{4.19c}$$

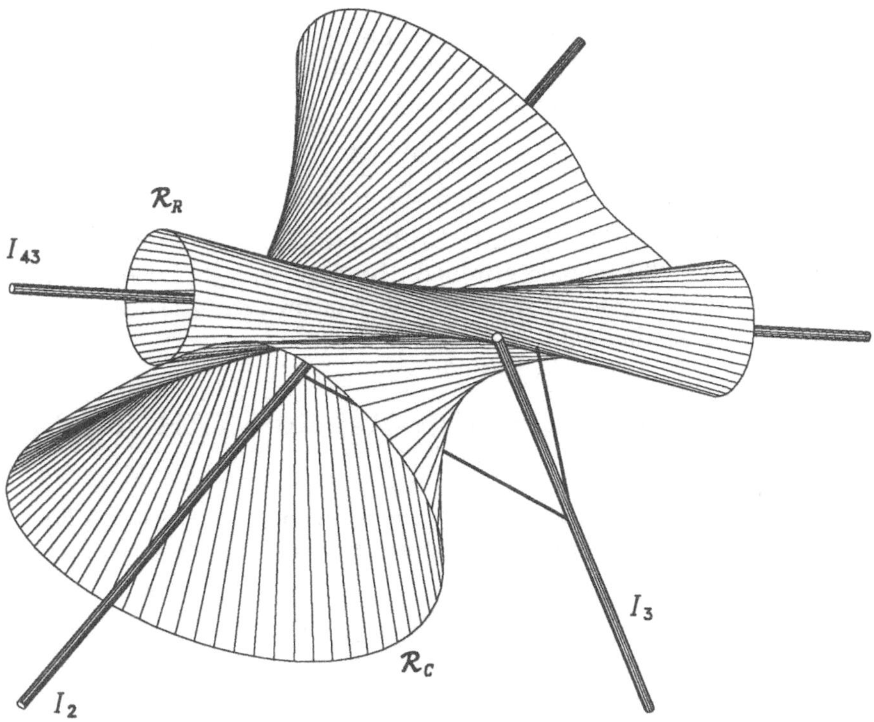

Figure 4.3 Ruled surfaces of the cam ($\mathcal{R}_C$) and roller ($\mathcal{R}_R$) of an RHCR cam mechanism.

Upon substituting eqs.(4.19) into eq.(4.18) and multiplying both sides by $\hat{\mathbf{Q}}^T(\hat{\theta}_2)$, we obtain

$$\hat{\omega}_{32}\mathbf{k} = \hat{\omega}_{42}\hat{\mathbf{Q}}(\hat{\beta})\hat{\mathbf{S}}(\hat{\phi})\hat{\mathbf{Q}}(\hat{\alpha}_3)\hat{\mathbf{S}}(\hat{\nu})\hat{\mathbf{Q}}(\hat{\alpha}_4)\mathbf{k} + \hat{\omega}_{43}\hat{\mathbf{Q}}(\hat{\beta})\hat{\mathbf{S}}(\hat{\phi})\hat{\mathbf{Q}}(\hat{\alpha}_3)\mathbf{k} \tag{4.20}$$

Notice that $\hat{\mathbf{Q}}^T(\hat{\theta}_2)\hat{\mathbf{Q}}(\hat{\alpha}_1) = \hat{\mathbf{Q}}(\hat{\beta})$ and $\hat{\beta} = \hat{\alpha}_1 - \hat{\theta}_2$.

From the eq.(4.20), one can readily obtain the relations below:

$$\hat{\omega}_{42} \sin\hat{\alpha}_4 \sin\hat{\nu} = -\hat{\omega}_{32} \sin\hat{\beta} \sin\hat{\phi} \tag{4.21a}$$

$$\hat{\omega}_{42} \sin\hat{\alpha}_4 \cos\hat{\nu} = -\hat{\omega}_{32}(\cos\hat{\beta}\sin\hat{\alpha}_3 + \cos\hat{\alpha}_3\cos\hat{\phi}\sin\hat{\beta}) \tag{4.21b}$$

$$\hat{\omega}_{42}\cos\hat{\alpha}_4 = \hat{\omega}_{43} - \hat{\omega}_{32}(\cos\hat{\beta}\cos\hat{\alpha}_3 - \sin\hat{\beta}\sin\hat{\alpha}_3\cos\hat{\phi}) \tag{4.21c}$$

Thus, from eqs.(4.21a & 4.21c),

$$\hat{\omega}_{42} = -\frac{\sin\hat{\beta}\sin\hat{\phi}}{\sin\hat{\alpha}_4\sin\hat{\nu}}\hat{\omega}_{32} \tag{4.22a}$$

$$\hat{\omega}_{43} = \hat{\omega}_{42}\cos\hat{\alpha}_4 + \hat{\omega}_{32}(\cos\hat{\beta}\cos\hat{\alpha}_3 - \sin\hat{\beta}\sin\hat{\alpha}_3\cos\hat{\phi}) \tag{4.22b}$$

Moreover, the dual angle $\hat{\nu}$ can be computed following two approaches, one of which is based on eq.(4.16), while the other consists of dividing each side of eq.(4.21a) by the corresponding side of eq.(4.21b), namely,

$$\tan\hat{\nu} = -\frac{\sin\hat{\beta}\sin\hat{\phi}}{\cos\hat{\beta}\sin\hat{\alpha}_3 + \cos\hat{\alpha}_3\cos\hat{\phi}\sin\hat{\beta}} \tag{4.23}$$

It can be proven that both eqs.(4.16) and (4.23) lead to the same result. However, for planar mechanisms, eq.(4.23) is undetermined. Nevertheless, if planar mechanisms are considered as particular cases of spherical mechanisms, eq.(4.23) can be applied as well, as shown in Subsection 4.2.1.

4.2.1 Revolute-Higher-Cylindric-Revolute (RHCR) Mechanisms

For *revolute-higher-cylindric-revolute* (RHCR) mechanisms, the cam and follower are coupled to the frame via revolute pairs; therefore, $z_2 = z_3 = 0$. Moreover, $\phi = \phi(\psi)$, and eq.(4.2) can be written as

$$\mathbf{r}_p(\psi, \lambda) = \mathbf{p}_{43}(\psi) + \lambda\mathbf{e}_{43}(\psi) \tag{4.24}$$

with

$$\mathbf{e}_{43} = \begin{bmatrix} -h_1 \\ -h_2 \\ k_1 \end{bmatrix} \quad \text{and} \quad \mathbf{p}_{43} = \begin{bmatrix} (h_2k_3 + k_1^2c\psi)a_1 + (h_2k_4 + h_4k_1)a_3 \\ -(h_1k_3 + k_1^2s\psi)a_1 - (h_1k_4 + h_3k_1)a_3 \\ (h_1c\psi - h_2s\psi)k_1a_1 + (h_1h_4 - h_2h_3)a_3 \end{bmatrix} \tag{4.25}$$

Similarly, eq.(4.12) is expressed as

$$\mathbf{r}_C(\psi, \lambda) = \mathbf{p}_{42}(\psi) + \lambda\mathbf{e}_{42}(\psi) \tag{4.26}$$

with

$$\mathbf{e}_{42} = \begin{bmatrix} -\tilde{h}_1 \\ -\tilde{h}_2 \\ \tilde{k}_1 \end{bmatrix} \quad \text{and} \quad \mathbf{p}_{42} = \begin{bmatrix} (\tilde{h}_2\tilde{k}_3 + \tilde{k}_1^2c\psi)b_2 + (\tilde{h}_2\tilde{k}_4 + \tilde{h}_4\tilde{k}_1)(b_3 - a_4) \\ -(\tilde{h}_1\tilde{k}_3 + \tilde{k}_1^2s\psi)b_2 - (\tilde{h}_1\tilde{k}_4 + \tilde{h}_3\tilde{k}_1)(b_3 - a_4) \\ (\tilde{h}_1c\psi - \tilde{h}_2s\psi)\tilde{k}_1b_2 + (\tilde{h}_1\tilde{h}_4 - \tilde{h}_2\tilde{h}_3)(b_3 - a_4) \end{bmatrix} \tag{4.27}$$

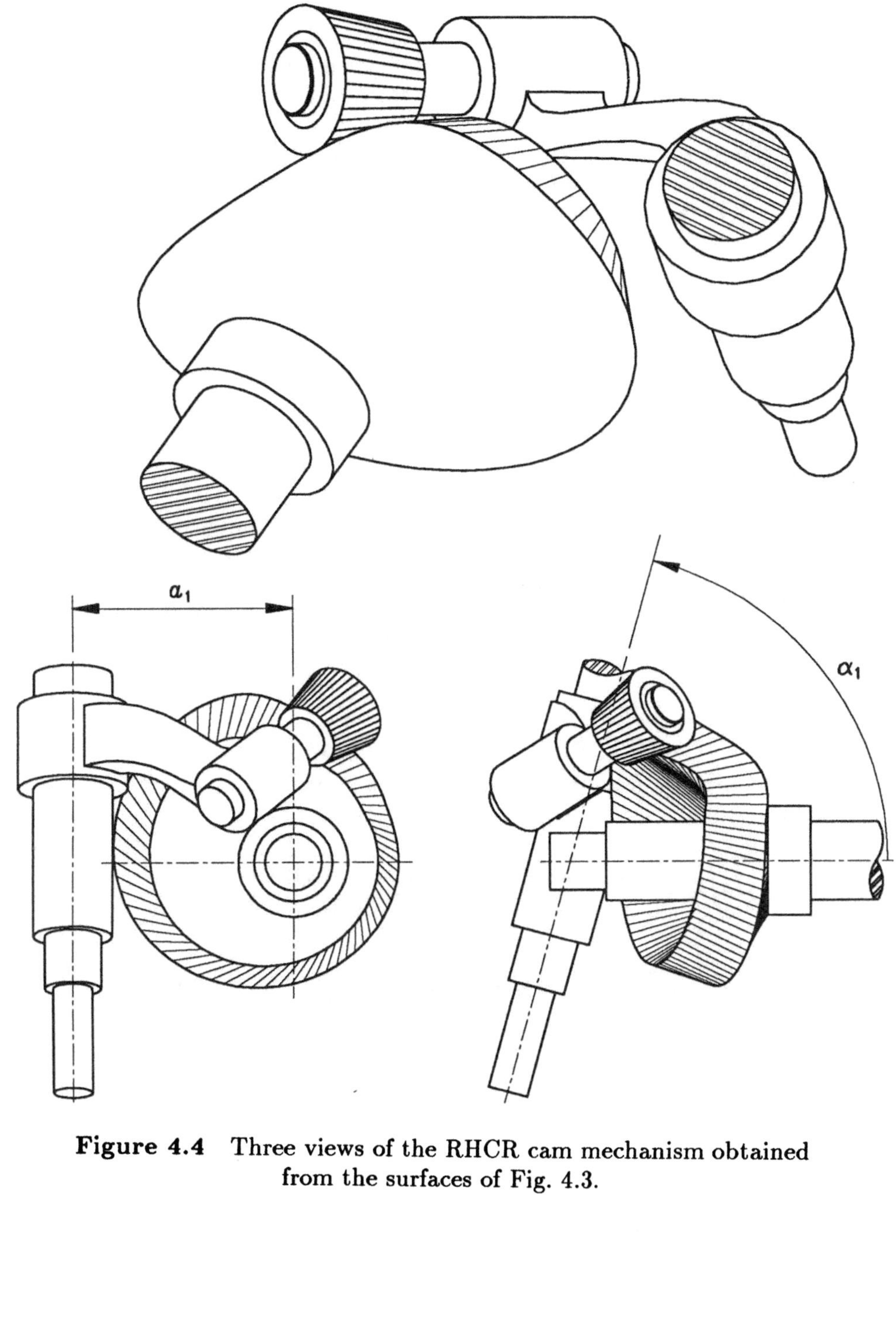

Figure 4.4 Three views of the RHCR cam mechanism obtained from the surfaces of Fig. 4.3.

where $\hat{\theta}_2$, $\hat{\eta}$ and $\hat{\delta}$ are evaluated with the aid of eqs.(3.16c, 3.16d, 4.6 & 4.10).

A geometric representation of the poles of Fig. 4.1 is shown in Fig. 4.2, where $\hat{P}_{ij} \rightarrow I_{ij}$ and $\mathcal{C}_{ijk} \rightarrow \mathcal{L}_{ijk}$. Since I_{41} and $\mathcal{L}_{412}$ are not needed for this derivation, they do not appear in Fig. 4.2. Moreover, if $\hat{\alpha}_4$ is constant, the surface of the roller $\mathcal{R}_R$, will be in general a hyperboloid. Under the assumption that $-\lambda_0 \leq \lambda \leq \lambda_0$ in eq.(4.26), where λ_0 is a design parameter, one obtains the surfaces shown in Fig. 4.3. On the other hand, if we bound λ as: $0.7\lambda_0 \leq \lambda \leq \lambda_0$, one can obtain the cam mechanism shown in Fig. 4.4 in various views.

For the cases when all axes are parallel or intersecting, z_{43} is always zero, and hence, the follower-roller pair becomes a revolute. These two cases are called RHRR mechanisms. The resulting surface geometry for RHRR mechanisms is rather simple and can be readily derived from the general formulation given above. However, a special treatment to compute dual cross products is needed for planar RHRR mechanisms, as discussed in Chapter 2.1 for the analysis pertaining to parallel dual vectors.

Spherical RHRR Mechanisms

For spherical RHRR mechanisms, the dual terms of the dual angles shown in Table 4.1 are all zero and, from eqs.(4.24 & 4.26), one can see that the pitch and contact surfaces of the cam reduce to conical surfaces, namely,

$$\mathbf{r}_P(\psi, \lambda) = \lambda \mathbf{e}_{43} \tag{4.28}$$

$$\mathbf{r}_C(\psi, \lambda) = \lambda \mathbf{e}_{42} \tag{4.29}$$

where $\mathbf{e}_{43}$ and $\mathbf{e}_{42}$ are defined in eqs.(4.25 & 4.27), respectively.

Now, θ_2 is computed from eq.(3.16c) and, from eqs.(4.7, 4.10 & 4.16), one can readily obtain

$$\tan\theta_3 = \frac{\sqrt{[c(\alpha_1-\theta_2)c\phi s\alpha_3 + c\alpha_3 s(\alpha_1-\theta_2)]^2 + s^2\alpha_3 s^2\phi}}{c\alpha_3 c(\alpha_1-\theta_2) - c\phi s\alpha_3 s(\alpha_1-\theta_2)} \tag{4.30}$$

$$\tan\delta = \frac{s\alpha_3 s\phi}{s\alpha_3 c(\alpha_1-\theta_2)c\phi + c\alpha_3 s(\alpha_1-\theta_2)} \tag{4.31}$$

$$\tan\nu = \frac{-s(\alpha_1-\theta_2)s\phi}{c(\alpha_1-\theta_2)s\alpha_3 + c\alpha_3 c\phi s(\alpha_1-\theta_2)} \tag{4.32}$$

A typical design of a spherical RHRR cam mechanism is shown in Fig. 4.5.

Example 4.2.1: *Given the synthesis parameters of three types of spherical cam mechanisms as shown in table 4.2 and the displacement program defined as in Fig. 4.6,*

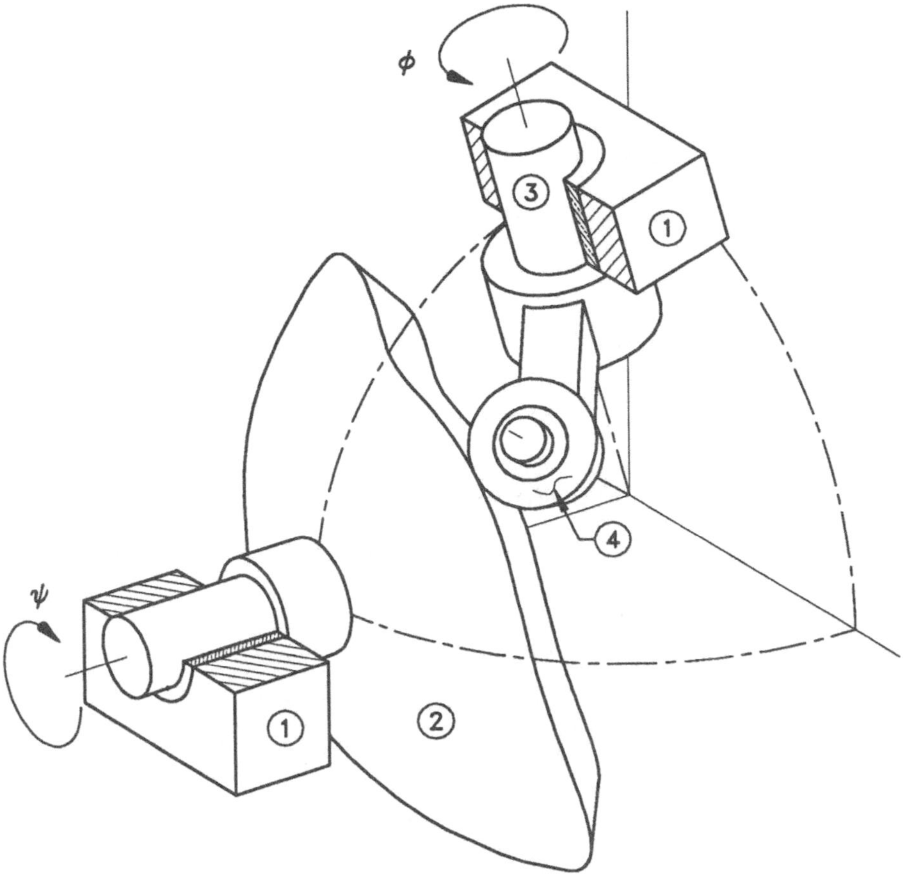

Figure 4.5 Spherical RHRR cam mechanism.

Table 4.2 Synthesis parameters of Example 4.2.1

Case	α_1	α_3	α_4
	Degrees		
1	90	90	8
2	90	70	8
3	70	45	8

where the intervals of rise and return are defined by the 3-4-5 polynomial motion introduced in eqs.(B.4), plot the curves of the profiles projected onto a unit sphere.

Solution:

The surface of the cam is represented by eq.(4.29), the curves of the cam profiles projected onto the unit sphere thus being defined as

$$\mathbf{r}_c(\psi, 1) = \mathbf{e}_{42} \equiv [-\tilde{h}_1, -\tilde{h}_2, \tilde{k}_1]^T$$

But,

$$\tilde{h}_1 = [s\theta_2 c(\theta_3 - \alpha_4) + c\theta_2 s(\theta_3 - \alpha_4)c\delta]s\psi - s(\theta_3 - \alpha_4)s\delta c\psi$$

$$\tilde{h}_2 = [s\theta_2 c(\theta_3 - \alpha_4) + c\theta_2 s(\theta_3 - \alpha_4)c\delta]c\psi + s(\theta_3 - \alpha_4)s\delta s\psi$$

$$\tilde{k}_1 = c\theta_2 c(\theta_3 - \alpha_4) - s\theta_2 s(\theta_3 - \alpha_4)c\delta$$

Thus, θ_2, θ_3 and δ are computed with the aid of eqs.(3.16c, 4.30 & 4.31), namely,

$$\theta_2 = \tan^{-1}\left(\frac{\phi' \sin\alpha_1}{\phi' \cos\alpha_1 - 1}\right)$$

$$\theta_3 = \tan^{-1}\left(\frac{\sqrt{[c(\alpha_1 - \theta_2)c\phi s\alpha_3 + c\alpha_3 s(\alpha_1 - \theta_2)]^2 + s^2\alpha_3 s^2\phi}}{c\alpha_3 c(\alpha_1 - \theta_2) - c\phi s\alpha_3 s(\alpha_1 - \theta_2)}\right)$$

$$\delta = \tan^{-1}\left(\frac{s\alpha_3 s\phi}{s\alpha_3 c(\alpha_1 - \theta_2)c\phi + c\alpha_3 s(\alpha_1 - \theta_2)}\right)$$

Now, considering intervals given in Fig. 4.6, ϕ and ϕ' are defined as follows,

$$\phi = \begin{cases} \pi/4 + h(10x_1^3 - 15x_1^4 + 6x_1^5) & 0 \le \psi \le \pi \\ \pi/4 + h & \pi < \psi < 7\pi/5 \\ \pi/4 + h(10x_2^3 - 15x_2^4 + 6x_2^5) & 7\pi/5 \le \psi \le 2\pi \end{cases} \tag{4.33a}$$

$$\phi' = \begin{cases} 30h(x_1^2 - 2x_1^3 + x_1^4)/\pi & 0 \le \psi \le \pi \\ 0 & \pi < \psi < 7\pi/5 \\ -150h(x_2^2 - 2x_2^3 + x_2^4)/3\pi & 7\pi/5 \le \psi \le 2\pi \end{cases} \tag{4.33b}$$

where

$$h = \frac{\pi}{6}, \quad x_1 = \frac{\psi}{\pi}, \quad x_2 = \frac{10\pi - 5\psi}{3\pi} \tag{4.33c}$$

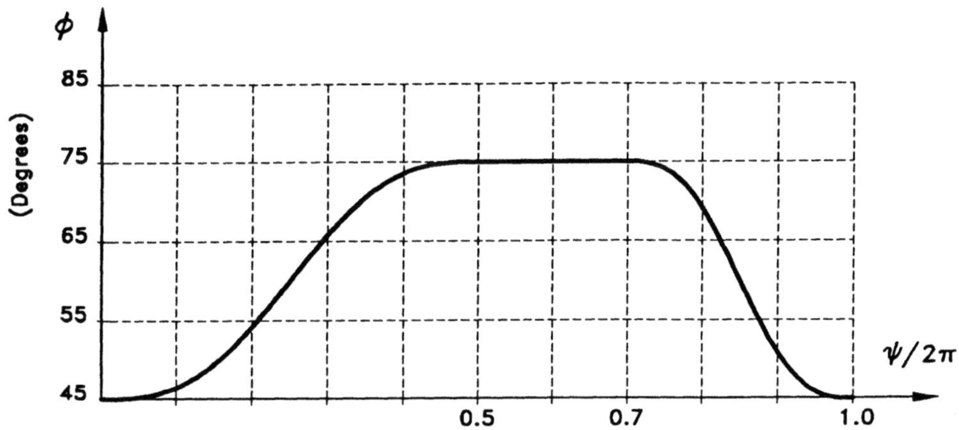

Figure 4.6 Follower displacement program of Example 4.2.1.

Thus, with the aid of a CAD system, we can plot the curves shown in Fig. 4.7, where the follower for each case is shown at the lowest position, i.e., at $\psi = 45°$. Comparing the results obtained here with those reported in (González-Palacios and Angeles, 1991*a*), where we solved this example by using cycloidal motion, it is apparent that the 3-4-5 motion yields better results from the viewpoint of the maximum values of first and second derivatives of the displacement program, these values are 5.77% and 8.11% smaller than those obtained with a cycloidal motion, respectively. However, from the viewpoint of the maximum value of the third derivative, the cycloidal motion is better, since its maximum value is 35.2% smaller. ∎

Example 4.2.2: *Derive an expression for the top and bottom surfaces, $\mathcal{U}$ and $\mathcal{B}$, of a spherical cam as per Fig. 4.8, showing $\mathcal{B}$.*

Solution:

If we define $\mathcal{U}$ and $\mathcal{B}$ as *spherical surfaces*, we can then generate them by rotating the arc AB of Fig. 4.9, about the Z-axis, while varying γ as described below. Plane OAB makes an angle σ with the XZ plane, this angle being determined as

$$\sigma = \tan^{-1}\left(\frac{\mathbf{e}_{42}^T\mathbf{e}_y}{\mathbf{e}_{42}^T\mathbf{e}_x}\right) \tag{4.34}$$

while γ varies as a function of $\mathbf{e}_{23}$, namely,

$$\gamma = \cos^{-1}(\mathbf{e}_{42}^T\mathbf{k}), \qquad 0 \le \gamma \le \pi \tag{4.35}$$

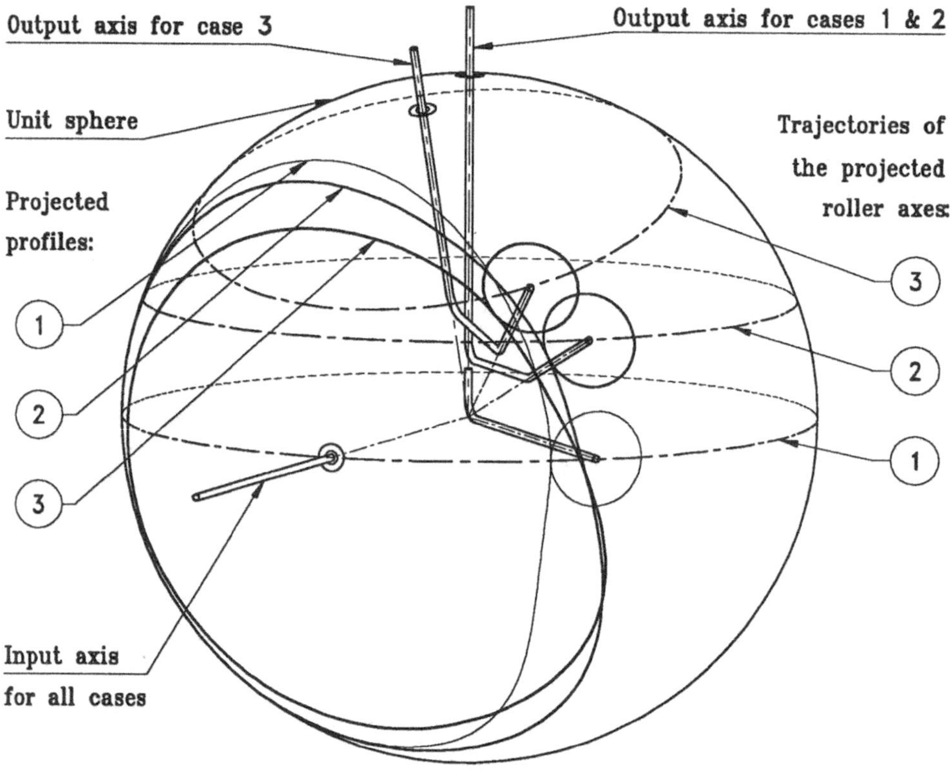

Figure 4.7 Curves of the cam profiles of Example 4.2.1.

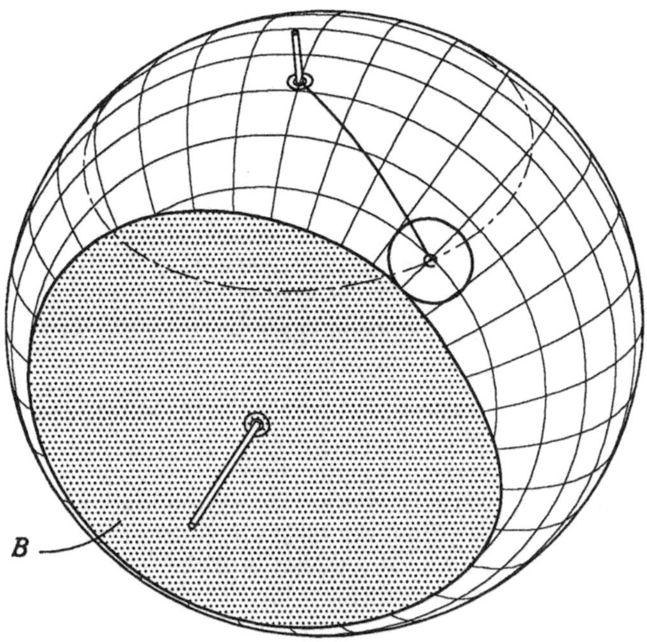

Figure 4.8 Boundaries of a cam profile.

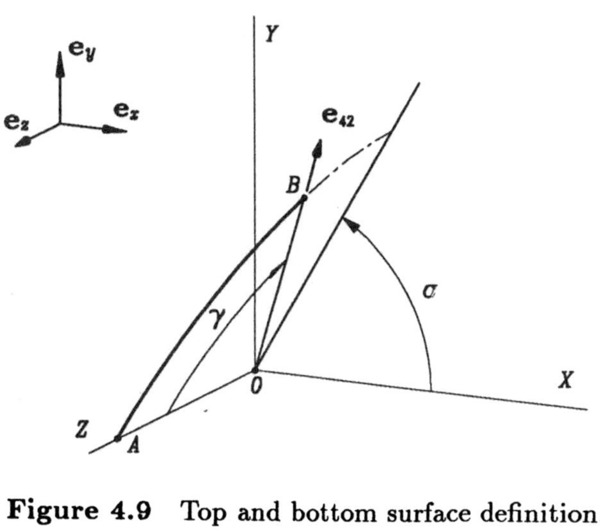

Figure 4.9 Top and bottom surface definition.

Now, a curve on the unit sphere is defined in parametric form, namely,

$$\mathbf{e}(\nu) = \mathbf{S}(\sigma)\mathbf{R}(\nu\gamma)\mathbf{k}, \qquad 0 \le \nu \le 1 \tag{4.36}$$

Thus, $\mathcal{U}$ and $\mathcal{B}$ are generated by the position vectors $\mathbf{u}(\psi,\nu)$ and $\mathbf{b}(\psi,\nu)$, respectively, of arbitrary points on these surfaces. These vectors are calculated as indicated below:

$$\mathbf{u}(\psi,\nu) = \lambda_{max}\mathbf{S}^T(\psi)\mathbf{e}(\nu) \tag{4.37a}$$

$$\mathbf{b}(\psi,\nu) = \lambda_{min}\mathbf{S}^T(\psi)\mathbf{e}(\nu) \tag{4.37b}$$

Based on these results, the *solid model* of the cams synthesized in Example 4.2.1 can be obtained as shown in Figs. 4.10–4.12. ■

Planar RHRR Mechanisms

The ISAs of planar RHRR mechanisms are all parallel, as shown in Fig. 4.13; consequently, the primal terms of $\hat{\alpha}_i$ and $\hat{\theta}_i$, shown in Table 4.1, are all zero and the parameters k_i and h_i of eqs.(4.4) take on the values

$$k_1 = 1, \; k_2 = \cos\phi$$

$$k_3 = k_4 = k_7 = h_1 = h_2 = 0$$

$$k_5 = h_4 = \cos(\psi - \phi)$$

$$k_6 = h_3 = \sin(\psi - \phi)$$

Similarly, parameters $\tilde{k}_i$ and $\tilde{h}_i$ of eqs.(4.14), take on the values

$$\tilde{k}_1 = 1, \; \tilde{k}_2 = \cos\delta$$

$$\tilde{k}_3 = \tilde{k}_4 = \tilde{k}_7 = \tilde{h}_1 = \tilde{h}_2 = 0$$

$$\tilde{k}_5 = \tilde{h}_4 = \cos(\psi - \delta)$$

$$\tilde{k}_6 = \tilde{h}_3 = \sin(\psi - \delta)$$

Now, the pitch and contact surfaces are defined as

$$\mathbf{r}_P(\psi,\lambda) = \mathbf{p}_{43} + \lambda\mathbf{e}_{43} \tag{4.38}$$

$$\mathbf{r}_C(\psi,\lambda) = \mathbf{p}_{42} + \lambda\mathbf{e}_{42} \tag{4.39}$$

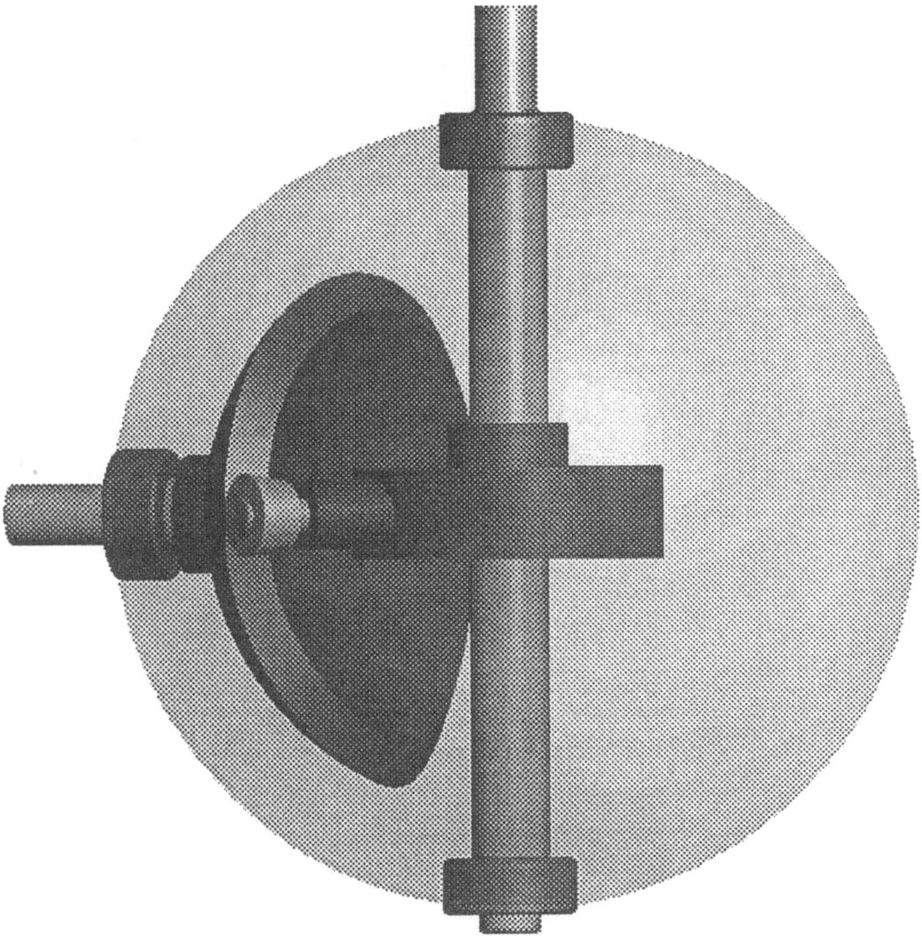

Figure 4.10 Spherical oscillating cam mechanism. Case 1 of Example 4.2.1.

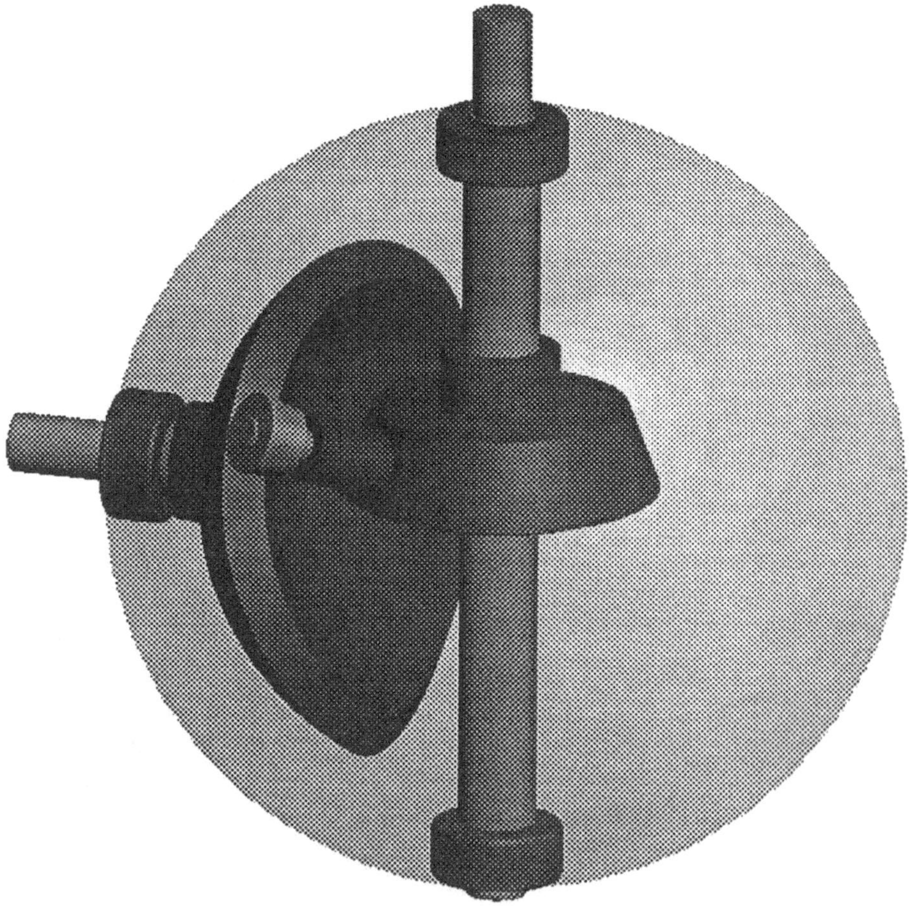

Figure 4.11 Spherical oscillating cam mechanism. Case 2 of Example 4.2.1.

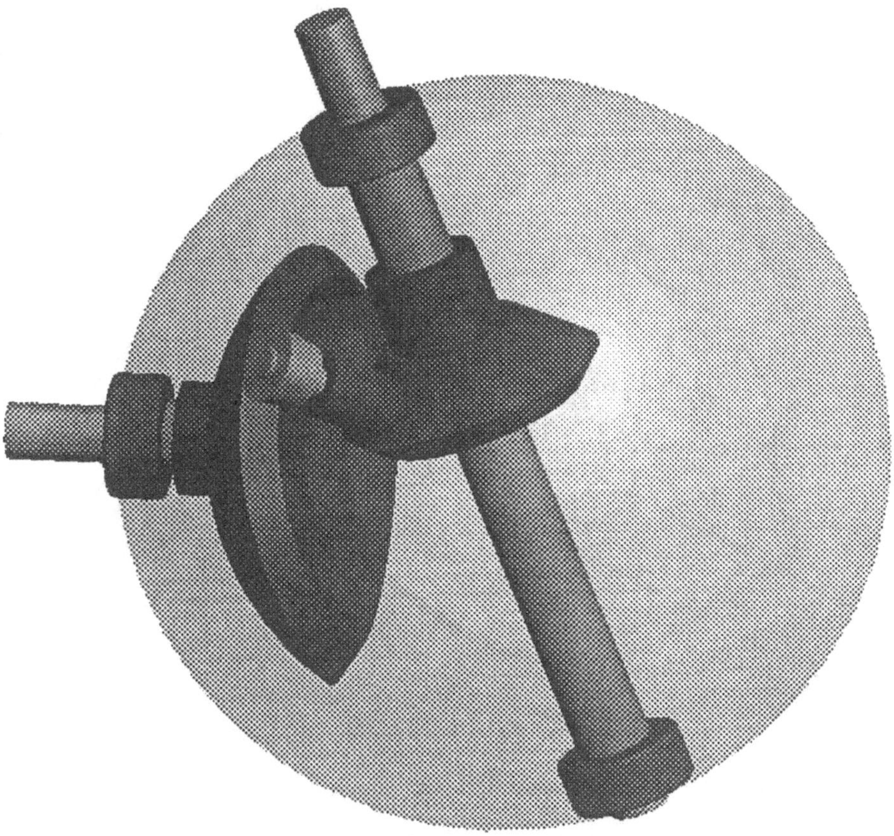

Figure 4.12 Spherical oscillating cam mechanism. Case 3 of Example 4.2.1.

with

$$\mathbf{e}_{43} = \mathbf{e}_{42} = [0, \quad 0, \quad 1]^T \tag{4.40}$$

and

$$\mathbf{p}_{43} = \begin{bmatrix} c\psi a_1 + c(\psi - \phi)a_3 \\ -s\psi a_1 - s(\psi - \phi)a_3 \\ 0 \end{bmatrix} \tag{4.41}$$

$$\mathbf{p}_{42} = \begin{bmatrix} c\psi b_2 + c(\psi - \delta)(b_3 - a_4) \\ -s\psi b_2 - s(\psi - \delta)(b_3 - a_4) \\ 0 \end{bmatrix} \tag{4.42}$$

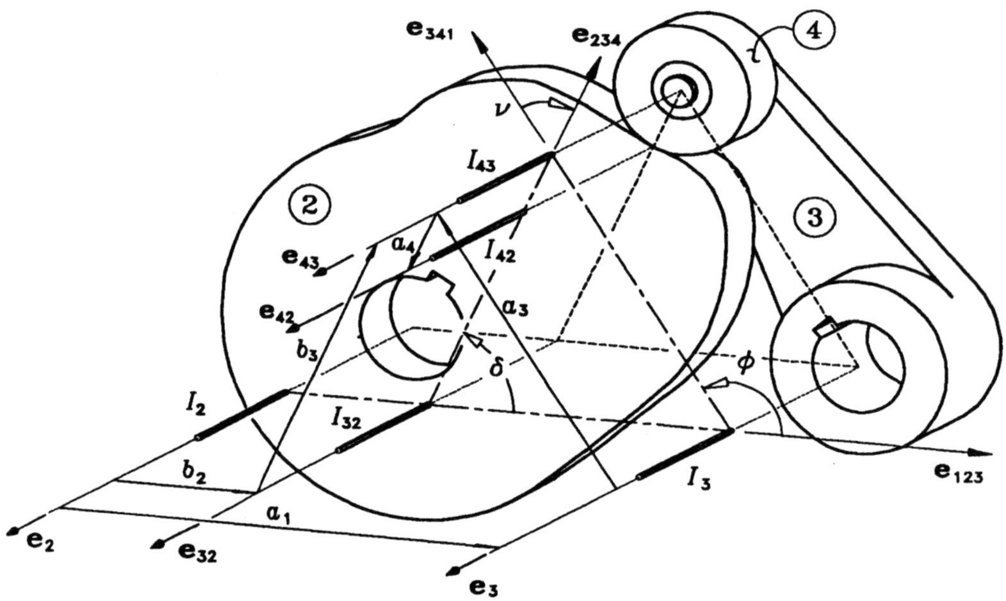

Figure 4.13 Planar RHRR cam mechanism.

Here, the special theory for parallel dual vectors introduced in Section 2.1 is applied to derive b_3 and δ. From eqs.(4.8),

$$[\hat{\mathbf{e}}_{32}]_1 \equiv [\mathbf{e}_{32}]_1 + \epsilon[\mathbf{m}_{32}]_1 = \begin{bmatrix} 0 \\ 0 \\ 1 \end{bmatrix} + \epsilon \begin{bmatrix} 0 \\ -b_2 \\ 0 \end{bmatrix} \tag{4.43a}$$

$$[\hat{\mathbf{e}}_{43}]_1 \equiv [\mathbf{e}_{43}]_1 + \epsilon[\mathbf{m}_{43}]_1 = \begin{bmatrix} 0 \\ 0 \\ 1 \end{bmatrix} + \epsilon \begin{bmatrix} a_3 s\phi \\ -a_3 c\phi - a_1 \\ 0 \end{bmatrix} \tag{4.43b}$$

Thus, the dual unit vector $\hat{\mathbf{e}}_{234}$ is defined by means of eq.(2.44), namely,

$$[\hat{\mathbf{e}}_{234}]_1 = \frac{[\mathbf{p}_{43}]_1 - [\mathbf{p}_{32}]_1}{b_3} + \epsilon\frac{[\mathbf{p}_{32}]_1 \times [\mathbf{p}_{43}]_1}{b_3} \tag{4.44}$$

where

$$[\mathbf{p}_{32}]_1 \equiv [\mathbf{e}_{32}]_1 \times [\mathbf{m}_{32}]_1 = \begin{bmatrix} b_2 \\ 0 \\ 0 \end{bmatrix} \tag{4.45a}$$

$$[\mathbf{p}_{43}]_1 \equiv [\mathbf{e}_{43}]_1 \times [\mathbf{m}_{43}]_1 = \begin{bmatrix} a_3 c\phi + a_1 \\ a_3 c\phi \\ 0 \end{bmatrix} \tag{4.45b}$$

and

$$b_3 \equiv \| [\mathbf{p}_{43}]_1 - [\mathbf{p}_{32}]_1 \| = \sqrt{(a_3 c\phi + a_1 - b_2)^2 + {a_3}^2 s^2\phi} \tag{4.46}$$

Combining eqs.(4.44–4.46) and substituting the dual vectors $\mathbf{p}_{32}$ and $\hat{\mathbf{e}}_{234}$ into eq.(4.10), one can readily obtain

$$\tan\delta = \frac{a_3 s\phi}{a_3 c\phi + a_1 - b_2} \tag{4.47}$$

A similar procedure can be followed to derive the dual unit vector $[\hat{\mathbf{e}}_{341}]_1$ and evaluate ν from eq.(4.16) as

$$\tan\nu = \frac{-(a_1 - b_2)s\phi}{a_3 + (a_1 - b_2)c\phi} \tag{4.48}$$

Furthermore, b_2 is obtained from eq.(3.16d), namely,

$$b_2 = \frac{\phi'}{\phi' - 1}a_1 \tag{4.49}$$

It is apparent that eqs.(4.47, 4.48 & 4.49) can be derived directly from eqs.(4.31, 4.32 & 3.16c), respectively, under the assumption that, for small arcs on the unit sphere, $\sin\alpha_i \rightarrow a_i$; $\cos\alpha_i \rightarrow 1$; $\sin\theta_i \rightarrow b_i$ and $\cos\theta_i \rightarrow 1$. We then conclude that planar mechanisms can be regarded as a special case of spherical mechanisms.

4.2.2 Revolute-Higher-Cylindric-Prismatic (RHCP) Mechanisms

Revolute-higher-cylindric-prismatic (RHCP) mechanisms have revolute and prismatic pairs as input and output, respectively, and an intermediate cylindrical pair, as shown in Fig. 4.14. Thus, $z_2 = \phi = 0$. The input-output function is given in eq.(3.19) as

$$z_3 = z_3(\psi) \tag{3.19}$$

Moreover, expanding eq.(4.21a) and knowing from eq.(3.21c) that $\theta_2 = 0$, one obtains

$$\omega_{42} s\alpha_4 s\nu = 0 \tag{4.50a}$$

$$\omega_{42}(z_{43} s\alpha_4 c\nu + a_4 c\alpha_4 s\nu) + v_{42} s\alpha_4 s\nu = -\omega_{32} z_3 s\beta \tag{4.50b}$$

Now, considering that $\omega_{42} \neq 0$ and $\sin\nu \neq 0$, from eq.(4.50a), one concludes that α_4 must be zero and, hence, from eq.(4.50b),

$$\omega_{42} = -\frac{z_3 \sin\alpha_1}{a_4 \sin\nu}\omega_{32} \tag{4.51}$$

Furthermore, expanding eq.(4.21b) with $\alpha_4 = 0$, one obtains from its primal part the relationship

$$0 = \omega_{32}\sin(\alpha_3 + \alpha_1) \tag{4.52}$$

Moreover, from eq.(3.21a), $\omega_{32} \neq 0$, and hence, $\alpha_3 = -\alpha_1$.

Now, substituting the corresponding values of α_3 and ϕ into eqs.(4.4), one obtains

$$k_1 = k_2 = 1$$

$$k_3 = k_4 = k_7 = h_1 = h_2 = 0$$

$$k_5 = h_4 = \cos\psi$$

$$k_6 = h_3 = \sin\psi$$

With these values, the vectors of eqs.(4.3a & 4.3c) take on the form

$$\mathbf{e}_{43} = \begin{bmatrix} 0 \\ 0 \\ 1 \end{bmatrix}, \qquad \mathbf{p}_{43} = \begin{bmatrix} (a_1 + a_3)\cos\psi - z_3 \sin\alpha_1 \sin\psi \\ -(a_1 + a_3)\sin\psi - z_3 \sin\alpha_1 \cos\psi \\ 0 \end{bmatrix} \tag{4.53}$$

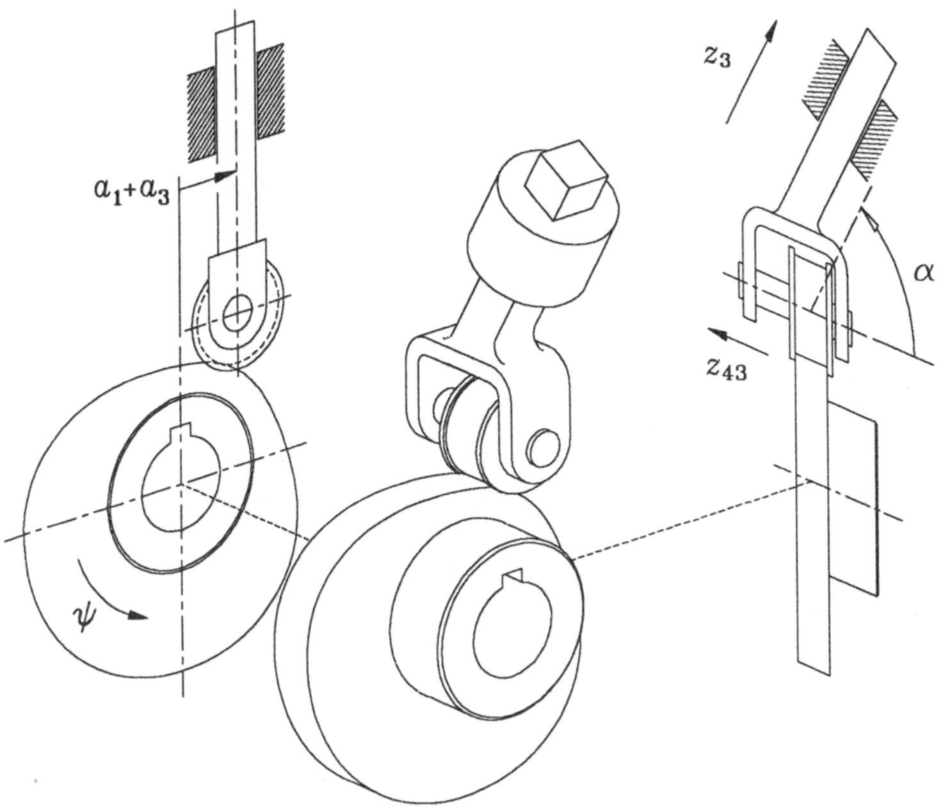

Figure 4.14 RHCP cam mechanism.

In order to derive the surface of the cam, the dual unit vectors $[\hat{\mathbf{e}}_{32}]_1$ and $[\hat{\mathbf{e}}_{43}]_1$ are computed from eqs.(4.8), namely,

$$[\hat{\mathbf{e}}_{32}]_1 = \begin{bmatrix} 0 \\ 0 \\ 1 \end{bmatrix} + \epsilon \begin{bmatrix} 0 \\ -b_2 \\ 0 \end{bmatrix} \tag{4.54a}$$

$$[\hat{\mathbf{e}}_{43}]_1 = \begin{bmatrix} 0 \\ 0 \\ 1 \end{bmatrix} + \epsilon \begin{bmatrix} -z_3 \sin\alpha_1 \\ -a_1 - a_3 \\ 0 \end{bmatrix} \tag{4.54b}$$

It is clear that $[\hat{\mathbf{e}}_{32}]_1$ and $[\hat{\mathbf{e}}_{43}]_1$ are parallel, and hence, the theory of parallel dual vectors introduced in Chapter 2.1 is applied to obtain the expressions below:

$$\hat{\theta}_3 \equiv \theta_3 + \epsilon b_3 = 0 + \epsilon\sqrt{(a_3 + a_1 + z_3' \sin\alpha_1)^2 + z_3^2 \sin^2\alpha_1} \tag{4.55a}$$

$$\hat{\delta} \equiv \delta + \epsilon z_{32} = \arctan \frac{-z_3 \sin \alpha_1}{a_1 + a_3 + z_3' \sin \alpha_1} + \epsilon 0 \tag{4.55b}$$

$$\hat{\nu} \equiv \nu + \epsilon z_{43} = \arctan \frac{z_3 \sin \alpha_1}{-(a_1 + a_3 + z_3' \sin \alpha_1)} + \epsilon(-z_3 \cos \alpha_1) \tag{4.55c}$$

Now, substituting $\theta_2 = 0$ and $\eta \equiv \theta_3 - \alpha_4 = 0$ into eqs.(4.14), one obtains

$$\tilde{k}_1 = 1, \quad \tilde{k}_2 = \cos \delta$$

$$\tilde{k}_3 = \tilde{k}_4 = \tilde{k}_7 = \tilde{h}_1 = \tilde{h}_2 = 0$$

$$\tilde{k}_5 = \tilde{h}_4 = \cos(\psi - \delta)$$

$$\tilde{k}_6 = \tilde{h}_3 = \sin(\psi - \delta)$$

Thus, vectors $\mathbf{e}_{42}$ and $\mathbf{p}_{42}$ of eqs.(4.13b & 4.13c) take on the forms

$$\mathbf{e}_{42} = \begin{bmatrix} 0 \\ 0 \\ 1 \end{bmatrix} \quad \text{and} \quad \mathbf{p}_{42} = \begin{bmatrix} b_2 \cos \psi + (b_3 - a_4) \cos(\delta - \psi) \\ -b_2 \sin \psi + (b_3 - a_4) \sin(\delta - \psi) \\ 0 \end{bmatrix} \tag{4.56}$$

RHRP Mechanisms

It was mentioned before that the condition of having a revolute pair between follower and roller is that $z_{43} = 0$. Moreover, from eq.(4.55c),

$$z_{43} = -z_3 \cos \alpha_1$$

Therefore, RHRP mechanisms can be obtained if $\alpha_1 = \pm\pi/2$.

Additionally, expanding and separating the primal and dual parts of eq.(4.21c), one obtains

$$\omega_{42} = \omega_{43} - \omega_{32} \tag{4.57a}$$

$$v_{42} = v_{43} - v_{32} \tag{4.57b}$$

Now, with $\alpha_1 = \pm\pi/2$, $v_{43} \equiv dz_{43}/dt = 0$ and, from eq.(3.21b), $v_{32} = 0$. Substitution of these two values into eq.(4.57b) reveals that v_{42} is zero and pure rolling between cam and roller is achieved.

4.2.3 Prismatic-Higher-Cylindric-Revolute (PHCR) Mechanisms

Shown in Fig. 4.15 is a ***prismatic-higher-cylindric-revolute*** (PHCR) mechanism, with prismatic and revolute pairs providing its input and output motions, respectively, and an intermediate cylindrical pair. Thus, $\psi = 0$, $z_3 = 0$, and the input-output function is given as in eq.(3.25), namely,

$$\phi = \phi(z_2) \tag{3.25}$$

Furthermore, it is known from eq.(3.28c) that $\theta_2 = \alpha_1$. Substituting this value into eq.(4.21a), one can readily obtain the primal and dual parts of the expression thus resulting, namely,

$$\omega_{42} s\alpha_4 s\nu = 0 \tag{4.58a}$$

$$\omega_{42}(z_{43} s\alpha_4 c\nu + a_4 s\nu c\alpha_4) + v_{42} s\alpha_4 s\nu = -\omega_{32}(a_1 - b_2)s\phi \tag{4.58b}$$

In general, $\omega_{42} \neq 0$ and $\nu \neq 0$, and hence, from eq.(4.58a), $\sin\alpha_4$ must be zero. Moreover, with $\sin\alpha_4 = 0$ the primal part of eq.(4.21b) takes on the form

$$0 = -\omega_{32}\sin\alpha_3 \tag{4.59}$$

Thus, from eq.(3.28a), $\omega_{32} = \pm\phi' v_2$, and one concludes that $\sin\alpha_3 = 0$.

Now, with $\psi = \sin\alpha_3 = 0$, eqs.(4.4) reduce to

$$k_1 = \cos\alpha_1$$

$$k_2 = h_4 = \cos\alpha_1\cos\phi$$

$$k_3 = h_2 = \sin\alpha_1$$

$$k_4 = \sin\alpha_1\cos\phi$$

$$k_5 = \cos\phi$$

$$k_6 = -\sin\phi\cos\alpha_1$$

$$k_7 = h_1 = 0$$

$$h_3 = -\sin\phi$$

while vectors $\mathbf{e}_{43}$ and $\mathbf{p}_{43}$ are obtained from eqs.(4.3a & 4.3c), namely,

$$\mathbf{e}_{43} = \begin{bmatrix} 0 \\ -\sin\alpha_1 \\ \cos\alpha_1 \end{bmatrix} \quad \text{and} \quad \mathbf{p}_{43} = \begin{bmatrix} a_1 + a_3\cos\phi \\ \cos\alpha_1(a_3\sin\phi - z_2\sin\alpha_1) \\ \sin\alpha_1(a_3\sin\phi - z_2\sin\alpha_1) \end{bmatrix} \tag{4.60}$$

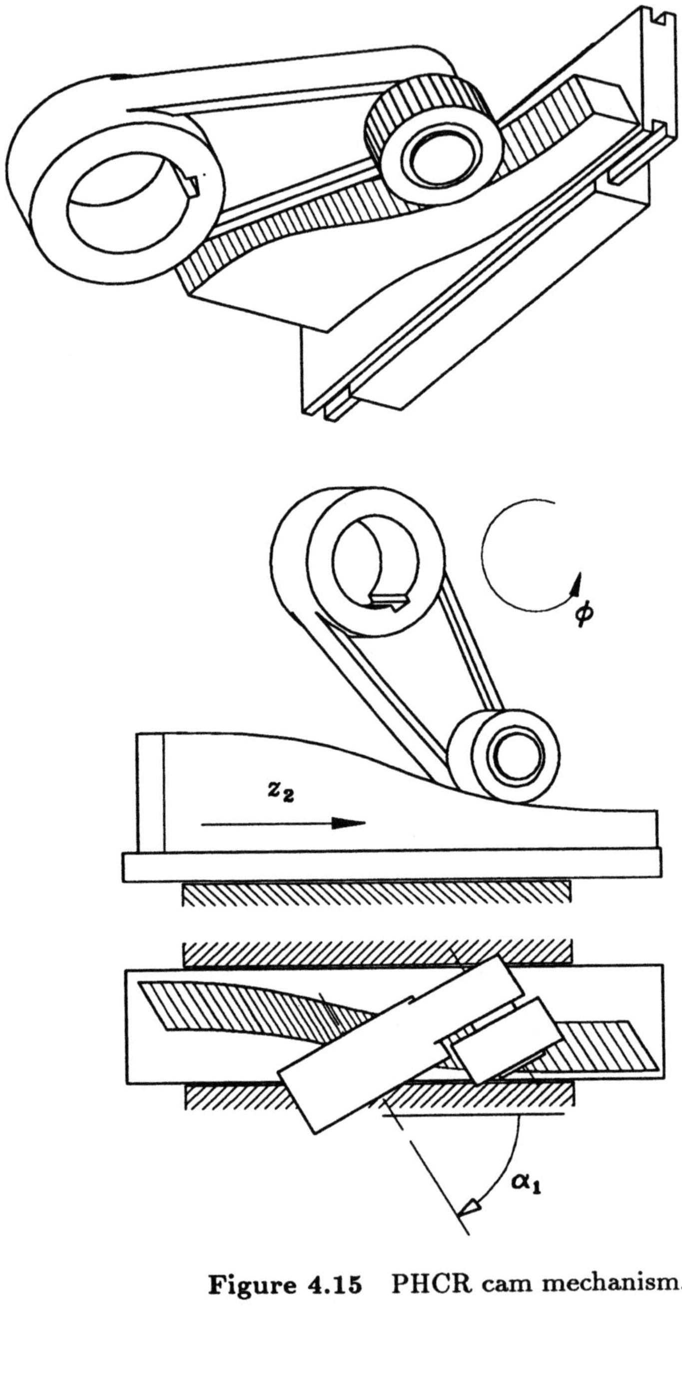

Figure 4.15 PHCR cam mechanism.

Next, the dual vectors $[\hat{\mathbf{e}}_{32}]_1$ and $[\hat{\mathbf{e}}_{43}]_1$ are computed from eqs.(4.8) as

$$[\hat{\mathbf{e}}_{32}]_1 = \begin{bmatrix} 0 \\ -\sin\alpha_1 \\ \cos\alpha_1 \end{bmatrix} + \epsilon b_2 \begin{bmatrix} 0 \\ -\cos\alpha_1 \\ \sin\alpha_1 \end{bmatrix} \tag{4.61a}$$

$$[\hat{\mathbf{e}}_{43}]_1 = \begin{bmatrix} 0 \\ -\sin\alpha_1 \\ \cos\alpha_1 \end{bmatrix} + \epsilon \begin{bmatrix} a_3\sin\phi \\ -(a_1 + a_3\cos\phi)\cos\alpha_1 \\ -(a_1 + a_3\cos\phi)\sin\alpha_1 \end{bmatrix} \tag{4.61b}$$

With application of the theory of parallel dual vectors of Section 2.1, one can obtain, from eqs.(4.7, 4.10 & 4.16), the expressions below:

$$\hat{\theta}_3 \equiv \theta_3 + \epsilon b_3 = 0 + \epsilon \frac{\sqrt{(a_3\phi'\cos\phi - \sin\alpha_1)^2 + (a_3\phi'\sin\phi)^2}}{\phi'} \tag{4.62a}$$

$$\hat{\delta} \equiv \delta + \epsilon z_{32} = \arctan\frac{a_3\phi'\sin\phi}{a_3\phi'\cos\phi - \sin\alpha_1} + \epsilon 0 \tag{4.62b}$$

$$\hat{\nu} \equiv \nu + \epsilon z_{43} = \arctan\frac{\sin\alpha_1\sin\phi}{a_3\phi' + \sin\alpha_1\cos\phi} + \epsilon 0 \tag{4.62c}$$

According to eq.(4.62b), $z_{43} = 0$, which means that the follower-roller pair is a revolute; therefore, a PHCR mechanism is, in fact, a PHRR mechanism.

In order to find expressions for $\mathbf{e}_{42}$ and $\mathbf{p}_{42}$, values of $\theta_2 = \alpha_1$ and $\eta \equiv \theta_3 - \alpha_4 = 0$ are substituted into eqs.(4.14) to obtain

$$\tilde{k}_1 = \cos\alpha_1$$
$$\tilde{k}_2 = \tilde{h}_4 = \cos\alpha_1\cos\delta$$
$$\tilde{k}_3 = \tilde{h}_2 = \sin\alpha_1$$
$$\tilde{k}_4 = \sin\alpha_1\cos\delta$$
$$\tilde{k}_5 = \cos\delta$$
$$\tilde{k}_6 = -\sin\delta\cos\alpha_1$$
$$\tilde{k}_7 = \tilde{h}_1 = 0$$
$$\tilde{h}_3 = -\sin\delta$$

Thus, substituting these values into eqs.(4.13b & 4.13c),

$$\mathbf{e}_{42} = \begin{bmatrix} 0 \\ -\sin\alpha_1 \\ \cos\alpha_1 \end{bmatrix} \quad \text{and} \quad \mathbf{p}_{42} = \begin{bmatrix} a_1 + \sin\alpha_1/\phi' + (b_3 - a_4)\cos\delta \\ [(b_3 - a_4)\sin\delta - z_2\sin\alpha_1]\cos\alpha_1 \\ [(b_3 - a_4)\sin\delta - z_2\sin\alpha_1]\sin\alpha_1 \end{bmatrix} \tag{4.63}$$

Further analysis of eq.(4.21c) leads to the relations

$$\omega_{42} = \omega_{43} - \omega_{32} \tag{4.64a}$$

$$v_{42} = v_{32} \tag{4.64b}$$

Now, from eq.(3.28b) $v_{32} = \pm v_2 \cos \alpha_1$, and hence, pure rolling between cam and roller is achieved if $\alpha_1 = \pm\pi/2$.

According to the results pertaining to RHCP and PHRR mechanisms, similar theorems to those for RHP and PHR mechanisms can be stated, namely,

Theorem 4.1: *All ISAs of RHCP and PHRR mechanisms, except for the one associated with the prismatic pair, are parallel.*

Theorem 4.2: *Pure rolling between cam and roller on RHCP and PHRR mechanisms is achieved if the axis of the prismatic pair is perpendicular to all other ISAs.*

4.2.4 Prismatic-Higher-Cylindric-Prismatic (PHCP) Mechanisms

A *prismatic-higher-cylindric-prismatic* (PHCP) mechanism is defined with prismatic pairs in both its input and output motions, and an intermediate cylindrical pair. Thus, $\psi = 0$ and ϕ is constant, and the input-output function is given as in eq.(3.30), namely,

$$z_3 = z_3(z_2) \tag{3.30}$$

Furthermore, it is known from Subsection 3.2.4 that $a_1 = b_2 = 0$. Upon substitution of this value into eq.(4.21a), one can readily obtain the primal and dual parts of this equation as

$$\omega_{42} s\alpha_4 s\nu = 0 \tag{4.65a}$$

$$\omega_{42}(z_{43} s\alpha_4 c\nu + a_4 s\nu c\alpha_4) + v_{42} s\alpha_4 s\nu = -v_{32} s\beta s\phi \tag{4.65b}$$

In general, $\omega_{42} \neq 0$ and $\nu \neq 0$; therefore, from eq.(4.65a), $\sin \alpha_4$ must be zero. Moreover, with $\sin \alpha_4 = 0$, the primal and dual parts of eq.(4.21c) take on the form

$$\omega_{42} = \omega_{43} \tag{4.66a}$$

$$v_{42} = v_{43} - v_{32}(c\beta c\alpha_3 - s\beta s\alpha_3 c\phi) \tag{4.66b}$$

The desirable PHCP mechanism is one with pure rolling between cam and roller. Thus, v_{42} must be zero, and from eq.(4.66b), v_{43} is given as

$$v_{43} = v_{32}(c\beta c\alpha_3 - s\beta s\alpha_3 c\phi) \tag{4.67}$$

Pitch and contact surfaces are derived with application of eqs.(4.3 & 4.13).

PHRP mechanisms

The cylindrical pair becomes a revolute pair if v_{43} vanishes. Thus, from eq.(4.67), the coefficient of v_{32} must be zero, i.e.,

$$c\beta c\alpha_3 - s\beta s\alpha_3 c\phi = 0 \tag{4.68}$$

Since β is a variable and α_3 and ϕ are constants, the only solution of eq.(4.68) is that in which each term vanishes independently, i.e., $\alpha_3 = \pm\pi/2$ and $\phi = \pm\pi/2$.

Now, with $\psi = 0$, $\alpha_3 = \pi/2$ and $\phi = \pi/2$, eqs.(4.4) reduce to

$$k_1 = k_3 = h_2 = h_3 = 0$$

$$k_2 = -k_7 = h_4 = -\sin\alpha_1$$

$$k_4 = -k_6 = \cos\alpha_1$$

and vectors $\mathbf{e}_{43}$ and $\mathbf{p}_{43}$ are obtained from eqs.(4.3a & 4.3c), namely,

$$\mathbf{e}_{43} = \begin{bmatrix} 1 \\ 0 \\ 0 \end{bmatrix} \quad \text{and} \quad \mathbf{p}_{43} = \begin{bmatrix} 0 \\ a_3\cos\alpha_1 - z_3\sin\alpha_1 \\ a_3\sin\alpha_1 + z_3\cos\alpha_1 - z_2 \end{bmatrix} \tag{4.69}$$

Moreover, the dual vectors $[\hat{\mathbf{e}}_{32}]_1$ and $[\hat{\mathbf{e}}_{43}]_1$ are computed from eqs.(4.8), namely,

$$[\hat{\mathbf{e}}_{32}]_1 = \begin{bmatrix} 0 \\ -\sin\theta_2 \\ \cos\theta_2 \end{bmatrix} + \epsilon 0 \tag{4.70a}$$

$$[\hat{\mathbf{e}}_{43}]_1 = \begin{bmatrix} 1 \\ 0 \\ 0 \end{bmatrix} + \epsilon \begin{bmatrix} 0 \\ z_3\cos\alpha_1 + a_3\sin\alpha_1 \\ z_3\sin\alpha_1 - a_3\cos\alpha_1 \end{bmatrix} \tag{4.70b}$$

Now, from eqs.(4.7, 4.10 & 4.16), one can readily obtain the expressions below:

$$\hat{\theta}_3 \equiv \theta_3 + \epsilon b_3 = \pi/2 + \epsilon[a_3\cos(\alpha_1 - \theta_2) - z_3\sin(\alpha_1 - \theta_2)] \tag{4.71a}$$

$$\hat{\delta} \equiv \delta + \epsilon z_{32} = \pi/2 + \epsilon[a_3\sin(\alpha_1 - \theta_2) + z_3\cos(\alpha_1 - \theta_2)] \tag{4.71b}$$

$$\hat{\nu} \equiv \nu + \epsilon z_{43} = \arctan\frac{-\sin(\alpha_1 - \theta_2)}{\cos(\alpha_1 - \theta_2)} + \epsilon 0 \tag{4.71c}$$

Moreover, combining eqs.(3.35b & 4.71c),

$$\tan\nu = \frac{\sin\alpha_1}{z_3' - \cos\alpha_1} \tag{4.71d}$$

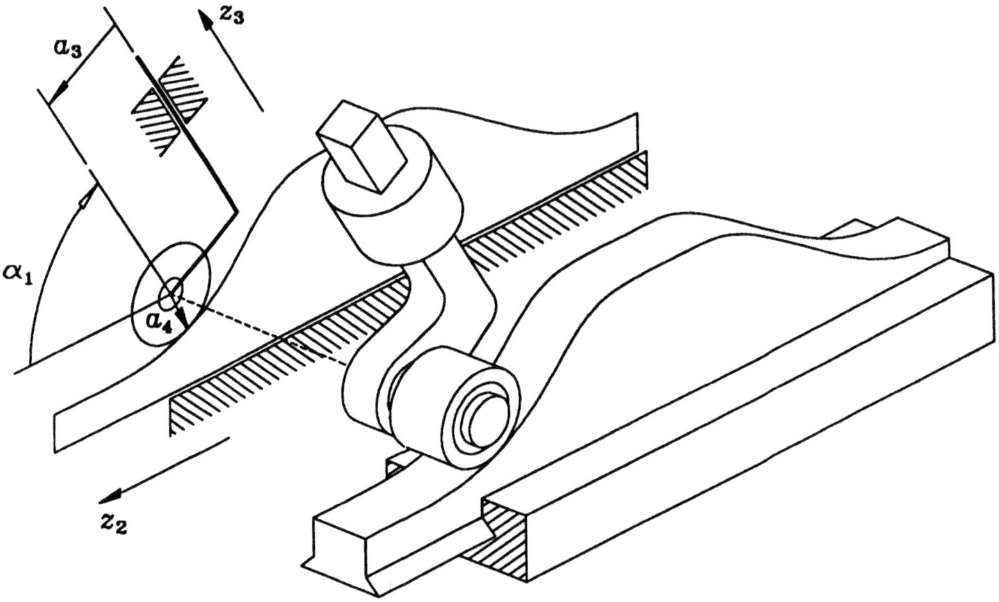

Figure 4.16 PHRP cam mechanism.

In order to find expressions for $\mathbf{e}_{42}$ and $\mathbf{p}_{42}$, values $\delta_2 = \pi/2$, $\eta \equiv \theta_3 - \alpha_4 = \pi/2$ and $\psi = 0$ are substituted into eqs.(4.14) to obtain

$$\tilde{k}_1 = \tilde{k}_3 = \tilde{h}_2 = \tilde{h}_3 = 0$$

$$\tilde{k}_2 = -\tilde{k}_7 = \tilde{h}_4 = -\sin\theta_2$$

$$\tilde{k}_4 = -\tilde{k}_6 = \cos\theta_2 \cos\delta$$

Thus, substituting these values into eqs.(4.13b & 4.13c), one obtains

$$\mathbf{e}_{42} = \begin{bmatrix} 1 \\ 0 \\ 0 \end{bmatrix} \quad \text{and} \quad \mathbf{p}_{42} = \begin{bmatrix} 0 \\ (b_3 - a_4)\cos\theta_2 - z_{32}\sin\theta_2 \\ (b_3 - a_4)\sin\theta_2 + z_{32}\cos\theta_2 - z_2 \end{bmatrix} \tag{4.72}$$

4.3 Pressure Angle

The pressure angle is defined as that comprised between the direction of the unit normal to the pitch surface $\mathcal{R}_p$ and the direction of the velocity of the follower at the contact point. Thus, the unit normal defined in eq. (2.57) is written as

$$\mathbf{n} = \frac{\mathbf{u}}{\| \mathbf{u} \|}; \qquad \mathbf{u} \equiv \frac{\partial \mathbf{r}_p}{\partial \psi} \times \frac{\partial \mathbf{r}_p}{\partial \lambda} \tag{4.73}$$

On the other hand, the unit vector $\mathbf{w}$ parallel to the follower velocity at points on I_{43} is obtained as

$$\mathbf{w} = \frac{\boldsymbol{\xi}}{\| \boldsymbol{\xi} \|}; \qquad \boldsymbol{\xi} \equiv \boldsymbol{\omega} \times \mathbf{r}_p \tag{4.74}$$

where $\boldsymbol{\omega}$ is the angular velocity of the follower.

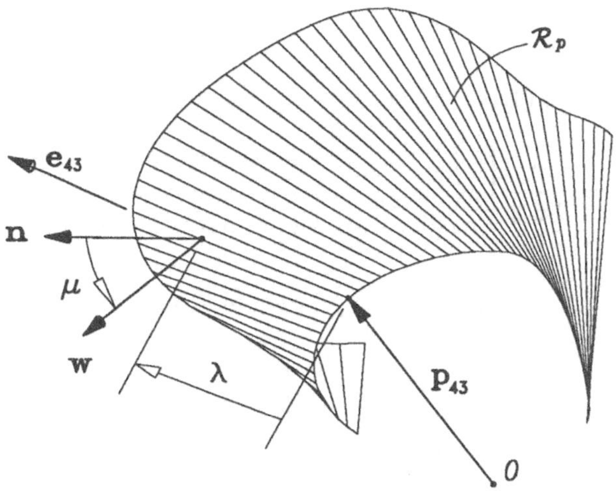

Figure 4.17 Graphical representation of the pressure angle.

Thus, the pressure angle is derived as

$$\tan \mu = \frac{\| \mathbf{n} \times \mathbf{w} \|}{\mathbf{n} \cdot \mathbf{w}} \tag{4.75}$$

A graphical representation of this definition is shown in Figure 4.17.

4.3.1 Pressure Angle of RHCR Mechanisms

The unit normal of the contact surfaces of RHCR mechanisms is computed from application of eq.(2.57), with the aid of eq.(4.24). Thus, the three components of $\mathbf{u}$

of eq.(4.73) in frame $\mathcal{F}_1$ are given as

$$u_1 = a_1 k_1 - a_3 \cos\phi h_5 - \lambda \sin\alpha_3 \sin\phi h_5$$

$$u_2 = -a_3 \cos\alpha_1 \sin\phi h_5 + \lambda[\cos\alpha_1 \cos\phi \sin\alpha_3 h_5 + \sin\alpha_1(\phi' + \cos\alpha_3 h_5)]$$

$$u_3 = \sin\phi(a_1 \sin\alpha_3 - a_3 \sin\alpha_1 h_5) + \lambda(1 - k_1 h_5 - \phi' \cos\alpha_1)$$

and

$$h_5 = k_1 - \phi' \cos\alpha_3 \tag{4.76a}$$

$$h_6 = k_4 - \phi' \sin\alpha_3 \tag{4.76b}$$

Furthermore, from eq.(4.74), $\mathbf{w}$ can be expressed as

$$\mathbf{w} = \frac{1}{\|\boldsymbol{\xi}\|} \begin{bmatrix} -a_3 \sin\phi + \lambda \sin\alpha_3 \cos\phi \\ a_3 \cos\alpha_1 \cos\phi + \lambda \sin\alpha_3 \cos\alpha_1 \sin\phi \\ a_3 \sin\alpha_1 \cos\phi + \lambda \sin\alpha_1 \sin\phi \sin\alpha_3 \end{bmatrix} \tag{4.77}$$

Moreover, the cross product appearing in eq. (4.75) is computed as

$$\mathbf{n} \times \mathbf{w} = \frac{\lambda^2 \mathbf{d} + \lambda \mathbf{f} + \mathbf{g}}{\|\boldsymbol{\xi}\| \, \|\mathbf{u}\|} \tag{4.78}$$

where the components of vectors $\mathbf{d}$, $\mathbf{f}$ and $\mathbf{g}$, in frame $\mathcal{F}_1$, are given below:

$$\begin{aligned}
d_1 &= -\sin^2\alpha_3 \sin\phi h_6 \\
d_2 &= \sin\alpha_3[\sin\alpha_1 \sin\alpha_3 h_5 + \cos\phi(\cos\alpha_1 \sin\alpha_3 h_6 + \sin^2\alpha_1)] \\
d_3 &= \sin\alpha_3[\sin\alpha_3(\cos\phi \sin\alpha_1 h_6 - \cos\alpha_1 h_5) - \cos\alpha_1 \cos\phi \sin\alpha_1] \\
f_1 &= -\sin\alpha_3(a_1 \cos\alpha_1 \sin\alpha_3 \sin^2\phi + a_3 \cos\phi h_6) \\
f_2 &= \sin\phi[a_1 \cos\alpha_1 \sin\alpha_3 k_3 - a_3(1 - \phi' \cos\alpha_1 - \cos\alpha_1 \cos\alpha_3 h_5)] \\
f_3 &= -\sin\phi[a_1 \cos\alpha_1 \sin\alpha_3 k_1 - a_3(\phi' \sin\alpha_1 + \sin\alpha_1 \cos\alpha_3 h_5)] \\
g_1 &= -a_1 a_3 \cos\alpha_1 \cos\phi \sin\alpha_3 \sin\phi \\
g_2 &= a_3[-a_1(\sin\alpha_3 - \cos\phi \cos\alpha_1 k_3) + a_3 \sin\alpha_1 h_5] \\
g_3 &= a_3 \cos\alpha_1(-a_1 \cos\phi k_1 - a_3 h_5)
\end{aligned} \tag{4.79}$$

Then, $\| \mathbf{n} \times \mathbf{w} \|$ can be written as

$$\| \mathbf{n} \times \mathbf{w} \| = \frac{\sqrt{A\lambda^4 + B\lambda^3 + C\lambda^2 + D\lambda + E}}{\| \boldsymbol{\xi} \| \| \mathbf{u} \|} \tag{4.80}$$

where

$$\begin{aligned} A &= \|\mathbf{d}\|^2 \\ B &= 2\mathbf{d} \cdot \mathbf{f} \\ C &= \|\mathbf{f}\|^2 + 2\mathbf{d} \cdot \mathbf{g} \\ D &= 2\mathbf{f} \cdot \mathbf{g} \\ E &= \|\mathbf{g}\|^2 \end{aligned} \tag{4.81}$$

Furthermore, the denominator of the right-hand side of eq. (4.75) can be expressed as

$$\mathbf{n} \cdot \mathbf{w} = \frac{F\lambda^2 + G\lambda + H}{\| \boldsymbol{\xi} \| \| \mathbf{u} \|} \tag{4.82}$$

where

$$\begin{aligned} F &= \sin\alpha_1 \sin\alpha_3 \sin\phi \\ G &= a_1 \sin\alpha_3(\sin\alpha_1 \sin\alpha_3 - \cos\alpha_1 \cos\alpha_3 \cos\phi) + a_3 \cos\phi \sin\alpha_1 \\ H &= a_1 a_3 \cos\alpha_1 \cos\alpha_3 \sin\phi \end{aligned} \tag{4.83}$$

Thus, the general expression for the pressure angle is given by

$$\tan\mu = \frac{\sqrt{A\lambda^4 + B\lambda^3 + C\lambda^2 + D\lambda + E}}{F\lambda^2 + G\lambda + H} \tag{4.84}$$

Spherical Cam Mechanisms

The pressure angle for spherical cams is derived from eq. (4.78) by taking the limit of $\tan\mu$ as $\lambda \to \infty$ (González-Palacios and Angeles, 1991*a*), i.e.,

$$\tan\mu = \lim_{\lambda\to\infty} \frac{\sqrt{A\lambda^4 + B\lambda^3 + C\lambda^2 + D\lambda + E}}{F\lambda^2 + G\lambda + H} = \frac{\sqrt{A}}{F} \tag{4.85}$$

Thus, from eqs. (4.79, 4.81 & 4.83), one can readily obtain the desired general expression for spherical cam mechanisms, namely,

$$\tan\mu = \frac{(\phi' - \cos\alpha_1)\sin\alpha_3 - \sin\alpha_1 \cos\alpha_3 \cos\phi}{\sin\alpha_1 \sin\phi} \tag{4.86}$$

Comparing eq.(4.86) with eq.(4.32) one can find a similitude. In fact, one can eliminate θ_2 from eq.(4.32) with the aid of eq.(3.16c). Thus, upon reduction, eq.(4.32) takes on the form

$$\tan\nu = \frac{-(\sin\alpha_1 - \cos\alpha_1 \tan\theta_2)\sin\phi}{(\cos\alpha_1 - \sin\alpha_1\tan\theta_2)\sin\alpha_3 + (\sin\alpha_1 - \cos\alpha_1\tan\theta_2)\cos\alpha_3\cos\phi} \tag{4.87}$$

which, with the aid of eq.(3.16c), becomes

$$\tan\nu = \frac{\sin\alpha_1\sin\phi}{(\phi' - 1)\sin\alpha_3 - \sin\alpha_1\cos\alpha_3\cos\phi} \tag{4.88}$$

Consequently,

$$\tan\mu = \frac{1}{\tan\nu} \tag{4.89}$$

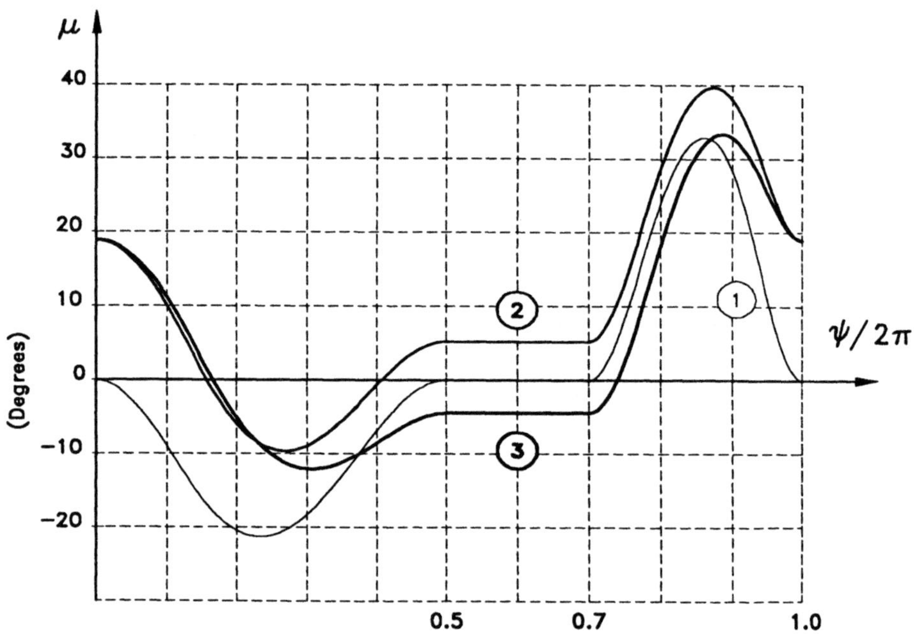

Figure 4.18 Pressure angle distribution for the mechanisms of Example 4.3.1.

Example 4.3.1: *Plot the pressure angle distribution for each of the cases of Example 4.2.1.1.*

Solution:

With the aid of eqs.(4.33), we obtain the plots shown in Fig. 4.18 upon substitution of the corresponding values of α_1 and α_3 into eq.(4.86). Furthermore, comparing the three examples, we conclude that the third one reports better results than the others, since the cam size is smallest, and the pressure angle remains within smaller bounds. It is then apparent that, in an optimization procedure, α_1 and α_3 can be regarded as design variables. ■

Planar Cam Mechanisms

The pressure angle for planar cam mechanisms is obtained from eq. (4.84) as indicated below. Since all the ISAs are parallel, $\alpha_1 = \alpha_3 = 0$. Moreover, variables k_1, k_3, h_5 and h_6, appearing in eqs. (4.4a), (4.4c), (4.76a) and (4.76b), respectively, take on the values

$$k_1 = 1$$

$$k_3 = 0$$

$$h_5 = 1 - \phi'$$

$$h_6 = 0$$

Furthermore, all components appearing in eq. (4.79) become zero, except for g_3, which reduces to

$$g_3 = a_3[a_3(\phi' - 1) - a_1 \cos\phi] \tag{4.90}$$

Consequently,

$$A = B = C = D = 0$$

$$E = {a_3}^2[a_3(\phi' - 1) - a_1 \cos\phi]^2$$

$$F = G = 0$$

$$H = a_1 a_3 \sin\phi$$

and the expression for the pressure angle takes on the form

$$\tan\mu = \frac{a_3(\phi' - 1) - a_1 \cos\phi}{a_1 \sin\phi} \tag{4.91}$$

in agreement with results available in the literature (Rothbart, 1956; Tesar, 1976; Angeles and López-Cajún, 1991).

Here, it can also be proven that $\tan\mu = 1/\tan\nu$ if eq.(4.49) is substituted into eq.(4.48).

Spatial Cam Mechanisms

We have shown that eq. (4.84) represents the pressure angle for all planar, spherical and spatial cam mechanisms. In fact, planar and spherical cam mechanisms are particular cases in which the expressions for the pressure angle are independent of λ. Here we present two more cases as examples of spatial cam mechanisms, namely, *cylindrical* and *globoidal* cam mechanisms in order to show the influence of λ on the pressure angle.

The pressure angle of cylindrical cam mechanisms is derived by taking into account that $\alpha_1 = \pi/2$ and $\alpha_3 = 0$. Consequently, the expression for the pressure angle can be readily obtained from eq. (4.84) as

$$\tan\mu = \frac{\lambda\sin\phi + a_3\phi'}{\lambda\cos\phi} \tag{4.92}$$

On the other hand, $\alpha_1 = \alpha_3 = \pi/2$ and $a_3 = 0$ for globoidal cam mechanisms, eq. (4.84) thus reducing to

$$\tan\mu = \frac{\lambda\phi'}{a_1 + \lambda\sin\phi} \tag{4.93}$$

4.3.2 Pressure Angle of RHCP Mechanisms

The unit normal of the cam surface of RHCP mechanisms is computed from application of eq.(2.57) and with the aid of eqs.(4.53). Thus, $\mathbf{e}'_{43} = \mathbf{0}$ and $\mathbf{p}'_{43}$ is expressed as

$$\mathbf{p}'_{43} = \begin{bmatrix} -(a_1+a_3)\sin\psi - z_3\sin\alpha_1\cos\psi - z'_3\sin\alpha_1\cos\psi \\ -(a_1+a_3)\cos\psi + z_3\sin\alpha_1\sin\psi - z'_3\sin\alpha_1\cos\psi \\ 0 \end{bmatrix} \tag{4.94}$$

where the prime denotes differentiation with respect to ψ.

Now, $\mathbf{n}$ is given as in eq.(4.73), with $\mathbf{u}$ defined as

$$\mathbf{u} \equiv [\,-(a_1+a_3) - z'_3\sin\alpha_1, \quad z_3\sin\alpha_1, \quad 0\,]^T \tag{4.95}$$

Moreover, the unit vector in the direction of the velocity of the follower on I_{43} is given as

$$\mathbf{w} = [\,0, \quad 1, \quad 0\,]^T$$

From eq.(4.75), one can readily obtain

$$\tan\mu = -\frac{(a_1 + a_3 + z'_3)}{z_3\sin\alpha_1} \tag{4.96}$$

4.3.3 Pressure Angle of PHRR Mechanisms

The unit normal of the contact surface of PHRR mechanisms is computed from application of eq.(2.57), with the aid of eqs.(4.60). Thus, $\mathbf{e}'_{43} = \mathbf{0}$ and $\mathbf{p}'_{43}$ is expressed as

$$\mathbf{p}'_{43} = \begin{bmatrix} -\phi' a_3 \sin\phi \\ \cos\alpha_1(a_3\phi'\cos\phi - \sin\alpha_1) \\ \sin\alpha_1(a_3\phi'\cos\phi - \sin\alpha_1) \end{bmatrix} \tag{4.97}$$

where the prime denotes differentiation with respect to z_2. Now, $\mathbf{n}$ is given as in eq.(4.73) with $\mathbf{u}$ defined as

$$\mathbf{u} \equiv \begin{bmatrix} a_3\phi'\cos\phi - \sin\alpha_1 \\ a_3\phi'\sin\phi\cos\alpha_1 \\ a_3\phi'\sin\phi\sin\alpha_1 \end{bmatrix} \tag{4.98}$$

Moreover, the unit vector in the direction of the velocity of the follower on I_{43} is given as

$$\mathbf{w} = \begin{bmatrix} -\sin\phi \\ \cos\alpha_1\cos\phi \\ \sin\alpha_1\cos\phi \end{bmatrix}$$

From eq.(4.75) one obtains

$$\tan\mu = \frac{a_3\phi' - \sin\alpha_1\cos\phi}{\sin\phi\sin\alpha_1} \tag{4.99}$$

4.3.4 Pressure Angle of PHRP Mechanisms

The unit normal of the cam surface of PHRP mechanisms is computed from application of eq.(2.57), with the aid of eqs.(4.69). Thus, $\mathbf{e}'_{43} = \mathbf{0}$ and $\mathbf{p}'_{43}$ is expressed as

$$\mathbf{p}'_{43} = \begin{bmatrix} 0 \\ -z'_3\sin\alpha_1 \\ z'_3\cos\alpha_1 - 1 \end{bmatrix} \tag{4.100}$$

where the prime denotes differentiation with respect to z_2.

Now, $\mathbf{n}$ is given, again, as in eq.(4.73), with $\mathbf{u}$ defined as

$$\mathbf{u} \equiv \begin{bmatrix} 0 \\ z'_3\cos\alpha_1 - 1 \\ z'_3\sin\alpha_1 \end{bmatrix} \tag{4.101}$$

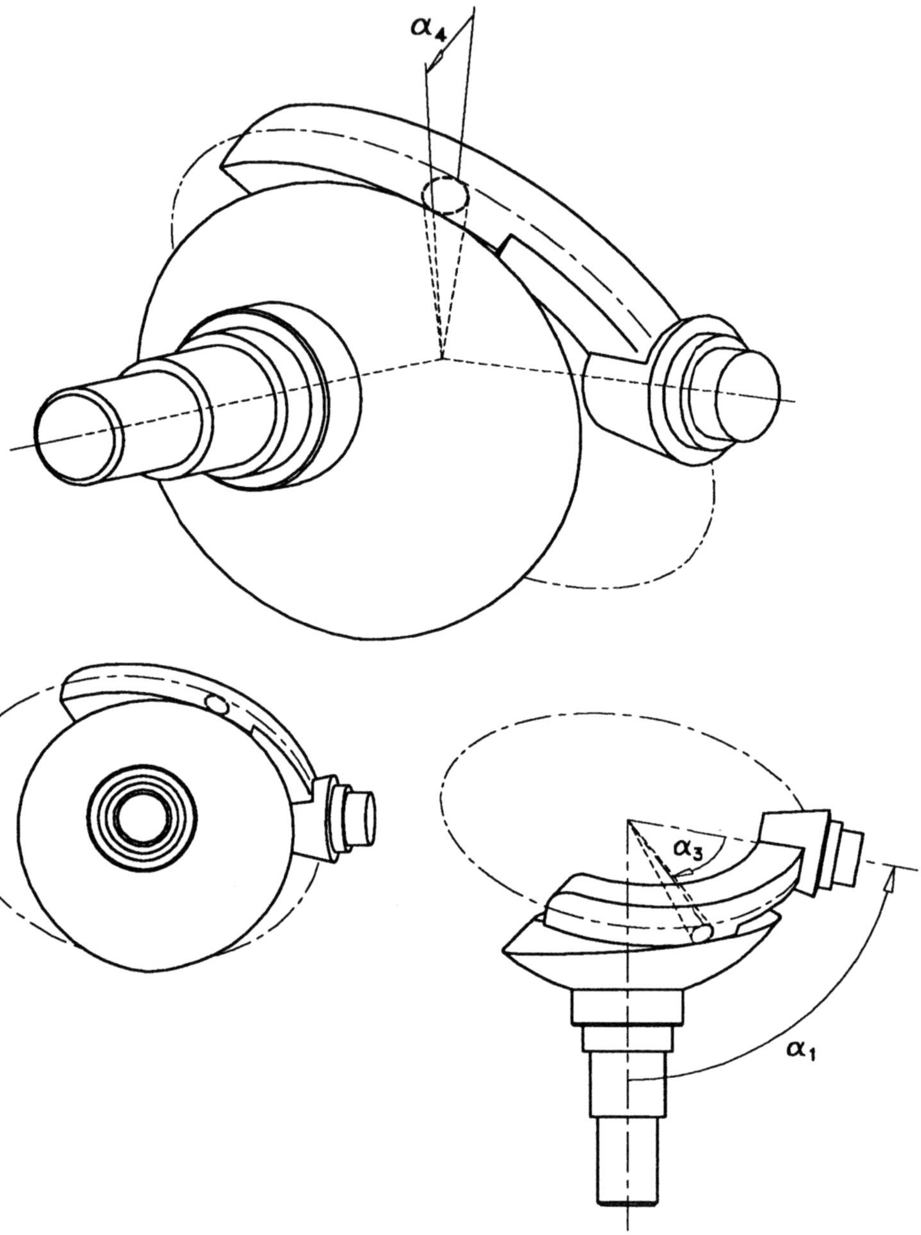

Figure 4.19 Spherical R$\overline{\text{HH}}$R cam mechanism.

Furthermore, the unit vector in the direction of the velocity of the follower on I_{43} is given as

$$\mathbf{w} = \begin{bmatrix} 0 \\ -\sin\alpha_1 \\ \cos\alpha_1 \end{bmatrix}$$

From eq.(4.75), one obtains

$$\tan\mu = \frac{z_3' - \cos\alpha_1}{\sin\alpha_1} \tag{4.102}$$

4.4 Three-Link Mechanisms with Constant Pressure Angle

Cam mechanisms with a *constant pressure angle* are attractive because they involve rather simple follower shapes. Planar cam mechanisms with a flat-face follower are good examples of this kind, their followers containing planar surfaces at the cam-follower interface. Here we study the synthesis of these mechanisms and their spherical counterparts. However, when dealing with the latter, one cannot speak in general of flat-face but rather of *conical-face followers*, as we will show presently.

The approach introduced here for the synthesis of the mechanisms under study is based on an auxiliary roller in contact with the cam and the follower, which gives rise to a four-link mechanism, although the mechanism under study is, in fact of the three-link type. The lines of contact of the roller with the cam and the follower are, respectively, the ISAs I_{42} and I_{43}, and hence, two higher pairs arise at the cam-follower interface. A convention is adopted here to distinguish the aforementioned four-link mechanisms from the three-link mechanisms analyzed in Chapter 2, namely, the two higher pairs of the roller are indicated as $\overline{HH}$ in order to emphasize that these two higher pairs replace a single one. The synthesis of cam mechanisms with a constant pressure angle is thus rendered similar to the synthesis of cam mechanisms with roller-followers, the difference here being that the dual angle $\hat{\alpha}_3$ is unknown. On the other hand, it was proven in Section 4.3 that $\tan\nu = 1/\tan\mu$, i.e., $\nu = \pi/2 - \mu$. Thus, the dual curve of the cam is given as

$$\mathbf{s}_c \equiv \mathbf{e}_{42} + \epsilon\mathbf{m}_{42} = \hat{\mathbf{S}}^T(\hat{\psi})\hat{\mathbf{Q}}(\hat{\alpha}_1)\hat{\mathbf{S}}(\hat{\phi})\hat{\mathbf{Q}}(\hat{\alpha}_3)\hat{\mathbf{S}}(\hat{\nu})\hat{\mathbf{Q}}(\hat{\alpha}_4)\mathbf{k} \tag{4.103}$$

and the surface of the cam is derived from

$$\mathbf{r}_c \equiv \mathbf{e}_{42} \times \mathbf{m}_{42} + \lambda\mathbf{e}_{42} = \mathbf{p}_{42} + \lambda\mathbf{e}_{42} \tag{4.104}$$

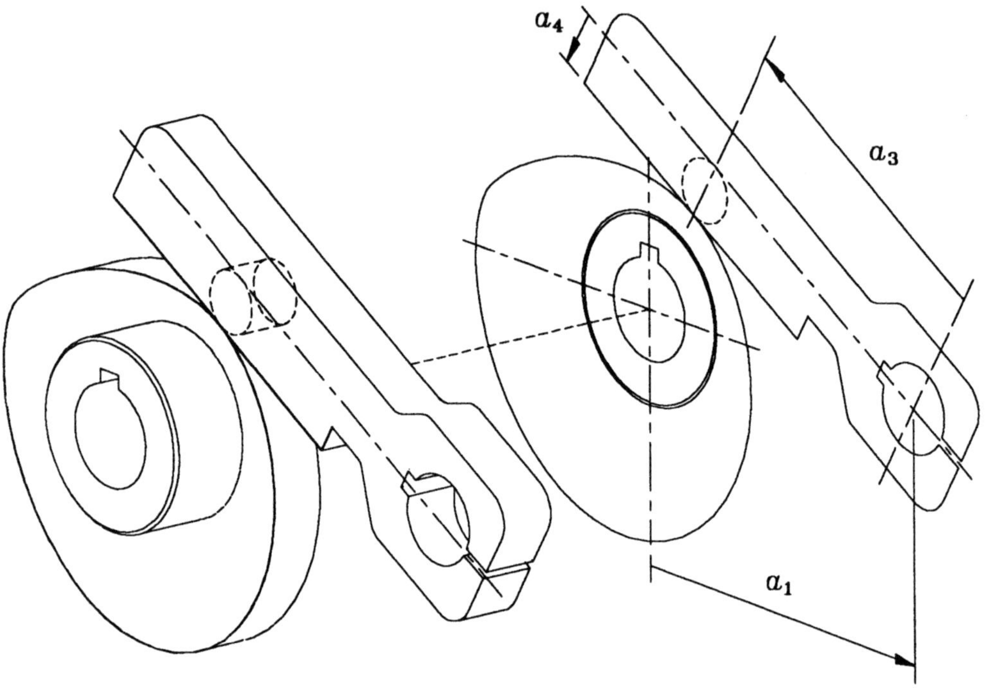

Figure 4.20 Planar $R\overline{HH}R$ cam mechanism.

This approach is applied to those cases in which pure rolling between the cam and the roller is achieved, and hence, conical and cylindrical surfaces are considered. Nevertheless, the condition that the pressure angle be constant eliminates the pure-rolling motion in mechanisms of this kind, and sliding occurs. The envelopes of the roller on the follower give rise to a regular cone for spherical $R\overline{HH}R$ mechanisms and a plane for planar $R\overline{HH}R$, $R\overline{HH}P$, $P\overline{HH}R$ and $P\overline{HH}P$ mechanisms. The latter are the well-known planar mechanisms with a flat-face follower. The input-output functions are the same as for the related RHR, RHP, PHR and PPP mechanisms discussed in Chapter 3.

4.4.1 $R\overline{HH}R$ Mechanisms

Figures 4.19 and 4.20 show the two cases of $R\overline{HH}R$ mechanisms with $\mu = 0$. For the spherical $R\overline{HH}R$, the dual terms of the angles of eq.(4.103) are zero, while, for planar $R\overline{HH}R$, the primal terms of the angles $\hat{\alpha}_i$ and the dual terms of the angles $\hat{\psi}$, $\hat{\phi}$ and $\hat{\nu}$ are zero.

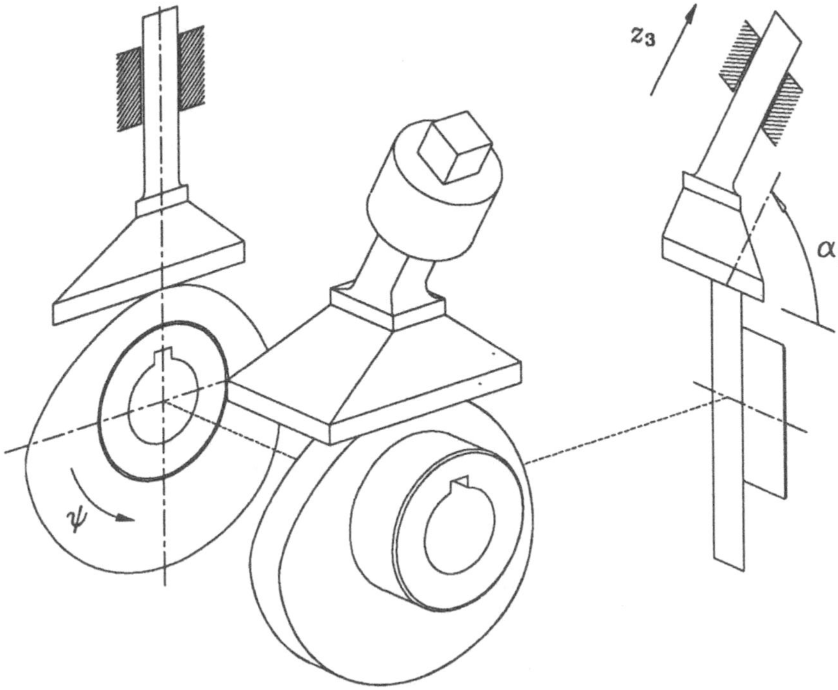

Figure 4.21 R$\overline{\text{HH}}$P cam mechanism.

Spherical R$\overline{\text{HH}}$R mechanisms

With application of eq.(4.103), the surface of the cam is given via vector $\mathbf{r}_c$ defined as

$$\mathbf{r}_c \equiv \lambda \mathbf{e}_{42} = \lambda \begin{bmatrix} k_5 \sin\alpha_4 - h_1 \cos\alpha_4 \\ -k_6 \sin\alpha_4 - h_2 \cos\alpha_4 \\ \sin\alpha_1 \sin\phi \sin\alpha_4 + k_1 \cos\alpha_4 \end{bmatrix} \tag{4.105}$$

where k_i and h_i are defined in eqs.(4.4) and, from eq.(4.85), it is clear that $\mu = 0$ if α_3 takes the value

$$\tan\alpha_3 = \frac{\sin\alpha_1 \cos\phi}{\phi' - \cos\alpha_1} \tag{4.106}$$

Note that spherical mechanisms with a constant pressure angle and zero offset, i.e., with $\alpha_4 = 0$, give rise to a flat-face follower.

Planar R$\overline{\text{HH}}$R mechanisms

Vectors $\mathbf{e}_{42}$ and $\mathbf{p}_{42}$ for planar R$\overline{\text{HH}}$R mechanisms are readily obtained from eqs.(4.103 & 4.104) as

$$\mathbf{e}_{42} = \begin{bmatrix} 0 \\ 0 \\ 1 \end{bmatrix} \quad \text{and} \quad \mathbf{p}_{42} = \begin{bmatrix} a_1 \cos\psi + a_3 \cos(\psi - \phi) + a_4 \sin(\psi - \phi) \\ -a_1 \sin\psi - a_3 \sin(\psi - \phi) + a_4 \cos(\psi - \phi) \\ 0 \end{bmatrix} \tag{4.107}$$

Now, from eq.(4.91) one obtains the value of a_3 in which $\mu = 0$, namely,

$$a_3 = \frac{a_1 \cos\phi}{\phi' - 1} \tag{4.108}$$

4.4.2 R$\overline{\text{HH}}$P Mechanisms

Considering that, from eq.(4.96), $a_3 = -a_1 - z_3' \sin\alpha_1$, vectors $\mathbf{e}_{42}$ and $\mathbf{p}_{42}$ of the surface of the cam are given as

$$\mathbf{e}_{42} = \begin{bmatrix} 0 \\ 0 \\ 1 \end{bmatrix} \quad \text{and} \quad \mathbf{p}_{42} = \begin{bmatrix} -z_3' \sin\alpha_1 \cos\psi + (a_4 - z_3 \sin\alpha_1) \sin\psi \\ z_3' \sin\alpha_1 \sin\psi + (a_4 - z_3 \sin\alpha_1) \cos\psi \\ 0 \end{bmatrix} \tag{4.109}$$

Figure 4.21 shows a general case of an R$\overline{\text{HH}}$P mechanism.

Note that a cam obtained with $\alpha_1 \neq \pi/2$ is of a smaller size than that obtained with $\alpha_1 = \pi/2$, for the same displacement program. Hence, angle α_1 can be considered as a design parameter in minimizing the cam size. However, as angle α_1 decreases, the sliding between cam and follower increases.

4.4.3 P$\overline{\text{HH}}$R Mechanisms

Figure 4.22 shows a P$\overline{\text{HH}}$R mechanism. Vectors $\mathbf{e}_{42}$ and $\mathbf{p}_{42}$, for the cam of this mechanism, are obtained from eqs.(4.103 & 4.104), namely,

$$\mathbf{e}_{42} = \begin{bmatrix} 0 \\ -\sin\alpha_1 \\ \cos\alpha_1 \end{bmatrix} \tag{4.110a}$$

and

$$\mathbf{p}_{42} = \begin{bmatrix} a_1 + a_3 \cos\phi - a_4 \sin\phi \\ a_3 \cos\alpha_1 \sin\phi + a_4 \cos\alpha_1 \cos\phi - z_2 \cos\alpha_1 \sin\alpha_1 \\ a_3 \sin\alpha_1 \sin\phi + a_4 \cos\phi \sin\alpha_1 - z_2 \sin^2\alpha_1 \end{bmatrix} \tag{4.110b}$$

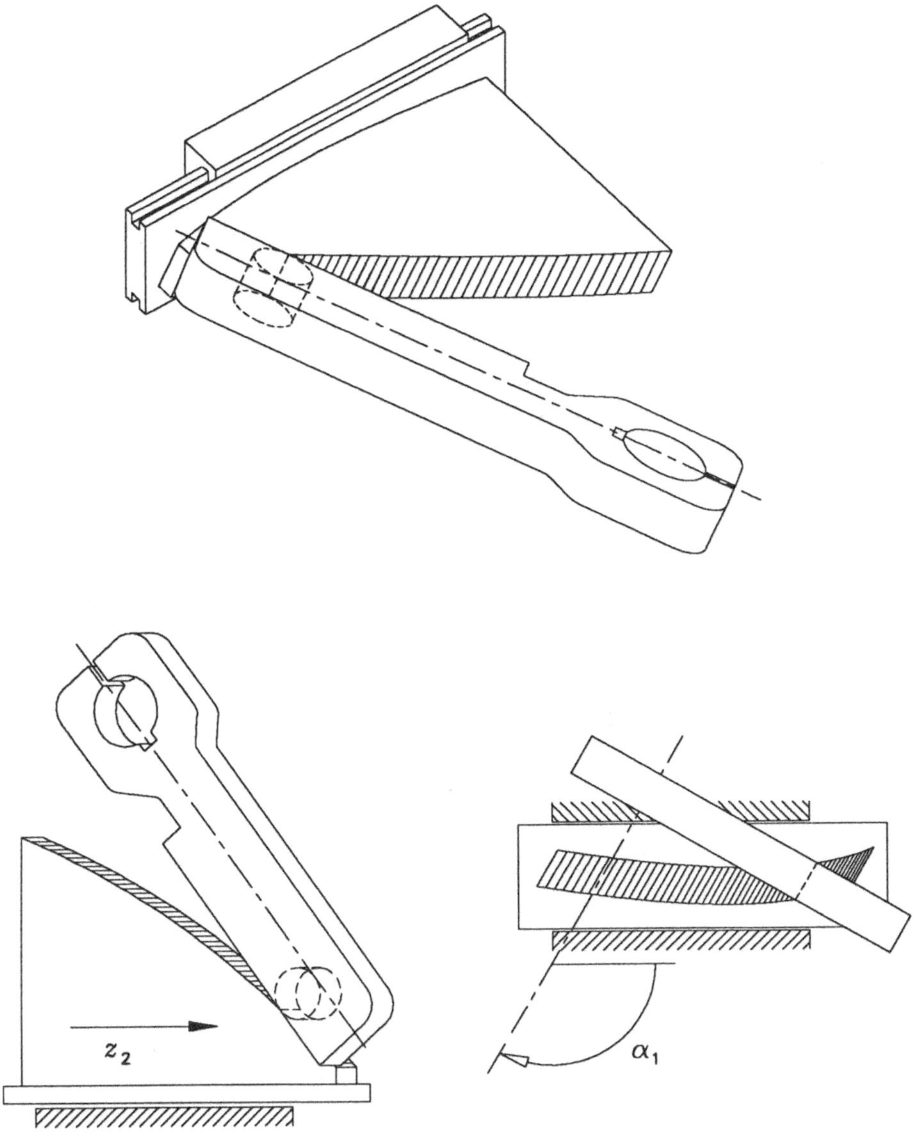

Figure 4.22 P$\overline{\text{HH}}$R cam mechanism.

with a_3 obtained from eq.(4.99) as

$$a_3 = \frac{\sin\alpha_1 \cos\phi}{\phi'} \tag{4.111}$$

4.4.4 $\mathrm{P\overline{HH}P}$ Mechanisms

For this mechanism, μ is independent of a_3. Moreover, one can consider a given value of μ, namely, $\mu = \mu_0$, and hence, from eq.(4.102) one obtains

$$\tan\mu_0 = \frac{z_3' - \cos\alpha_1}{\sin\alpha_1} \tag{4.112}$$

Therefore, z_3' must be constant and, consequently, both surfaces will have flat faces, which thus gives rise to wedge cam mechanisms.

Considering that $\nu = \pi/2 - \mu_0$, one can readily obtain the vectors below:

$$\mathbf{e}_{42} = \begin{bmatrix} 1 \\ 0 \\ 0 \end{bmatrix} \quad \text{and} \quad \mathbf{p}_{42} = \begin{bmatrix} 0 \\ a_3\cos\alpha_1 - a_4\sin(\alpha_1 - \mu_0) - z_3\sin\alpha_1 \\ a_3\sin\alpha_1 + a_4\cos(\alpha_1 - \mu_0) + z_3\cos\alpha_1 - z_2 \end{bmatrix} \tag{4.113}$$

Figure 4.23 illustrates a $\mathrm{P\overline{HH}P}$ mechanism. Notice that the contact between cam and follower is a plane, and hence, the $\overline{\mathrm{HH}}$ pair becomes a prismatic pair, a PPP mechanism thus being obtained.

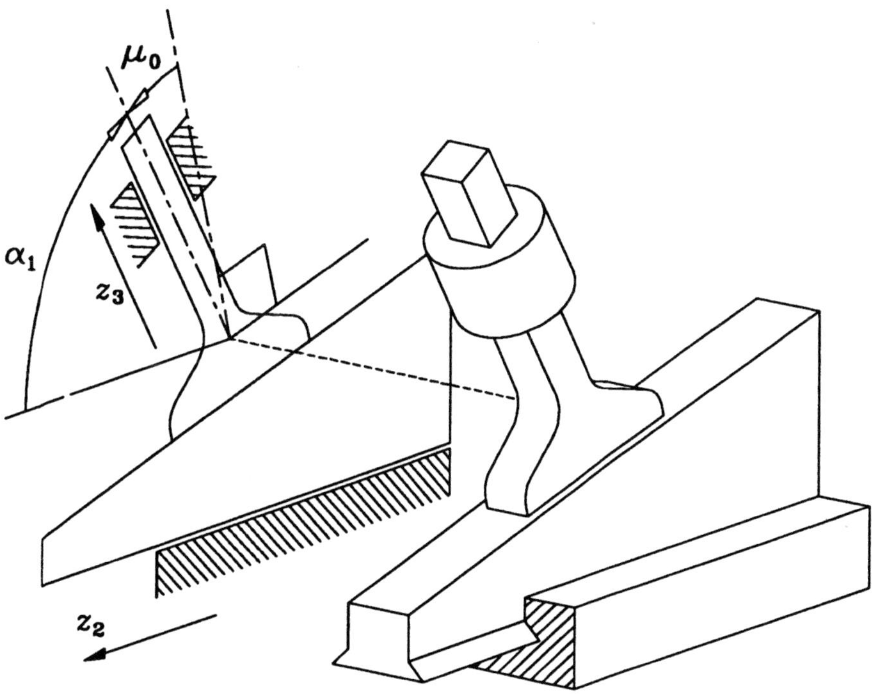

Figure 4.23 $\mathrm{P\overline{HH}P}$ cam mechanism.

Chapter 5

INDEXING MECHANISMS

5.1 Introduction

Mechanisms for intermittent motion and force transmission, termed *indexing mechanisms*, find extensive applications in industry. Mechanisms of this kind for motion transmission between parallel axes, such as the Geneva mechanism and external and internal indexing cam mechanisms, have been studied in the past (Guoxun, Zhengyang and Huimin 1988).

Indexing cam mechanisms (ICM) for skew axes, like the *Ferguson indexing cam mechanism*, have been used extensively in industry, but the optimization of their contact surfaces for minimum friction losses, under various functionality conditions, has not been given due attention. Although the kinematics of spatial cams has been studied in the past (Jensen 1965; Chakraborty and Dhande 1977; Koloc and Václavík 1988), the minimization of *power losses* in the synthesis of spatial ICM was addressed only recently (González-Palacios and Angeles, 1990; González-Palacios and Angeles, 1992*a*). In these references, a unified approach to the synthesis of cam and follower profiles was introduced that is applicable when motion is transmitted either through *direct contact* (RHR mechanisms) or through an intermediate roller (RHCR mechanisms), while power losses due to sliding are minimized.

The contact surfaces involved are generated with the application of the theory presented in Chapters 3 and 4, since this approach satisfies the condition that the relative velocity at the contact points be of minimum magnitude.

The method introduced in the above-mentioned references is recalled here, within the general framework developed in Chapters 2, 3 and 4. This method can be applied to the synthesis of the contact surfaces of both the cam and the follower for a prescribed indexing output motion of the latter. We assume here that the input cam rotates at a constant speed, while the relative layout of the axes of both the cam and the follower is given.

Moreover, indexing cam mechanisms with ***helical springs*** to maintain contact have been designed in the past (Johnson, 1958). However, the use of springs in cam mechanisms leads to dynamic problems of vibration. Consequently, the design of planar cam mechanisms with positive motion, i.e., without springs, has been proposed as an alternative (Jackowski and Dubil, 1967; Wunderlich, 1971; Hunt, 1973). This concept has been applied to the design of planar ICM (Makino, 1979; Jones and Tsang, 1987; Gouxun, Zhengyang and Huimin, 1988). In the latter reference, pure-rolling motion has been achieved. However, the mechanisms thus designed have the disadvantage that the pressure angle is large, especially at and near the *dwell* positions. Other indexing mechanisms like the Geneva mechanism have been studied (Fenton, 1965, 1975*a*, 1975*b*; Olędzki and Szydlowski, 1975), but these mechanisms have jerk discontinuities and wear problems. The elimination of jerk discontinuities in this context was reported by Sadek, Lloyd, and Smith (1990).

ICM with direct contact for spatial, spherical and planar motions were introduced by González-Palacios and Angeles (1990). A unified approach of ICM with roller-followers is presented in (González-Palacios and Angeles, 1991*a*,1992*b*), both approaches being discussed here for completeness. These formulations are combined in this chapter to obtain a new type of mechanism that is termed here *PRICAM* (*P*ure *R*olling *I*ndexing *C*am *M*echanism) with positive motion, pure rolling and improved pressure-angle distribution. The pressure angle vanishes at the beginning and at the middle of the rise, which makes PRICAM a unique design. The condition applied on the design of PRICAM is the elimination of wear, and hence, two versions are proposed here, namely, planar and spherical mechanisms.

5.2 Input-Output Function of Indexing Cam Mechanisms (ICM)

In the realm of indexing cam mechanisms, a full rotation of the cam is divided into two intervals, of lengths $\Delta\psi$ and $2\pi - \Delta\psi$, as shown in Fig. 5.1. When the cam rotates in the second interval, ϕ remains constant while, in the first interval, it is defined as

$$\phi(\psi) = \frac{2\pi}{N}\tau\left(\frac{\psi}{\Delta\psi}\right) \tag{5.1}$$

where $\tau(x)$ is a ***normal input-output function***, namely,

$$\tau = \tau(x), \quad 0 \le \tau \le 1, \quad 0 \le x \le 1 \tag{5.2}$$

and N is the number of ***indexing steps*** for one full rotation of the follower. Various kinds of functions $\tau(x)$ are discussed in Appendix B.

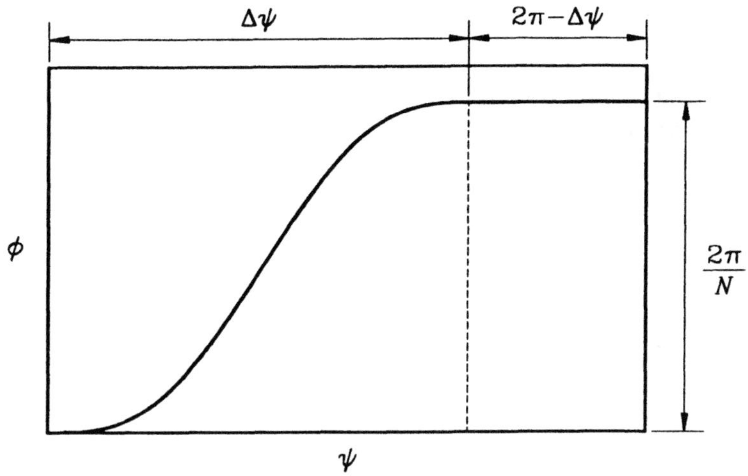

Figure 5.1 Motion function $\phi = \phi(\psi)$

5.3 ICM of RHR Type

It was shown in Chapter 2 that, with the application of eqs. (3.16 & 3.18), it is possible to generate the shapes of two rigid bodies in contact, when transmitting a motion given by the functional relation $\phi = \phi(\psi)$. Now, with $\phi(\psi)$ defined as in eq.(5.1), one can readily obtain the surfaces of an ICM of the RHR type (RHR-ICM) by prescribing the values N, $\Delta\psi$, a_1 and α_1.

The first and second derivatives of ϕ with respect to ψ are readily derived with the values of N and $\Delta\psi$, namely,

$$\phi' = \frac{2\pi}{N\Delta\psi}\tau'\left(\frac{\psi}{\Delta\psi}\right) \tag{5.3a}$$

$$\phi'' = \frac{4\pi^2}{N(\Delta\psi)^2}\tau''\left(\frac{\psi}{\Delta\psi}\right) \tag{5.3b}$$

Equations (5.3) are valid in the range $0 \le \psi \le \Delta\psi$. For $\Delta\psi < \psi \le 2\pi$, $\phi = 2\pi/N$ and $\phi' = \phi'' = 0$.

The cam and follower surfaces are now defined, as in eqs.(3.18), by the vectors $\mathbf{r}_2$ and $\mathbf{r}_3$, respectively, as indicated below:

$$\mathbf{r}_2 = b_2\begin{bmatrix}\cos\psi \\ -\sin\psi \\ 0\end{bmatrix} + \lambda\begin{bmatrix}-\sin\psi\sin\theta_2 \\ -\cos\psi\sin\theta_2 \\ \cos\theta_2\end{bmatrix}$$

$$\mathbf{r}_3 = (b_2 - a_1)\begin{bmatrix} \cos\phi \\ -\sin\phi \\ 0 \end{bmatrix} + \lambda \begin{bmatrix} -\sin\phi\sin(\theta_2 - \alpha_1) \\ -\cos\phi\sin(\theta_2 - \alpha_1) \\ \cos(\theta_2 - \alpha_1) \end{bmatrix}$$

where λ defines the thickness of the surfaces, while $\tan\theta_2$ and b_2 are computed as in eqs.(3.16c & 3.16d), namely,

$$\tan\theta_2 = \frac{\phi'\sin\alpha_1}{\phi'\cos\alpha_1 - 1}$$

$$b_2 = \frac{\phi'^2 - \phi'\cos\alpha_1}{\phi'^2 - 2\phi'\cos\alpha_1 + 1}a_1$$

A solid model of the surfaces defined above was implemented on a *Silicon Graphics Power Series Workstation (IRIS 4D/420VGX)*. Moreover, the motion is simulated for any value of the four parameters defining the RHR-ICM. Thus, fixing three of them, the changes of the profile can be appreciated according to the variation of the fourth parameter.

In order to give a better idea of the software implemented for the above-mentioned purpose, we present four figures with six still frames each, so that the changes can be appreciated. Thus, in Fig. 5.2, six RHR-ICM are shown with different values of N, while in Fig 5.3, $\Delta\psi$ is given different values. Furthermore, the transition from spatial to spherical RHR-ICM, when a_1 varies from 1 to 0, is shown in Fig. 5.4. Moreover, with $\alpha_1 = 0$ in Fig. 5.5, the cam and the follower have parallel axes and rotate in the same direction. In this configuration, the mechanism is an *internal* RHR-ICM. As α_1 increases, the input and output axes are skew until they become parallel again with $\alpha_1 = \pi$. At this value, the cam and the follower rotate in opposite directions, thus giving rise to an *external* RHR-ICM. The transition from internal to external RHR-ICM is shown with four intermediate configurations.

5.4 ICM of RHCR Type

The theory for the synthesis of RHCR mechanisms was already introduced in Section 4.2.1 and is applied here to the synthesis of mechanisms of interest. An ICM is defined by its input-output function, which is given as

$$\phi = \phi_m + \frac{2\pi}{N}\tau\left(\frac{\psi}{\Delta\psi}\right) \tag{5.4}$$

where ϕ_m is the value of ϕ upon engagement of the roller with the cam, as illustrated in Fig. 5.6. For the applications presented in Sections 5.5 and 5.6, ϕ_m is defined as

$$\phi_m = \pi\left(1 - \frac{1}{N}\right) \tag{5.5}$$

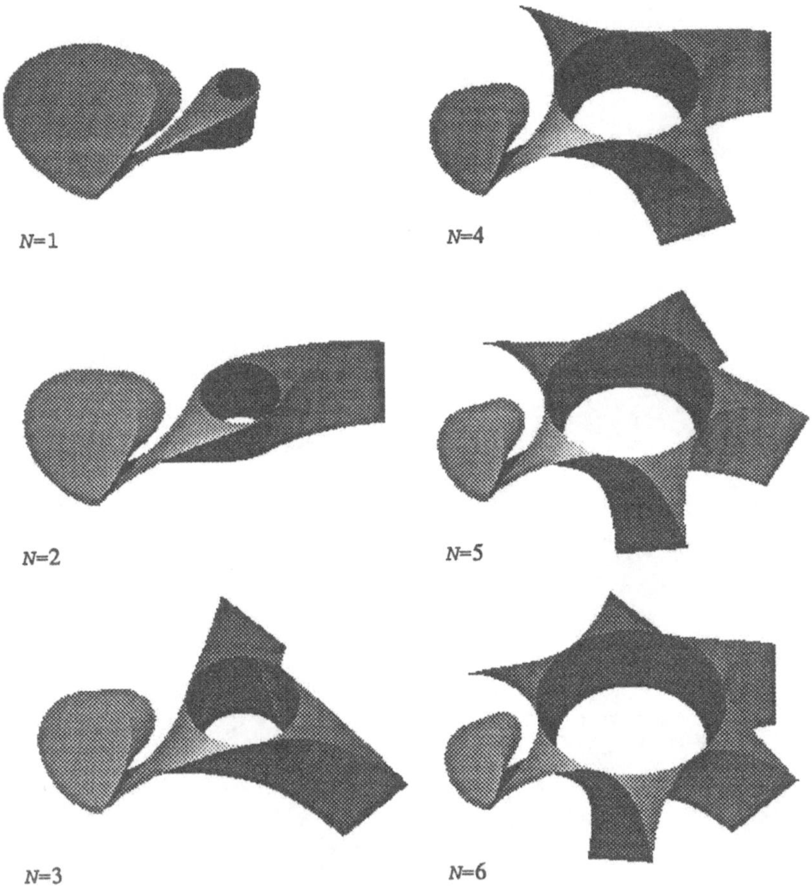

Figure 5.2 Spatial RHR-ICM with $\Delta\psi = 120°$, $a_1 = 1$, $\alpha_1 = 60°$ and $0.25 \leq \lambda \leq 0.85$ for six different values of N

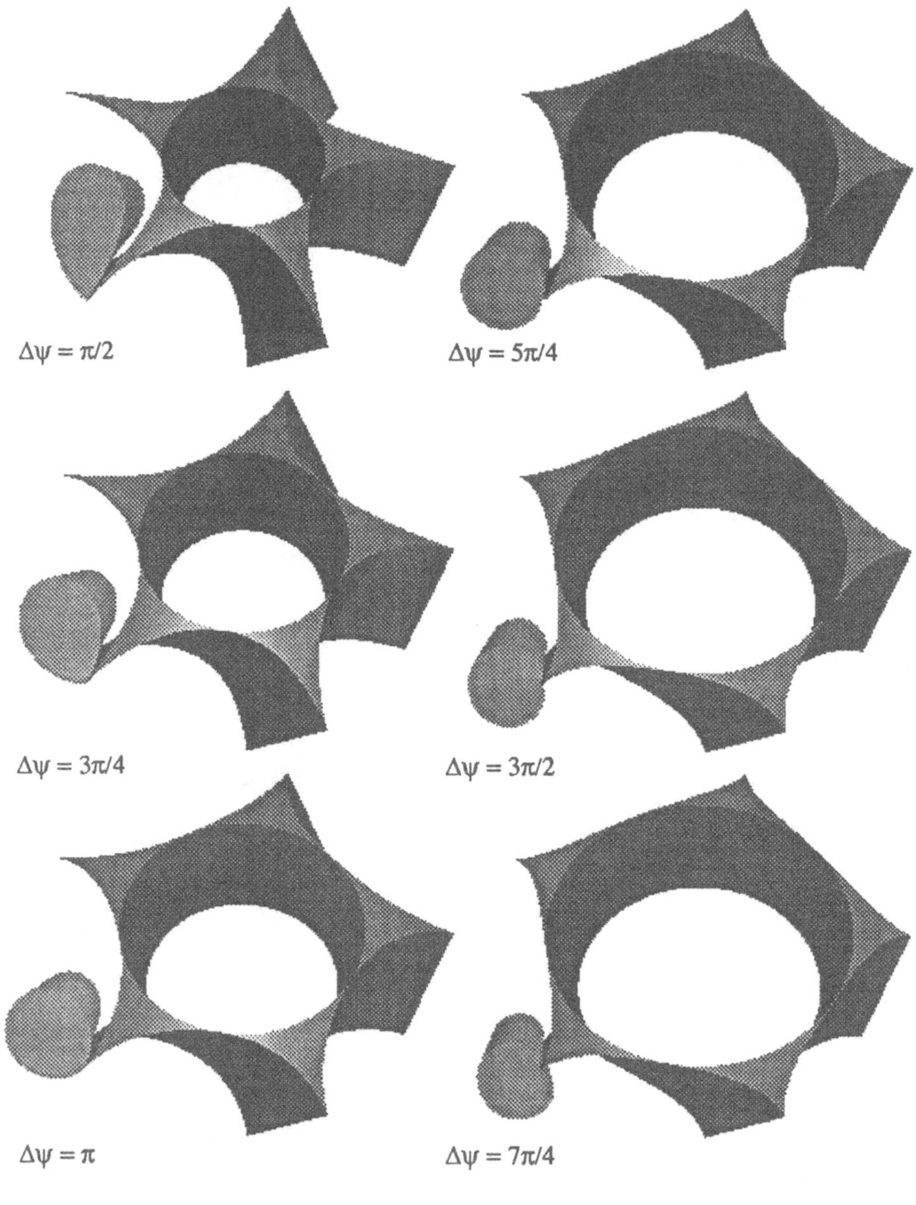

Figure 5.3 Spatial RHR-ICM with $N = 5$, $a_1 = 0.8$, $\alpha_1 = 45°$ and $0.4 \leq \lambda \leq 1.0$ for six different values of $\Delta\psi$

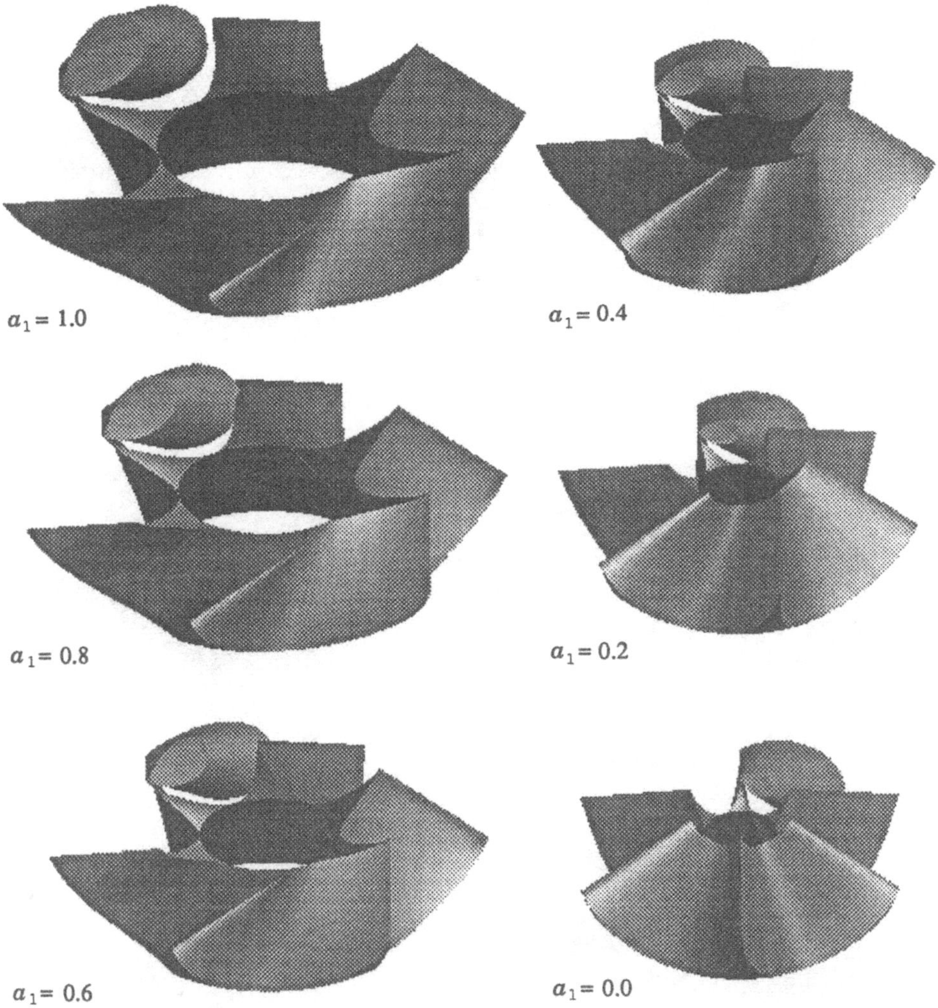

Figure 5.4 Transition from spatial to spherical RHR-ICM with $N = 6$, $\Delta\psi = 120°$, $\alpha_1 = 70°$ and $0.3 \leq \lambda \leq 0.95$

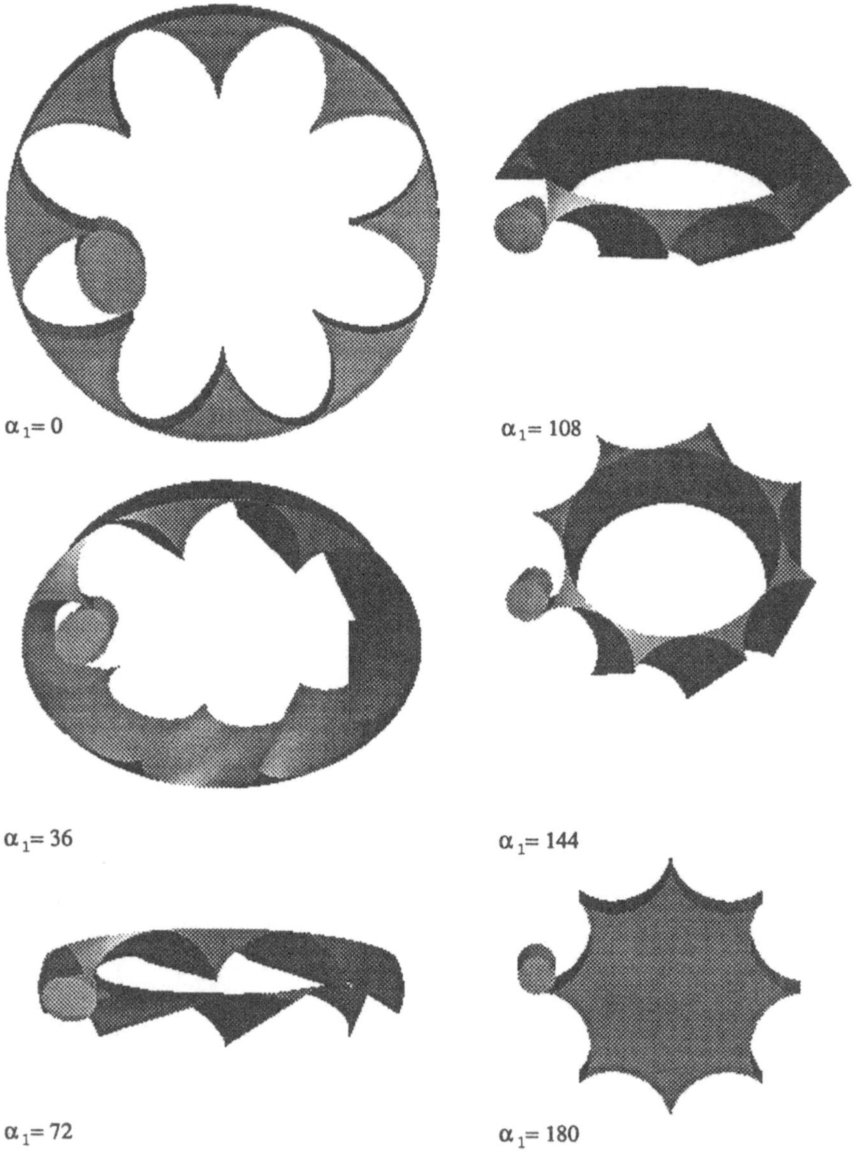

Figure 5.5 Transition from internal to external RHR-ICM with $N = 8$, $\Delta\psi = 220°$, $a_1 = 1$ and $0.4 \leq \lambda \leq 1.0$

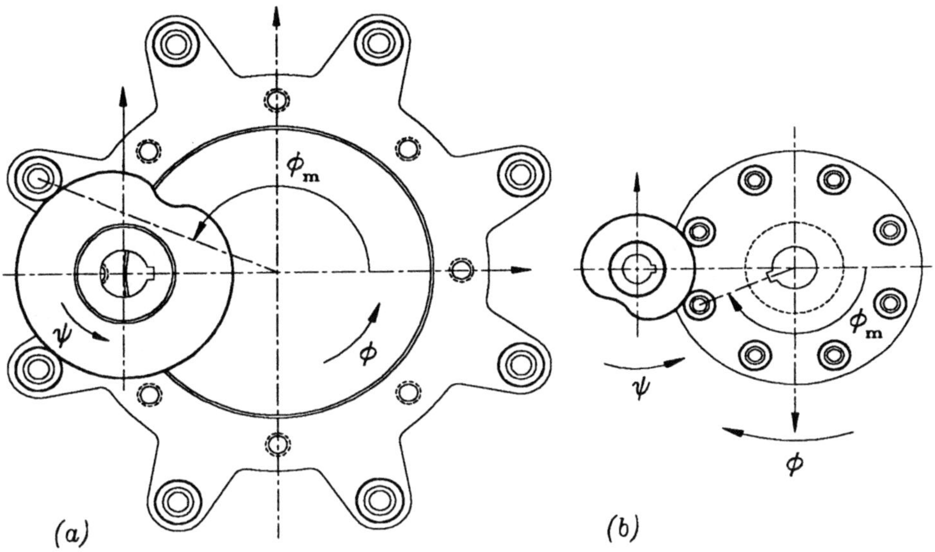

Figure 5.6 Definition of ϕ_m for RHCR-ICM: *a*) internal; *b*) external

In Fig. 5.6*a* ϕ_m is shown for an internal ICM, whereas, in Fig. 5.6*b*, ϕ_m is shown for an external ICM.

There are two approaches to determine the geometry of an RHCR-ICM. The first one consists of the application of eqs.(4.27); the second on the application of eqs.(4.103 & 4.104). The latter was found more reliable from the computational point of view; it was therefore, applied in *USYCAMS*—pronounced *you see cams*, for *U*nified *Sy*nthesis of *Ca*m *M*echanism*s*— the software written for the interactive synthesis of cam mechanisms, and outlined in Chapter 7.

Equation (4.103) is rewritten here for quick reference, namely,

$$\mathbf{s}_c \equiv \mathbf{e}_{42} + \epsilon\mathbf{m}_{42} = \hat{\mathbf{S}}^T(\psi)\hat{\mathbf{Q}}(\hat{\alpha}_1)\hat{\mathbf{S}}(\phi)\hat{\mathbf{Q}}(\hat{\alpha}_3)\hat{\mathbf{S}}(\hat{\nu})\hat{\mathbf{Q}}(\hat{\alpha}_4)$$

where $\hat{\nu}$ was defined in eq.(4.23) as

$$\tan\hat{\nu} \equiv \tan\nu + \epsilon z_{43}(1+\tan^2\nu) = \frac{-\sin\hat{\beta}\sin\hat{\phi}}{\cos\hat{\beta}\sin\hat{\alpha}_3 + \cos\hat{\alpha}_3\cos\hat{\phi}\sin\hat{\beta}}$$

which can be readily expanded in its primal and dual parts, namely,

$$\tan\nu = \frac{-\sin\beta\sin\phi}{\cos\beta\sin\alpha_3 + \cos\alpha_3\sin\beta\cos\phi} \tag{5.6a}$$

$$z_{43} = \frac{a_3\sin\beta(\cos\beta\cos\alpha_3 - \sin\alpha_3\sin\beta\cos\phi) - d\sin\alpha_3}{(\cos\beta\sin\alpha_3 + \cos\alpha_3\sin\beta\cos\phi)^2 + \sin^2\beta\sin^2\phi}\sin\phi \tag{5.6b}$$

where $\beta = \alpha_1 - \theta_2$ and $d = a_1 - b_2$.

Thus, given N, $\Delta\psi$, $\phi(\psi)$ and the dual angles $\hat{\alpha}_1$, $\hat{\alpha}_3$ and $\hat{\alpha}_4$, which are described in Table 4.1, the cam profile of an RHCR-ICM can be obtained with the aid of USYCAMS. However, the dual angle $\hat{\alpha}_3$ cannot be chosen arbitrarily as discussed in Section 5.5, where we show how to specify it. A solid model of this type of ICM was implemented using USYCAMS. The transition from internal to external RHCR-ICM is shown in Fig. 5.7 with $N = 8$ and $\Delta\psi = 240°$, while, in Fig. 5.8, the transition from spatial to spherical RHCR-ICM is displayed.

5.5 Pure Rolling with Indexing Cam Mechanisms

The last two sections of this chapter are devoted to the synthesis of novel mechanisms that produce indexing motion with pure rolling. Thus, two versions of these mechanisms are obtained, planar and spherical.

In fact, the mechanisms under study turn out to be the composition of two elementary mechanisms, namely, the *primary mechanism* (PM), which is an RHR ICM, and the *secondary mechanism* (SM), which is constituted by an RHRR ICM.

The generation of the higher-pair contact surface of both PM and SM is based on the minimization of the magnitude of the relative velocity at the contact points, i.e., at the cam-follower pair for the PM and at the cam-roller pair for the SM. As discussed in Section 3.2.1, for planar and spherical RHR mechanisms, the relative velocity v_{32} is zero. Similarly, for planar and spherical RHRR mechanisms, we showed in Section 4.2.1 that the velocity v_{42} between cam and roller is zero as well. Thus, as discussed in Chapters 2 and 3, those surfaces are generated as ruled surfaces and represented in the parametric form

$$\mathbf{r}(\psi,\lambda) = \mathbf{p}(\psi) + \lambda\mathbf{e}(\psi) \tag{5.7}$$

where $\mathbf{r}(\psi,\lambda)$ is the position vector of a point of the surface, ψ is the angle of rotation of the cam, λ is a real number, $\mathbf{p}$ is the position vector of the directrix, and $\mathbf{e}$ is a unit vector parallel to the generatrix.

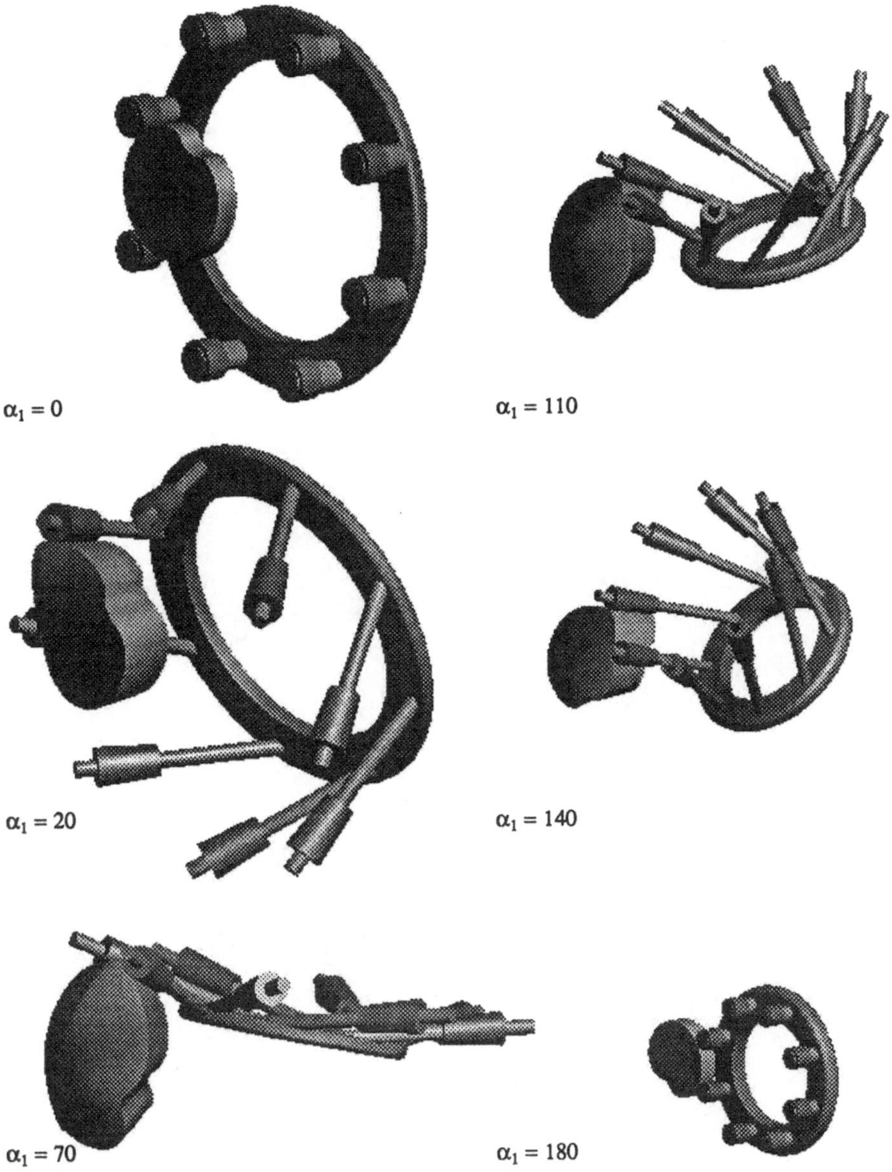

Figure 5.7 Transition from internal to external RHCR-ICM with $N = 8$, $\Delta\psi = 240°$

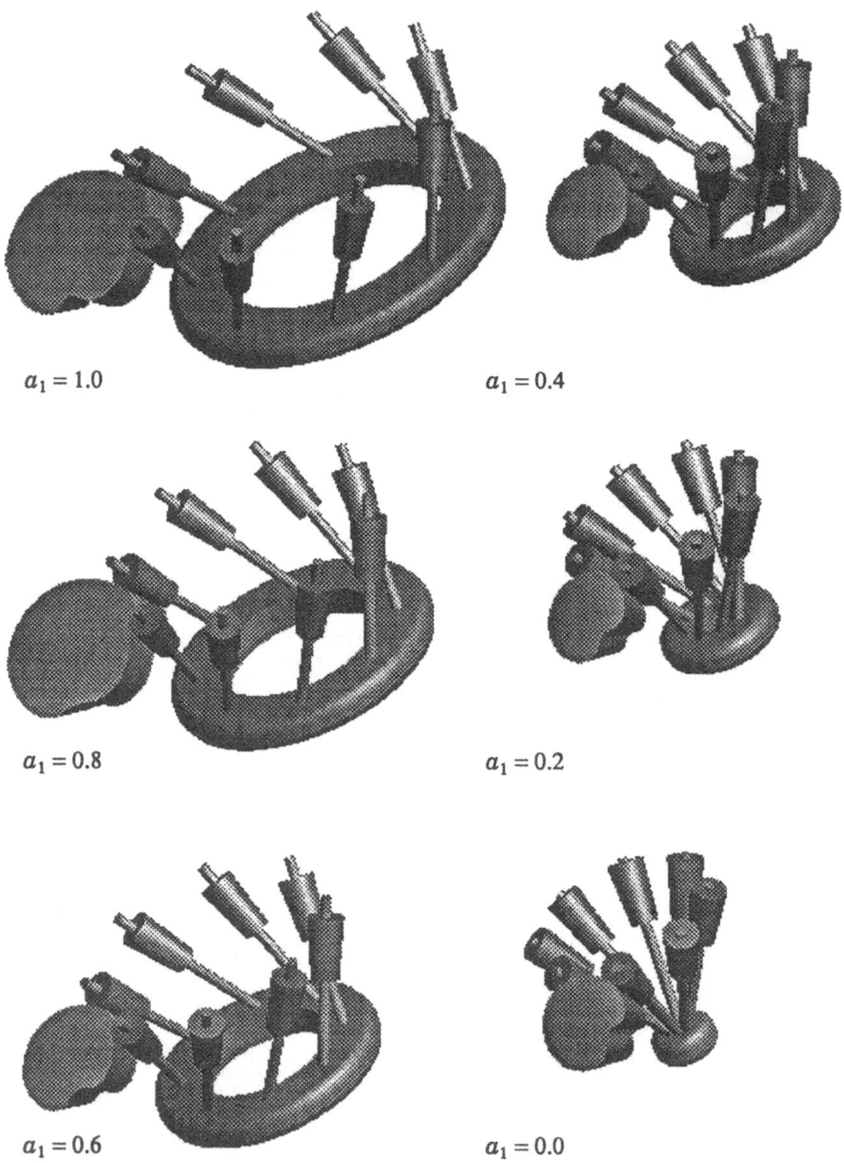

Figure 5.8 Transition from spatial to spherical RHCR-ICM with $N = 8$, $\Delta\psi = 240°$, $\alpha_1 = 135°$ and $0.3 \leq \lambda \leq 0.95$

5.5.1 Primary Mechanism

We regard here, as in Chapter 4, planar mechanisms as special cases of spherical mechanisms. Accordingly, we study the latter first, then the former.

Spherical PM

From the definition of the geometry of spherical RHR mechanisms with $a_1 = 0$, the geometry of the spherical PM is given as

$$\mathbf{r}_2 = \lambda \begin{bmatrix} -\sin\psi\sin\theta_2 \\ -\cos\psi\sin\theta_2 \\ \cos\theta_2 \end{bmatrix} \tag{5.8a}$$

$$\mathbf{r}_3 = \lambda \begin{bmatrix} -\sin\phi\sin(\theta_2 - \alpha_1) \\ -\cos\phi\sin(\theta_2 - \alpha_1) \\ \cos(\theta_2 - \alpha_1) \end{bmatrix} \tag{5.8b}$$

with θ_2 defined as in eq(3.16c), i.e.,

$$\tan\theta_2 = \frac{\phi'\sin\alpha_1}{\phi'\cos\alpha_1 - 1} \tag{5.9}$$

Moreover, the pressure angle can be computed from eq.(3.55) as,

$$\tan\mu = \frac{\phi'\sqrt{\phi'^2 - 2\phi'\cos\alpha_1 + 1}}{\phi''} \tag{5.10}$$

Planar PM

The geometry of the planar PM is defined by the geometry of planar RHR mechanisms with $\alpha_1 = 0, \pi$. Thus, the position vectors of both cam and follower are derived from eqs.(3.18a & 3.18b), namely,

$$\mathbf{r}_2 = \frac{\phi' a_1}{\phi' \pm 1} \begin{bmatrix} \cos\psi \\ -\sin\psi \\ 0 \end{bmatrix} + \lambda \begin{bmatrix} 0 \\ 0 \\ 1 \end{bmatrix} \tag{5.11a}$$

$$\mathbf{r}_3 = \frac{a_1}{\phi' \pm 1} \begin{bmatrix} \cos\phi \\ -\sin\phi \\ 0 \end{bmatrix} + \lambda \begin{bmatrix} 0 \\ 0 \\ 1 \end{bmatrix} \tag{5.11b}$$

The pressure angle, in turn, is as defined in eq.(3.61), i.e.,

$$\tan\mu = \frac{\phi'(\phi' \pm 1)}{\phi''} \tag{5.12}$$

the plus and minus signs corresponding to external and internal PM, respectively.

5.5.2 Secondary Mechanism

Spherical SM

The pitch and the cam surfaces for the spherical mechanism are defined as in eqs.(4.28 & 4.29), namely,

$$\mathbf{r}_P \equiv \lambda \mathbf{e}_P = \lambda \begin{bmatrix} s\alpha_3 s\phi c\psi - (s\alpha_1 c\alpha_3 + c\alpha_1 s\alpha_3 c\phi)s\psi \\ -s\alpha_3 s\phi s\psi - (s\alpha_1 c\alpha_3 + c\alpha_1 s\alpha_3 c\phi)c\psi \\ c\alpha_1 c\alpha_3 - s\alpha_1 s\alpha_3 c\phi \end{bmatrix} \tag{5.13}$$

$$\mathbf{r}_C \equiv \lambda \mathbf{e}_C = \lambda \begin{bmatrix} s(\theta_3 - \alpha_4)s\delta c\psi - [s\theta_2 c(\theta_3 - \alpha_4) + c\theta_2 s(\theta_3 - \alpha_4)c\delta]s\psi \\ -s(\theta_3 - \alpha_4)s\delta s\psi - [s\theta_2 c(\theta_3 - \alpha_4) + c\theta_2 s(\theta_3 - \alpha_4)c\delta]c\psi \\ c\theta_2 c(\theta_3 - \alpha_4) - s\theta_2 s(\theta_3 - \alpha_4)c\delta \end{bmatrix} \tag{5.14}$$

where θ_2, θ_3 and δ are given as

$$\tan\theta_2 = \frac{\phi' \sin\alpha_1}{\phi' \cos\alpha_1 - 1} \tag{5.15a}$$

$$\tan\theta_3 = \frac{\sqrt{[c(\alpha_1 - \theta_2)c\phi s\alpha_3 + c\alpha_3 s(\alpha_1 - \theta_2)]^2 + s^2\alpha_3 s^2\phi}}{c\alpha_3 c(\alpha_1 - \theta_2) - c\phi s\alpha_3 s(\alpha_1 - \theta_2)} \tag{5.15b}$$

$$\tan\delta = \frac{s\alpha_3 s\phi}{s\alpha_3 c(\alpha_1 - \theta_2)c\phi + c\alpha_3 s(\alpha_1 - \theta_2)} \tag{5.15c}$$

Furthermore, the pressure angle is as defined in eq.(4.86), i.e.,

$$\tan\mu = \frac{(\phi' - \cos\alpha_1)\sin\alpha_3 - \sin\alpha_1 \cos\alpha_3 \cos\phi}{\sin\alpha_1 \sin\phi} \tag{5.16}$$

In the design of spherical RHRR ICM, N, $\Delta\psi$ and α_1 are given as design specifications. However, α_3 cannot be assigned arbitrarily for, beyond a certain bound, undercutting will occur. The maximum allowable value of α_3, defined as $\bar{\alpha}_3$, can be derived by analyzing the spherical radius of curvature of the *pitch curve* Γ.

In order to generalize the solution, we write the radius of curvature as $\rho = \rho(\psi, \alpha_3, \alpha_1, N, \Delta\psi)$. Thus, in terms of $\mathbf{e}_P$, defined in eq.(5.13), and its first and second derivatives with respect to ψ, ρ is expressed as (Guggenheimer, 1977)

$$\rho = \arctan\left(\frac{(\mathbf{e}'_P \cdot \mathbf{e}'_P)^{3/2}}{\mathbf{e}_P \times \mathbf{e}'_P \cdot \mathbf{e}''_P}\right) \tag{5.17}$$

Now, we are interested in those points of Γ where cusps occur, i.e., where $\rho = 0$. First we search for the values ψ_0 where ρ attains a minimum value, i.e., we make $d\rho/d\psi = 0$, which readily leads to

$$f(\psi_0, \alpha_3, \alpha_1, N, \Delta\psi) \equiv \mathbf{e}_P \times \mathbf{e}'_P \cdot [3(\mathbf{e}'_P \cdot \mathbf{e}''_P)\mathbf{e}''_P - (\mathbf{e}'_P \cdot \mathbf{e}'_P)\mathbf{e}_P'''] = 0 \tag{5.18}$$

The analysis of the motion covers one indexing step and hence, it starts at $\phi(-\Delta\psi/2) = -\pi/N$ and finishes at $\phi(\Delta\psi/2) = \pi/N$. For example, for a 3-4-5 polynomial motion, ϕ and ϕ' are readily derived from eqs.(B.4), namely,

$$\phi = \Delta\phi(\frac{15}{8}x - 5x^3 + 6x^5), \quad x \equiv \frac{\psi}{\Delta\psi}, \quad \Delta\phi \equiv \frac{2\pi}{N} \tag{5.19}$$

$$\phi' = \frac{\Delta\phi}{\Delta\psi}(\frac{15}{8} - 15x^2 + 30x^5), \quad -\frac{1}{2} \leq x \leq \frac{1}{2} \tag{5.20}$$

With the aid of *symbolic algebra* (Wolfram, 1992), we found that $\psi_0 = 0$ always satisfies eq.(5.18), i.e., $f(0,\, \alpha_3,\, \alpha_1,\, N,\, \Delta\psi) \equiv 0$, as shown in Fig. 5.9. Moreover, we found graphically that f can have one or three real roots; however, the minimum absolute value of ρ is found at $\psi_0 = 0$.

Now, $\bar{\alpha}_3$ is obtained as a solution of $\rho(0,\, \bar{\alpha}_3,\, \alpha_1,\, N,\, \Delta\psi) = 0$. In view of eq.(5.17), ρ is zero if

$$\mathbf{e}'_P \cdot \mathbf{e}'_P = 0 \tag{5.21}$$

Furthermore, from eq.(5.13), $\mathbf{e}'_P$ at $\psi = 0$ is readily obtained as

$$\mathbf{e}'_P(0) = [-\sin\alpha_1 \cos\bar{\alpha}_3 + (\phi'_0 - \cos\alpha_1)\sin\bar{\alpha}_3, \quad 0, \quad 0]^T \tag{5.22}$$

where $\phi'_0 = 15\pi/4N\Delta\psi$ is computed from eq.(5.20) by setting $x = 0$.

Upon substitution of eq.(5.22) into eq.(5.21), we readily derive $\bar{\alpha}_3$, the maximum allowable value of α_3, as

$$\tan\bar{\alpha}_3 = \frac{\sin\alpha_1}{15\pi/4N\Delta\psi - \cos\alpha_1} \tag{5.23}$$

Figure 5.9 illustrates the pitch curve for three different values of α_3, with the corresponding plots of ρ and ρ'. Notice that, for $\alpha_3 < \bar{\alpha}_3$, ρ' has three roots but ρ attains its minimum always at $\psi = 0$.

These results are now summarized below:

Theorem 5.1: *Given α_1, N, $\Delta\psi$ and the input-output function defined as in eq.(5.1), the pitch curve of a spherical RHRR ICM reaches a cusp at $\psi = 0$ if $\alpha_3 = \bar{\theta}_2 - \alpha_1$, $\bar{\theta}_2$ being the value of θ_2 at $\psi = 0$.*

We prove this theorem showing that $\bar{\theta}_2 - \alpha_1 \equiv \bar{\alpha}_3$, with $\bar{\alpha}_3$ defined above as the maximum value attained by α_3. First, we expand $\tan(\bar{\theta}_2 - \alpha_1)$

$$\tan(\bar{\theta}_2 - \alpha_1) = \frac{\tan\bar{\theta}_2 - \tan\alpha_1}{1 + \tan\bar{\theta}_2 \tan\alpha_1} \tag{5.24}$$

Now, upon substitution of eq.(5.15a) into eq.(5.24), we obtain

$$\tan(\bar{\theta}_2 - \alpha_1) = \frac{\sin\alpha_1}{\phi'_0 - \cos\alpha_1} \equiv \tan\bar{\alpha}_3 \tag{5.25}$$

thereby completing the proof.

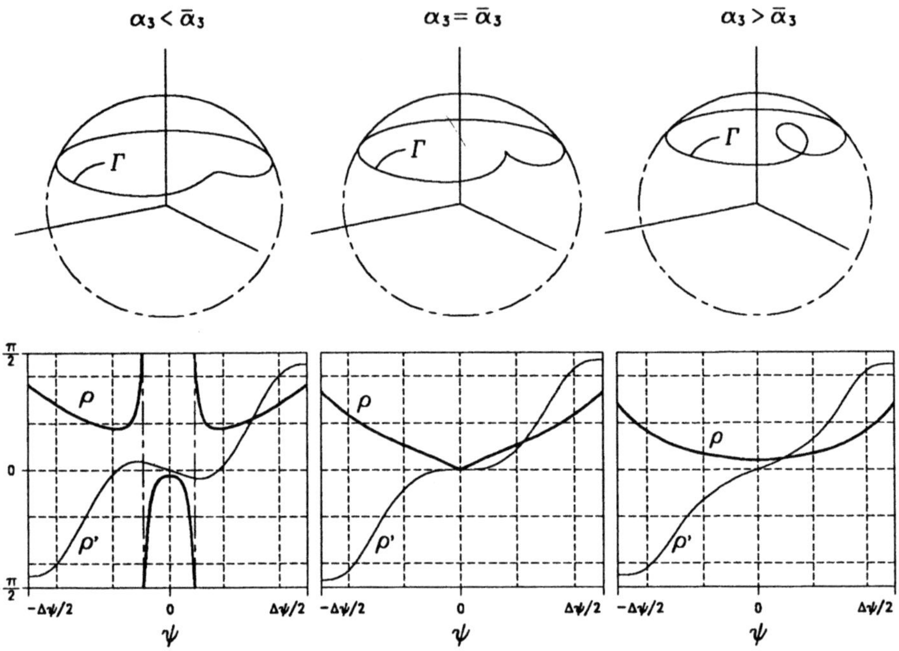

Figure 5.9 The pitch curve and the corresponding plots of ρ and ρ' for three values of α_3.

Planar SM

The geometry of planar SM is defined by the geometry of planar RHRR mechanisms with $\alpha_1 = \pi$ for external and $\alpha_1 = 0$ for internal SM. Thus, the pitch and the cam surfaces are defined according to eqs.(4.40–4.42) as

$$\mathbf{r}_P = \begin{bmatrix} c\psi a_1 + c(\psi \pm \phi)a_3 \\ -s\psi a_1 - s(\psi \pm \phi)a_3 \\ 0 \end{bmatrix} + \lambda \begin{bmatrix} 0 \\ 0 \\ 1 \end{bmatrix} \tag{5.26}$$

$$\mathbf{r}_C = \begin{bmatrix} c\psi b_2 + c(\psi - \delta)(b_3 - a_4) \\ -s\psi b_2 - s(\psi - \delta)(b_3 - a_4) \\ 0 \end{bmatrix} + \lambda \begin{bmatrix} 0 \\ 0 \\ 1 \end{bmatrix} \tag{5.27}$$

where

$$b_2 = \frac{\phi'}{\phi' \pm 1} a_1 \tag{5.28a}$$

$$b_3 = \sqrt{(a_3 c\phi + a_1 - b_2)^2 + a_3{}^2 s^2\phi} \tag{5.28b}$$

$$\tan\delta = \frac{a_3 s\phi}{a_3 c\phi + a_1 - b_2} \tag{5.28c}$$

The pressure angle is defined in turn as in eq.(4.91), namely,

$$\tan\mu = \frac{a_3(\phi' \pm 1) \pm a_1 \cos\phi}{a_1 \sin\phi} \tag{5.28d}$$

The plus sign of the doubly-signed terms in eqs.(5.28a & d) yields an external SM, whereas the minus, an internal SM.

As a counterpart of the spherical SM, the pitch curve of the planar SM can be analyzed to determine the maximum allowable value of a_3, defined as $\bar{a}_3$, so that undercutting can be avoided. Thus, the theorem below is established:

Theorem 5.2: *Given a_1, N, $\Delta\psi$ and the input-output function defined as in eq.(5.1), the pitch curve of a planar RHRR ICM reaches a cusp at $\psi = 0$ if $a_3 = \bar{b}_2 - a_1$, $\bar{b}_2$ being the value of b_2 at $\psi = 0$.*

5.6 Positive Action and Positive Motion

In the design of cam mechanisms we distinguish two kinds of actuating forces at the contact between the cam and the follower, namely, the force that transmits the motion to the follower and the force that opposes this motion. The action of each of these forces is termed here *positive action* (PA) and *negative action* (NA), respectively. Moreover, we call *positive motion* that in which both PA and NA are present in the transmission, and hence, there are at least two contact points or lines in a cam follower-system of this type, i.e., two or more cams attached to the input shaft interacting with two or more followers attached to the output shaft. Notice that cases like *constant-breadth* cam mechanisms use the same cam and follower interacting in two different points.

In order to identify the type of action that takes place on the follower, we resort to the value of the pressure angle. We assume in the pressure-angle analysis that friction forces are negligible, and hence, the direction of the contact force is parallel to the common normal. Furthermore, μ varies from 0° to 180° and PA and NA occur according to the rule given below:

$$\begin{array}{lll} \text{If} & 0^\circ \le \mu < 90^\circ, \text{ then} & \text{PA} \\ \text{If} & 90^\circ < \mu \le 180^\circ, \text{ then} & \text{NA} \end{array}$$

In Fig. 5.10 we plot the distribution of the pressure angle of both mechanisms for $N = 4$, $\Delta\psi = 120°$ and $\alpha_1 = 90°$, for an input-output function of the cycloidal type. Similar curves arise for other input-output functions. It is clear that the combination

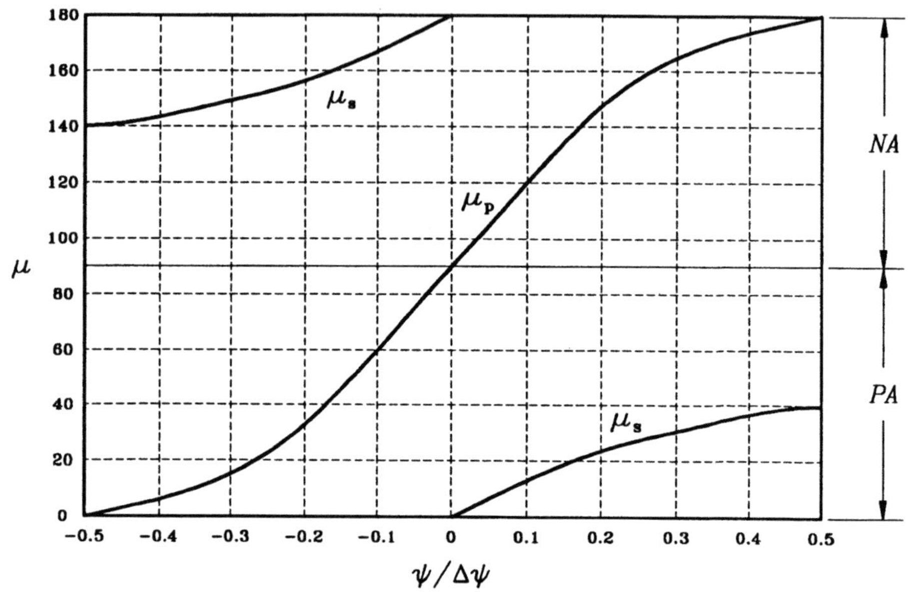

Figure 5.10 Pressure angle distribution of PRICAM.

of the two mechanisms satisfy the condition of positive motion. On the first half of the rise, the *primary mechanism* is under PA and starts with a zero pressure angle, while the *secondary mechanism* is under NA. In the second half, PA is present in the *secondary mechanism*, starting with a zero pressure angle. Shown in Fig. 5.11 is the pressure-angle distribution for three rotations of the cam. The encircled numbers indicate which roller is interacting with the cam. During the dwell phase, two rollers interact with the cam and lock the shaft of the follower.

Two prototypes, one planar and one spherical, were designed with the following characteristics:

- Four indexing steps, i.e., for a full rotation of the cam, the follower rotates 90° from one dwell to another.

- The dwell from one index to another is 2/3 of the rotation of the cam.

Prior to the manufacturing of the prototypes, solid models of both designs were created on a *Silicon Graphics Power Series Workstation (IRIS 4D/420VGX)* to animate the motion of PRICAM in its two versions. Three dimensional renderings of these designs are shown in Figs. 5.12 and 5.13, while photographs of the prototypes are shown in Figs. 5.14–5.17.

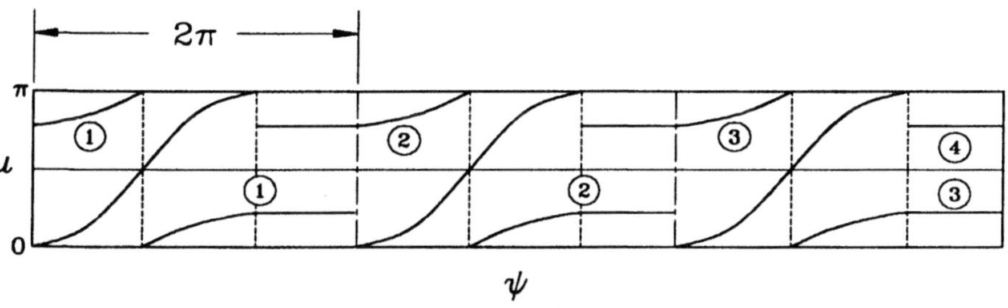

Figure 5.11 Three cycles of the cam motion

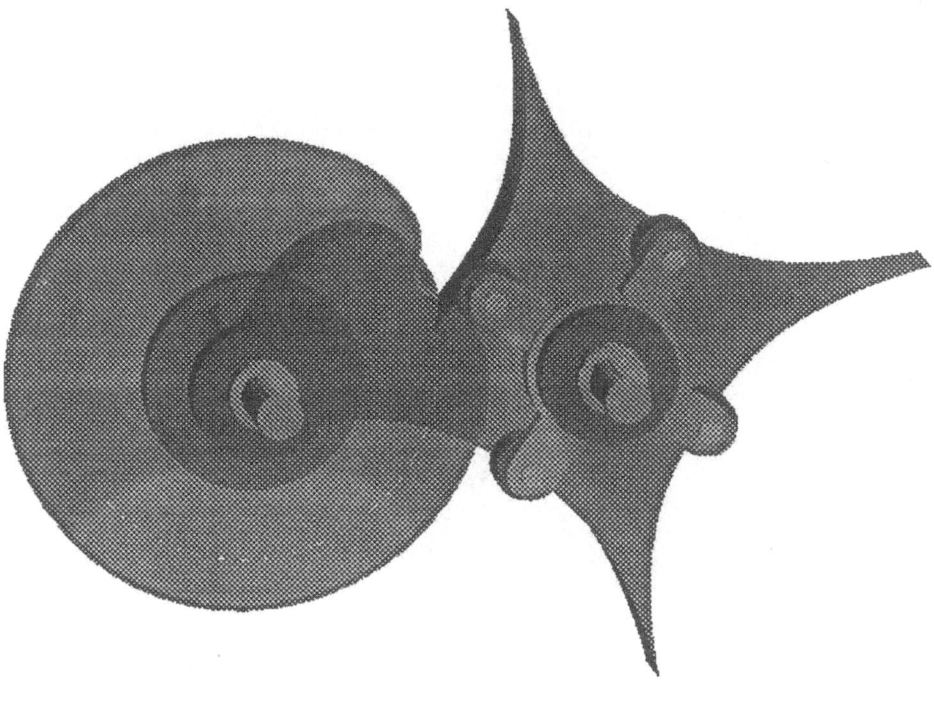

Figure 5.12 Solid model of a planar PRICAM.

Figure 5.13 Solid model of a spherical PRICAM.

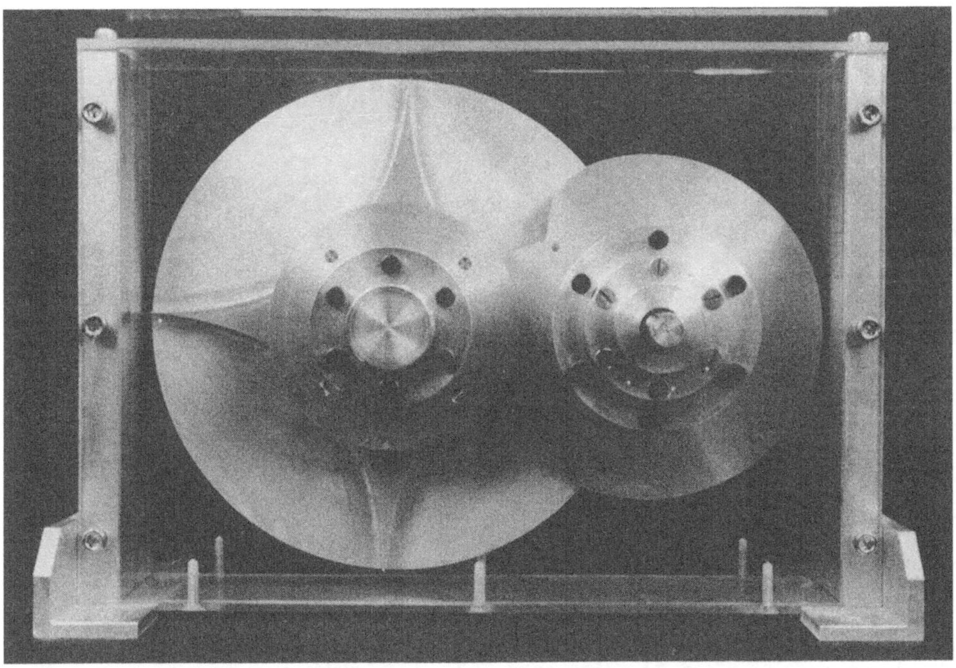

Figure 5.14 Front view of the planar PRICAM prototype

Figure 5.15 Top view of the planar PRICAM prototype

Figure 5.16 Front view of the spherical PRICAM prototype

Figure 5.17 Lateral view of the spherical PRICAM prototype

Chapter 6

THE VOLUMETRIC PROPERTIES OF CAMS

6.1 Introduction

The subject of this chapter is the *accurate* and *fast* calculation of the volumetric properties of solids of arbitrary shape, paradigms of which are the cams and followers synthesized in Chapters 3, 4 and 5. The need for accuracy needs no explanation, but the need for speed is less obvious. In fact cam synthesis, as discussed throughout this book, is only a stage of the whole mechanical design process, which includes both element and system optimization. In turn, optimization is unavoidably an iterative process, which requires a full synthesis of the overall mechanical system at each iteration. Since optimization may sometimes require hundreds of iterations, the relevance of the economy of computations within the synthesis stage is apparent.

Planar and spherical cam mechanisms contain special classes of solids, namely, cylinders bounded by planar parallel covers, and cones bounded by spherical concentric surfaces, respectively. In the spatial case, however, these solids are of a more general type, and cannot be produced with simple *primitives*. Primitives are elementary symmetrical shapes, whether planar or solid, which are fully described with a very limited amount of information. Planar primitives include the circle, the ellipse, the parabola, the hyperbola, the triangle and the quadrilateral; from these one can construct, by composition, a broad class of more general shapes. Primitive shapes in 3D include the sphere, the cylinder, the cone, the torus, the ellipsoid, the tetrahedron and the hexahedron; from these one may construct many other solids. A technique of solid modeling based on the composition of primitives is known as *Constructive Solid Geometry*, or CSG for brevity. We will refer to CSG later on in this chapter.

The shapes obtained when synthesizing cams and followers can often be described either as primitives or as combinations thereof. However, the contact surfaces belong to a more general class and cannot be composed by CSG using primitives. In this case, one must apply a more general technique of ***solid modeling***. This term is understood as the ***unambiguous*** representation of solids for computational purposes, as described in this chapter. Having a solid model of an object is of paramount importance because such a model allows one not only to render the object graphically on a monitor, but also to calculate all the geometric attributes of the object.

Of those attributes, we are interested here in the volumetric properties of an object, also known as its *moments*. Moreover, of these moments, we are interested in the first three, the simplest of which is the volume of the solid and is termed the *zeroth* moment. The first moment is the *weighted* volume of the solid, i.e., the product of the volume by the position vector of the centroid of the solid. The second moment is a matrix that gives the inertia matrix of the solid.

The importance of the first three moments of solids in mechanical design cannot be overstated. In fact, if an object is *homogeneous*, i.e., if the density of the material of which the object is made is constant throughout the object, then its volume is proportional to its weight, while its centroid coincides with its mass center and its geometric moments of inertia are proportional to its mass moments of inertia, the proportionality factor being the density. If the density is not constant, which happens when an object is made of various materials, we can divide the object into homogeneous parts. The overall moments are then calculated as the sum of the moments of the various homogeneous parts. Thus, when designing a moving mechanical element, we want to avoid inertia loads on the bearings that arise because of both *static* and *dynamic* unbalances of the element. The former refers to an offset of the mass center, while the latter to an offset of the principal axes of inertia, both with respect to the axis of rotation. Thus, knowing the volume and the first and the second moments of a solid we can determine these offsets and eliminate them by either adding or removing mass from the object; this operation is called *balancing*. Static balancing consists of eliminating the offset of the mass center, while dynamic balancing refers to eliminating the offset of the axis of rotation with respect to the *closest* principal axis of inertia.

In the discussion above we talked of the first and second moments of a solid, both of which involve the position vector of an element of volume. We assume that the position vector stems from a previously defined point. Whether this point belongs to the body or not is irrelevant. All that matters is that the same point is used to calculate the moments of all the points of the object. Usually, the second moment is defined with respect to the centroid, and hence, the first moment is computed before the second moment. Using the *parallel-axes theorem* (Meriam, 1993), the moment of inertia of a solid with respect to any other point fixed with respect to the body can be computed once the second moment about the centroid is known. The moments of inertia form a tensor quantity, and so, are represented by a matrix. This matrix is termed the *inertia matrix* of the object; it is of 2×2 for planar objects, 3×3 for solid objects. The inertia tensor is *symmetric* and *positive-definite*. The former property implies that the eigenvalues of the inertia matrix are all real and its eigenvectors are mutually orthogonal, the latter that the eigenvalues are positive. The three mutually orthogonal eigenvectors of the inertia matrix define three mutually orthogonal directions known as the *principal axes of inertia* of the object. In the planar case, there are two principal axes of inertia. When the moment of inertia is defined at the centroid of the object, the principal axes of inertia are lines passing through the centroid. Thus, in balancing an object that rotates about a given axis of rotation, we attempt *a*) to add or remove mass so that the mass center of the object lies on the axis of rotation so as to eliminate the centrifugal forces due to the motion of the mass center, and *b*) to add or remove mass so that the axis of rotation coincides with one of the principal axes of inertia, in order to eliminate the *gyroscopic* moments predicted by Euler's equations of motion about the mass center (Meriam, 1993). Obviously, the principal axis of inertia that we will try to make coincident with the axis of rotation is that making the smallest angle with the latter. We call this principal axis the closest to the axis of rotation.

In computing the volumetric properties of objects in this chapter, we will make extensive use of the *Gauss Divergence Theorem* (GDT), as described in (Al-Daccak and Angeles, 1993). The GDT offers the most economical and accurate way of calculating volume moments, because it reduces all volume integrals to surface integrals, and the latter to line integrals. Hence, by a repeated use of the GDT we can calculate the volumetric properties of solids as line integrals. Moreover, if the solid is approximated by a polyhedron, the line integrals are extremely simple to evaluate, for they entail integrations only along line segments.

We will start by recalling the GDT in its most general form, as applicable to ν-dimensional spaces, where ν can be any integer. Then, we specialize the associated expressions to two and three dimensions. The chapter includes a description of *MOMENT*, the software package that implements these calculations. In describing MOMENT, we outline the technique allowing us to produce the required database representing the solid. A few numerical examples are included at the end of the chapter. Where possible, we compare the results obtained by MOMENT with those obtained using AutoCAD^R AME^{TM}, one of the best known CAD systems with the capability of calculating volumetric properties. The accuracy of MOMENT is made apparent with these examples.

6.2 General Formulation

Let $\mathcal{E}^\nu$ be a ν-dimensional Euclidean space, in which a bounded region $\mathcal{R}$ is imbedded. The formulas presented here are valid for Euclidean spaces of arbitrary (finite) dimension. A few definitions are first introduced.

The kth moment of $\mathcal{R}$ is defined as the following integral:

$$\mathcal{I}_k = \int_{\mathcal{R}} f_k(\mathbf{r}) d\mathcal{R}, \qquad k = 0, 1, \cdots \tag{6.1}$$

Here $f_k(\mathbf{r})$ is a *homogeneous function* of kth degree of the position vector $\mathbf{r}$: for any scalar α, we have

$$f_k(\alpha \mathbf{r}) = \alpha^k f(\mathbf{r}) \tag{6.2}$$

Moreover, $f_k(\mathbf{r})$ is a kth-rank *tensor*, and hence, $\mathcal{I}_k$ also is a kth rank tensor. A kth-rank tensor is a kth-dimensional array of real numbers, the location of each individual entry within the array being indicated by an ordered set of k subscripts $\{i_j\}_{j=1}^{k}$, where each i_j runs from 1 to 3 in our case. For example, x_{3122} denotes the real number x located at the (3, 1, 2, 2) address in the array. First-rank tensors are the familiar Cartesian vectors and second-rank tensors are the familiar 3×3 matrices. The most familiar moments are the first three defined for $k = 0, 1$ and 2. The zeroth moment $\mathcal{I}_0$ is simply the volume of $\mathcal{R}$, while the first moment $\mathcal{I}_1$ is the first-rank tensor, i.e., the vector producing the position vector $\bar{\mathbf{r}}$ of the centroid of $\mathcal{R}$, namely,

$$\mathcal{I}_0 \bar{\mathbf{r}} = \mathcal{I}_1 \tag{6.3}$$

The second moment of $\mathcal{R}$, $\mathcal{I}_2$, a second-rank tensor, is the inertia matrix of $\mathcal{R}$.

Now let $\mathbf{\Phi}_m(\mathbf{r})$ be an mth-rank tensor function of $\mathbf{r}$ that is homogeneous of degree m in $\mathbf{r}$, $\text{div}(\mathbf{\Phi}_m)$ denoting its divergence. Let $(\cdot)$ denote the inner product of the tensor quantities beside it. We recall that the inner product between two tensors, one of rank m and one of rank n, is in turn a tensor of rank $m + n - 2$. The divergence operation involves an inner product, namely, the inner product of the ∇ operator (a first-rank tensor) by an mth-rank tensor. Hence, the divergence of an mth-rank tensor is an m'th-rank tensor, with $m' = m - 1$. The GDT states the relationship given below

$$\int_{\mathcal{R}} \text{div}(\mathbf{\Phi}_m) d\mathcal{R} = \int_{\mathcal{S}} \mathbf{\Phi}_m \cdot \mathbf{n} d\mathcal{S} \tag{6.4}$$

where $\mathcal{S}$ and $\mathbf{n}$ denote the boundary of $\mathcal{R}$ and the outward unit normal of this boundary, respectively. If the function $f_k(\mathbf{r})$ is such that

$$f_k(\mathbf{r}) = \text{div}(\mathbf{\Phi}_m) \equiv \nabla \cdot \mathbf{\Phi}_m \tag{6.5}$$

then eq.(6.1) reduces to an integral on $\mathcal{S}$ by application of the GDT. The relation between k and m is, clearly, $k = m - 1$. However, finding a function $\mathbf{\Phi}_m(\mathbf{r})$ whose divergence is a given function $f_k(\mathbf{r})$ can be, in general, a more difficult task than computing the volume integral of $f_k(\mathbf{r})$ directly. Nevertheless, the computation of the moments of regions, particularly the first three moments, involves the derivation of $\mathbf{\Phi}_m(\mathbf{r})$ functions that can be readily obtained, as described below. In fact, let V,

$\mathbf{q}$ and $\mathbf{I}$ denote the volume, the (vector) first moment and the (second-rank tensor) second moment of $\mathcal{R}$, the last two moments being taken with respect to a given point O. The computation of these quantities can be reduced to that of integrals on $\mathcal{S}$ by application of the GDT, as shown below.

6.2.1 Volume

In order to obtain the volume of $\mathcal{R}$ from eq.(6.1), $f_0(\mathbf{r})$ must be 1. Thus, we need to find the function Φ_1 whose divergence is unity. To this end, we first expand div($\mathbf{r}$) below:

$$\text{div}(\mathbf{r}) \equiv \nabla \cdot \mathbf{r} \equiv (\frac{\partial}{\partial r_1}\mathbf{e}_1 + \frac{\partial}{\partial r_2}\mathbf{e}_2 + \cdots + \frac{\partial}{\partial r_\nu}\mathbf{e}_\nu) \cdot (r_1\mathbf{e}_1 + r_2\mathbf{e}_2 + \cdots + r_\nu\mathbf{e}_\nu) \quad (6.6)$$

where r_i, for $i = 1, 2, \cdots, \nu$, denotes the ith component of vector $\mathbf{r}$ in the given reference frame, and hence,

$$\text{div}(\mathbf{r}) = \nu \quad (6.7)$$

from which we readily realize that

$$\Phi_1(\mathbf{r}) = \frac{\mathbf{r}}{\nu} \quad (6.8)$$

Hence, from eq.(6.4) the volume is expressed as

$$V = \frac{1}{\nu}\int_{\mathcal{S}} \mathbf{r} \cdot \mathbf{n} d\mathcal{S} \quad (6.9)$$

6.2.2 First Moment

Equation (6.1) represents the first moment of $\mathcal{R}$ when $f_1(\mathbf{r}) = \mathbf{r}$. Now we need a function $\Phi_2(\mathbf{r})$ whose divergence be the position vector $\mathbf{r}$; this function must be *a*) a second-rank tensor, and *b*) quadratic in $\mathbf{r}$. It is now apparent that there are two functions that can satisfy eq.(6.5), namely, $c_1(\mathbf{r} \cdot \mathbf{r})\mathbf{1}$ and $c_2\mathbf{r}\mathbf{r}^T$, where c_1 and c_2 are constant coefficients, as yet to be determined, and $\mathbf{1}$ denotes the $\nu \times \nu$ identity matrix. In order to determine c_1, we form

$$\begin{aligned} \text{div}[c_1((\mathbf{r} \cdot \mathbf{r})\mathbf{1}] &= c_1(\frac{\partial}{\partial r_1}\mathbf{e}_1 + \frac{\partial}{\partial r_2}\mathbf{e}_2 + \cdots + \frac{\partial}{\partial r_\nu}\mathbf{e}_\nu) \cdot (r_1^2 + r_2^2 + \cdots + r_\nu^2)\mathbf{1} \\ &\equiv 2c_1(r_1\mathbf{e}_1 + r_2\mathbf{e}_2 + \cdots + r_\nu\mathbf{e}_\nu) \cdot \mathbf{1} \equiv 2c_1\mathbf{r} \end{aligned} \quad (6.10)$$

and find

$$c_1 = \frac{1}{2}$$

This gives the result

$$(\Phi_2)_1 = \frac{1}{2}(\mathbf{r} \cdot \mathbf{r})\mathbf{1}$$

Thus,

$$\mathbf{q} = \frac{1}{2}\int_S (\mathbf{r}\cdot\mathbf{r})\mathbf{n}dS \tag{6.11}$$

Now we follow the second alternative. To find c_2, we first note that the product $\mathbf{r}\mathbf{r}^T$, termed the *tensor* product of vector $\mathbf{r}$ by itself, is a $\nu \times \nu$ matrix, namely,

$$\mathbf{r}\mathbf{r}^T \equiv \begin{bmatrix} r_1 \\ r_2 \\ \vdots \\ r_\nu \end{bmatrix} [r_1, r_2, \cdots, r_\nu] = \begin{bmatrix} r_1^2 & r_1 r_2 & \cdots & r_1 r_\nu \\ r_2 r_1 & r_2^2 & \cdots & r_2 r_\nu \\ \vdots & \vdots & \ddots & \vdots \\ r_\nu r_1 & r_\nu r_2 & \cdots & r_\nu^2 \end{bmatrix} \tag{6.12}$$

We need to find the divergence of a product, such as $\mathbf{r}\mathbf{r}^T$. Let $\mathbf{f}(\mathbf{r})$ and $\mathbf{g}(\mathbf{r})$ be two ν-dimensional vector functions of $\mathbf{r}$, the position vector of a point P in ξ^ν. Then, $\operatorname{div}(\mathbf{f}\mathbf{g}^T)$ is a ν-dimensional vector, whose jth component is found as

$$[\operatorname{div}(\mathbf{f}\mathbf{g}^T)]_j \equiv \frac{\partial}{\partial r_i}(f_i g_j) = \frac{\partial f_i}{\partial r_i} g_j + f_i \frac{\partial g_j}{\partial r_i}; \quad j = 1, \cdots, \nu \tag{6.13}$$

We note that

$$\frac{\partial f_i}{\partial r_i} g_j = [(\nabla\cdot\mathbf{f})\mathbf{g}]_j \tag{6.14}$$

$$f_i \frac{\partial g_j}{\partial r_i} = [\mathbf{f}\cdot\nabla\mathbf{g}]_j \tag{6.15}$$

and hence,

$$\operatorname{div}(\mathbf{f}\mathbf{g}^T) = (\nabla\cdot\mathbf{f})\mathbf{g} + \mathbf{f}\cdot\nabla\mathbf{g} \tag{6.16}$$

In particular, if $\mathbf{f} = \mathbf{g} = \mathbf{r}$, then we find

$$\operatorname{div}(\mathbf{r}\mathbf{r}^T) = (\nabla\cdot\mathbf{r})\mathbf{r} + \mathbf{r}\cdot\nabla\mathbf{r}, \tag{6.17}$$

but $(\nabla\cdot\mathbf{r}) = \nu$, from eq.(6.7), and $\nabla\mathbf{r} = \mathbf{1}$, so that

$$\operatorname{div}(c_2\mathbf{r}\mathbf{r}^T) = c_2(\nu+1)\mathbf{r} \tag{6.18}$$

from which we obtain

$$c_2 = \frac{1}{1+\nu} \tag{6.19}$$

$$(\Phi_2)_2 = \frac{1}{\nu+1}\mathbf{r}\mathbf{r}^T \tag{6.20}$$

This gives the alternative form

$$\mathbf{q} = \frac{1}{1+\nu}\int_S \mathbf{r}(\mathbf{r}\cdot\mathbf{n})dS \tag{6.21}$$

6.2.3 Second Moment

The second moment or the volumetric inertia matrix of $\mathcal{R}$ is represented as in eq.(6.1) if $f_2(\mathbf{r})$ is defined as

$$f_2(\mathbf{r}) = (\mathbf{r}\cdot\mathbf{r})\mathbf{1} - \mathbf{r}\mathbf{r}^T \tag{6.22}$$

Thus, $\Phi_3(\mathbf{r})$ should be a 3rd-rank tensor and cubic homogeneous in $\mathbf{r}$. The three linearly independent cubic forms in $\mathbf{r}$ are $(\mathbf{r}\cdot\mathbf{r})\mathbf{r}\mathbf{1}$, $(\mathbf{r}\cdot\mathbf{r})\mathbf{1}\mathbf{r}$ and $\mathbf{rrr}$, whose divergences can be found from the identities of tensor calculus (Brand, 1965), but are not derived here because they fall beyond the scope of this book. These relations are displayed below:

$$\text{div}[(\mathbf{r}\cdot\mathbf{r})\mathbf{r}\mathbf{1}] = 2\mathbf{r}\mathbf{r}^T + (\mathbf{r}\cdot\mathbf{r})\mathbf{1} \tag{6.23a}$$

$$\text{div}[(\mathbf{r}\cdot\mathbf{r})\mathbf{1}\mathbf{r}] = (\nu+2)(\mathbf{r}\cdot\mathbf{r})\mathbf{1} \tag{6.23b}$$

$$\text{div}[\mathbf{rrr}] = (\nu+2)\mathbf{r}\mathbf{r}^T \tag{6.23c}$$

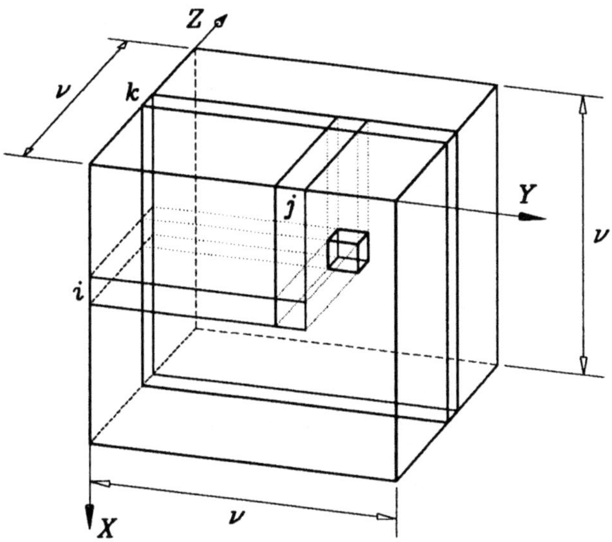

Figure 6.1 Pictorial representation of a 3rd-rank tensor.

The 3rd-rank tensor $\mathbf{rrr}$ can be interpreted as a direct generalization of its 2nd-rank counterpart $\mathbf{r}\mathbf{r}^T$. Thus, $\mathbf{rrr}$ can be regarded as a three-dimensional array, i.e., as an array of matrices. Such an array is depicted in Fig. 6.1, which shows a cube with ν slabs, each of which is an $\nu \times \nu$ matrix, a small cube representing the $(i,\, j)$

entry of the matrix in the kth slab. The kth slab of the aforementioned cube is thus the matrix $\mathbf{M}_k$ shown below:

$$\mathbf{M}_k = \begin{bmatrix} x_1y_1z_k & x_1y_2z_k & \cdots & x_1y_jz_k & \cdots & x_1y_\nu z_k \\ x_2y_1z_k & x_2y_2z_k & \cdots & x_2y_jz_k & \cdots & x_2y_\nu z_k \\ \vdots & \vdots & \ddots & \vdots & \ddots & \vdots \\ x_iy_1z_k & x_iy_2z_k & \cdots & x_iy_jz_k & \cdots & x_iy_\nu z_k \\ \vdots & \vdots & \ddots & \vdots & \ddots & \vdots \\ x_\nu y_1z_k & x_\nu y_2z_k & \cdots & x_\nu y_jz_k & \cdots & x_\nu y_\nu z_k \end{bmatrix} \tag{6.24}$$

Note that only two of the three eqs.(6.23) are linearly independent, so that we can express $\Phi_3(\mathbf{r})$ as a linear combination of $\Phi_1(\mathbf{r})$ and $\Phi_2\mathbf{r})$, namely,

$$\Phi_3(\mathbf{r}) = (\mathbf{r}\cdot\mathbf{r})[\frac{3}{2(\nu+2)}\mathbf{1}\mathbf{r} - \frac{1}{2}\mathbf{r}\mathbf{1}] \tag{6.25}$$

and, by means of eq.(6.4), $\mathbf{I}$ is readily derived as

$$\mathbf{I} = \int_{\mathcal{S}} \mathbf{r}\cdot\mathbf{r}[\frac{3}{2(\nu+2)}\mathbf{1}(\mathbf{r}\cdot\mathbf{n}) - \frac{1}{2}\mathbf{r}\mathbf{n}^T]d\mathcal{S} \tag{6.26}$$

In the sections below, these general equations are applied to two- and three-dimensional regions, under a piecewise linear approximation of their boundaries. It is pointed out that $\mathbf{q}$ can be computed by two alternative formulas, as in eq.(6.21). Moreover, the first is dimension invariant, whereas the second is more suitable for applications involving piecewise linear approximations, due to the simple forms which the $\mathbf{r}\cdot\mathbf{n}$ term produces in such cases. Both formulas will prove to be useful in deriving practical simple approximation procedures.

6.3 Planar Objects

In this section, eqs.(6.9, 6.21 & 6.26) are applied to 2-dimensional objects by defining $\nu \equiv 2$. Thus, we will derive the formulas to compute the area, the position vector of the centroid, and the inertia matrix of planar regions $\mathcal{R}$ bounded by a closed contour $\mathcal{C}$. In order to avoid confusion, we will rename V as A, the area of $\mathcal{R}$. The computation of these quantities can be reduced to the integration on the boundary $\mathcal{C}$ by application of the GDT, namely,

$$A = \frac{1}{2}\int_{\mathcal{C}} \mathbf{r}\cdot\mathbf{n}d\mathcal{C} \tag{6.27a}$$

$$\mathbf{q} = \frac{1}{2}\int_{\mathcal{C}} (\mathbf{r}\cdot\mathbf{r})\mathbf{n}d\mathcal{C} = \frac{1}{3}\int_{\mathcal{C}} \mathbf{r}(\mathbf{r}\cdot\mathbf{n})d\mathcal{C} \tag{6.27b}$$

$$\mathbf{I} = \int_{\mathcal{C}} \mathbf{r}\cdot\mathbf{r}[\frac{3}{8}\mathbf{1}(\mathbf{r}\cdot\mathbf{n}) - \frac{1}{2}\mathbf{r}\mathbf{n}^T]d\mathcal{C} \tag{6.27c}$$

Explicit formulas are now derived that apply to a piecewise linear approximation of the boundary. Thus, if $\mathcal{C}$ in eqs.(6.27a–c) is approximated by a closed n-sided polygon, then

$$\mathcal{C} \approx \bigcup_1^n \mathcal{C}_i \tag{6.28}$$

where $\mathcal{C}_i$ denotes the ith side of the polygon. Thus, eqs.(6.27) can be approximated as

$$A \approx \frac{1}{2}\sum_1^n \int_{\mathcal{C}_i} \mathbf{r}\cdot\mathbf{n}_i d\mathcal{C}_i \tag{6.29a}$$

$$\mathbf{q} \approx \frac{1}{2}\sum_1^n \int_{\mathcal{C}_i} (\mathbf{r}\cdot\mathbf{r})\mathbf{n}_i d\mathcal{C}_i = \frac{1}{3}\sum_1^n \int_{\mathcal{C}_i} \mathbf{r}(\mathbf{r}\cdot\mathbf{n}_i) d\mathcal{C}_i \tag{6.29b}$$

$$\mathbf{I} \approx \sum_1^n \int_{\mathcal{C}_i} \mathbf{r}\cdot\mathbf{r}[\frac{3}{8}\mathbf{1}(\mathbf{r}\cdot\mathbf{n}_i) - \frac{1}{2}\mathbf{r}\mathbf{n}_i^T]d\mathcal{C}_i \tag{6.29c}$$

where $\mathbf{n}_i$ is the unit outward normal vector of $\mathcal{C}_i$, and therefore constant along the ith side of the approximating polygon.

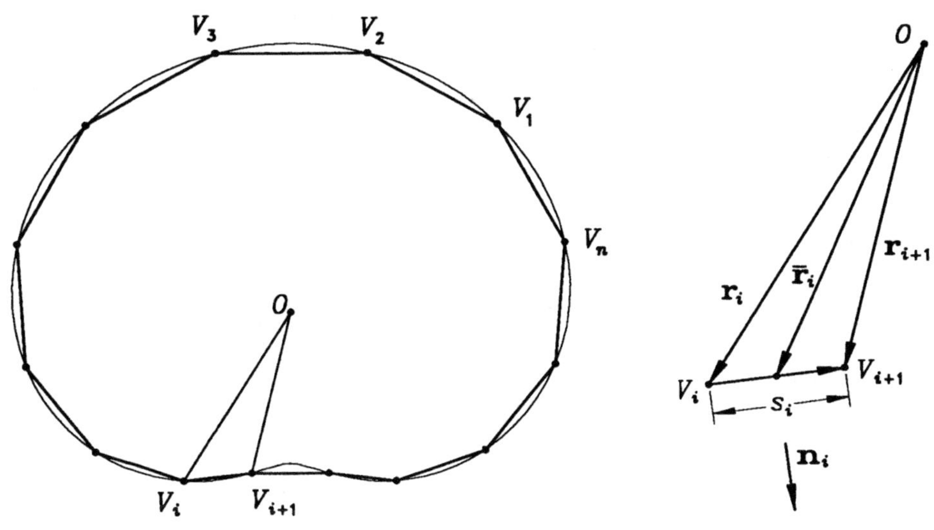

Figure 6.2 *a*) Polygon approximating the boundary $\mathcal{C}$ of a region $\mathcal{R}$ *b*) a segment of the polygon.

Furthermore, let A_i, $\mathbf{q}_i$ and $\mathbf{I}_i$ be the contributions of the ith side $\mathcal{C}_i$ of the polygon to the corresponding integral, and let s_i and $\bar{\mathbf{r}}_i$ denote the length and the position vector of its centroid, as shown in Fig. 6.2. From this figure, the reduced calculations are readily derived, namely,

$$A_i = \frac{1}{2}\mathbf{n}_i \cdot \int_{C_i} \mathbf{r} dC_i = \frac{1}{2}\mathbf{n}_i \cdot \bar{\mathbf{r}}_i s_i \tag{6.30}$$

Using each of the two formulas of eq.(6.29b), we obtain two alternative expressions for $\mathbf{q}_i$, namely,

$$\mathbf{q}_i = \frac{1}{2}(\int_C \mathbf{r} \cdot \mathbf{r} dC_i)\mathbf{n}_i \tag{6.31a}$$

$$\mathbf{q}_i = \frac{1}{3}(\int_{C_i} \mathbf{r}\mathbf{r}^T dC_i) \cdot \mathbf{n}_i \tag{6.31b}$$

Moreover, by subtracting two times both sides of eq.(6.31a) from three times both sides of eq.(6.31b), we obtain

$$\mathbf{q}_i = -\left\{\int_{C_i} [(\mathbf{r} \cdot \mathbf{r})\mathbf{1} - \mathbf{r}\mathbf{r}^T] dC_i\right\} \mathbf{n}_i \tag{6.32}$$

The right-hand side of eq.(6.32) is readily recognized to be the projection onto $-\mathbf{n}_i$ of the second moment of segment C_i with respect to O, represented here as $\mathbf{J}_i$. Thus,

$$\mathbf{q}_i = -\mathbf{J}_i \mathbf{n}_i \tag{6.33}$$

where $\mathbf{J}_i$ is defined as

$$\mathbf{J}_i \equiv \int_{C_i} [(\mathbf{r} \cdot \mathbf{r})\mathbf{1} - \mathbf{r}\mathbf{r}^T] dC_i \tag{6.34}$$

Furthermore,

$$\mathbf{I}_i = \int_{C_i} \mathbf{r} \cdot \mathbf{r}[\frac{3}{8}(\mathbf{r} \cdot \mathbf{n}_i)\mathbf{1} - \frac{1}{2}\mathbf{r}\mathbf{n}_i^T] dC_i = \frac{3}{8}\mathbf{1}(\mathbf{n}_i \cdot \mathbf{v}_i) - \frac{1}{2}\mathbf{v}_i \mathbf{n}_i^T \tag{6.35a}$$

with $\mathbf{v}_i$ defined as

$$\mathbf{v}_i \equiv \int_{C_i} (\mathbf{r} \cdot \mathbf{r})\mathbf{r} dC_i \tag{6.35b}$$

Now, $\mathbf{r}$ appearing in the integrand of eq.(6.35b) is expressed as

$$\mathbf{r} = \mathbf{r}_i + \sigma(\mathbf{r}_{i+1} - \mathbf{r}_i), \qquad 0 \leq \sigma \leq 1 \tag{6.35c}$$

where $\mathbf{r}_i$ and $\mathbf{r}_{i+1}$ denote the position vectors of the end points of C_i. Thus, the foregoing vectors are the position vectors of the ith and the $(i+1)$st vertices of the approximating polygon, which are assumed to be numbered in counterclockwise order, as shown in Fig. 6.2a. Moreover, since the polygon is closed, the sum in the subscript of $\mathbf{r}$ is to be understood modulo n, i.e., $\mathbf{r}_{n+1} = \mathbf{r}_1$ in eq.(6.35c). Substitution of eq.(6.35c) into eq.(6.35b) yields,

$$\mathbf{v}_i = \frac{s_i}{3}(A_i\mathbf{r}_i + B_i\mathbf{r}_{i+1}) \qquad \text{for } i = 1, \ldots, n \tag{6.35d}$$

where A_i and B_i are the scalars defined as

$$A_i \equiv \mathbf{r}_i \cdot \mathbf{r}_{i+1} + \frac{1}{2}\mathbf{r}_i \cdot \mathbf{r}_i + \frac{s_i^2}{4} \tag{6.35e}$$

$$B_i \equiv 2\,\mathbf{r}_i \cdot \mathbf{r}_{i+1} - \frac{1}{2}\mathbf{r}_i \cdot \mathbf{r}_i + \frac{3}{4}s_i^2 \tag{6.35f}$$

Equations (6.30), (6.33) and (6.35a–f) are the relations sought.

6.4 Solids

For solid objects, $\nu = 3$ and the general relations (6.9, 6.21 & 6.26) take on the forms

$$V = \frac{1}{3}\int_{\mathcal{S}} \mathbf{r} \cdot \mathbf{n} d\mathcal{S} \tag{6.36a}$$

$$\mathbf{q} = \frac{1}{2}\int_{\mathcal{S}} (\mathbf{r} \cdot \mathbf{r})\mathbf{n} d\mathcal{S} = \frac{1}{4}\int_{\mathcal{S}} \mathbf{r}(\mathbf{r} \cdot \mathbf{n}) d\mathcal{S} \tag{6.36b}$$

$$\mathbf{I} = \int_{\mathcal{S}} \mathbf{r} \cdot \mathbf{r}[\frac{3}{10}\mathbf{1}(\mathbf{r} \cdot \mathbf{n}) - \frac{1}{2}\mathbf{r}\mathbf{n}^T] d\mathcal{S} \tag{6.36c}$$

where $\mathcal{S}$ denotes the bounding surface. Explicit formulas are derived below that are applicable to piecewise-linear approximations of boundaries of solids of arbitrary shapes.

One simple approximation of the boundary can be obtained by means of a polyhedron formed by planar polygonal faces. The integrals appearing in eqs.(6.36a–c) then can be expressed as sums of integrals over the polyhedral faces, the whole boundary $\mathcal{S}$ thus being approximated as

$$\mathcal{S} \approx \bigcup_1^n \mathcal{S}_i \tag{6.37}$$

which is similar to the approximation appearing in eq.(6.28), except that now each part $\mathcal{S}_i$ is a polygonal portion of a plane. The integral formulas, thus, can be approximated as:

$$V \approx \frac{1}{3}\sum_{1}^{n}\int_{\mathcal{S}_i} \mathbf{r}\cdot\mathbf{n}_i d\mathcal{S}_i \tag{6.38a}$$

$$\mathbf{q} \approx \frac{1}{2}\sum_{1}^{n}\int_{\mathcal{S}_i} (\mathbf{r}\cdot\mathbf{r})\mathbf{n}_i d\mathcal{S}_i = \frac{1}{4}\sum_{1}^{n}\int_{\mathcal{S}_i} \mathbf{r}(\mathbf{r}\cdot\mathbf{n}_i) d\mathcal{S}_i \tag{6.38b}$$

$$\mathbf{I} \approx \sum_{1}^{n}\int_{\mathcal{S}_i} \mathbf{r}\cdot\mathbf{r}[\frac{3}{10}\mathbf{1}(\mathbf{r}\cdot\mathbf{n}_i) - \frac{1}{2}\mathbf{r}\mathbf{n}_i^T] d\mathcal{S}_i \tag{6.38c}$$

Now, let V_i, $\mathbf{q}_i$ and $\mathbf{I}_i$ be the contribution of the ith face, $\mathcal{S}_i$, of the polyhedron, to the corresponding integral, Δ_i, $\overline{\mathbf{r}}_i$ and $\mathbf{I}_{\Delta_i}$ being the area, the position vector of the centroid and the inertia matrix of the polygon $\mathcal{S}_i$, respectively, the last two quantities being taken with respect to O. Thus,

$$V_i = \frac{1}{3}\mathbf{n}_i \cdot \int_{\mathcal{S}_i} \mathbf{r} d\mathcal{S}_i = \frac{1}{3}\mathbf{n}_i \cdot \overline{\mathbf{r}}_i \Delta_i \tag{6.39}$$

The polygon area Δ_i and its centroid $\overline{\mathbf{r}}_i$ are calculated by using the method of planar regions presented in Section 6.3.

Furthermore, by subtracting two times the second part of (6.38b) from the first one, the relation shown below is obtained:

$$\begin{aligned} \mathbf{q}_i &= -\frac{1}{2}\int_{\mathcal{S}_i} [(\mathbf{r}\cdot\mathbf{r})\mathbf{n}_i - (\mathbf{r}\cdot\mathbf{n}_i)\mathbf{r}] d\mathcal{S}_i \\ &= -\frac{1}{2}\left\{\int_{\mathcal{S}_i} [(\mathbf{r}\cdot\mathbf{r})\mathbf{1} - \mathbf{r}\mathbf{r}^T] d\mathcal{S}_i\right\}\mathbf{n}_i \end{aligned} \tag{6.40}$$

The second integral of eq.(6.40) is readily identified as $\mathbf{I}_{\Delta_i}$, the second moment of polygon $\mathcal{S}_i$. Hence,

$$\mathbf{q}_i = -\frac{1}{2}\mathbf{I}_{\Delta_i}\mathbf{n}_i \tag{6.41}$$

which is a relation similar to that represented by eq.(6.33), i.e., the contribution of $\mathcal{S}_i$ to the first moment of $\mathcal{R}$ is recognized to be one-half the projection onto $-\mathbf{n}_i$ of the second moment of $\mathcal{S}_i$, both moments being taken, of course, with respect to the same point O. The centroidal inertia matrix of the polygon at hand is calculated using formulas developed for 2-D regions in a plane defined by the polygon. Using the parallel-axes theorem, we find $\mathbf{I}_{\Delta_i}$ from the calculated centroidal inertia matrix.

Matrix $\mathbf{I}_{\Delta_i}$ is thus defined as

$$\mathbf{I}_{\Delta_i} \equiv \int_{\mathcal{S}_i} [(\mathbf{r}\cdot\mathbf{r})\mathbf{1} - \mathbf{r}\mathbf{r}^T] d\mathcal{S}_i \tag{6.42}$$

Additionally, one has, for the contribution of the ith polygonal face to the second moment of the 3-D region under study,

$$\mathbf{I}_i = \frac{1}{2}\int_{\mathcal{S}_i} \mathbf{r}\cdot\mathbf{r}[\frac{3}{5}(\mathbf{r}\cdot\mathbf{n}_i)\mathbf{1} - \mathbf{r}\mathbf{n}_i^T]d\mathcal{S}_i$$

$$= \frac{3}{10}(\mathbf{n}_i\cdot\mathbf{w}_i)\mathbf{1} - \frac{1}{2}\mathbf{w}_i\mathbf{n}_i^T \tag{6.43a}$$

with $\mathbf{w}_i$ defined as:

$$\mathbf{w}_i \equiv \int_{\mathcal{S}_i} (\mathbf{r}\cdot\mathbf{r})\mathbf{r}d\mathcal{S}_i \tag{6.43b}$$

The integral appearing in eq.(6.43b) is evaluated below. To this end, $\mathbf{r}$ is expressed as

$$\mathbf{r} = \overline{\mathbf{r}}_i + \mathbf{p}_i \tag{6.44}$$

where $\mathbf{p}_i$ is a vector lying in the plane of the polygon $\mathcal{S}_i$ that stems from its centroid, as shown in Fig. 6.3a. Now $\mathbf{w}_i$ becomes

$$\mathbf{w}_i = \overline{\mathbf{r}}_i^2\Delta_i\mathbf{r} + \overline{\mathbf{r}}_i\cdot(\mathbf{A}_i + a_i\mathbf{1}) + \mathbf{b}_i \tag{6.45}$$

$\mathbf{A}_i$, a_i and $\mathbf{b}_i$ being the surface integrals over $\mathcal{S}_i$ defined as

$$\mathbf{A}_i = \int_{\mathcal{S}_i} \mathbf{p}_i\mathbf{p}_i^T d\mathcal{S}_i \tag{6.46a}$$

$$a_i = \int_{\mathcal{S}_i} \mathbf{p}_i^2 d\mathcal{S}_i \tag{6.46b}$$

$$\mathbf{b}_i = \int_{\mathcal{S}_i} \mathbf{p}_i^2\mathbf{p}_i d\mathcal{S}_i \tag{6.46c}$$

An exponent k over a vector quantity indicates the kth power of the magnitude of the vector, i.e., $\mathbf{r}^2 \equiv \mathbf{r}\cdot\mathbf{r} = ||\mathbf{r}||^2$, a notation that will be used throughout this section.

Since $\mathbf{p}_i$ of eqs.(6.46) is a vector lying entirely in the plane Π_i defined by the polygon $\mathcal{S}_i$, it can be represented uniquely in the 2-D subspace as a 2-D vector of that plane. Consequently, the GDT can be applied in this 2-D subspace to reduce the surface integrals (6.46a–c) to line integrals (Al-Daccak and Angeles, 1993), namely,

$$\mathbf{A}_i = \frac{1}{4}\int_{\mathcal{S}_i} \operatorname{div}(\mathbf{P}_i)d\mathcal{S}_i = \frac{1}{4}\int_{\mathcal{C}_i} \mathbf{p}_i\mathbf{p}_i^T(\mathbf{p}_i\cdot\overline{\mathbf{n}}_i)d\mathcal{C}_i \tag{6.47a}$$

$$a_i = \frac{1}{4}\int_{\mathcal{S}_i} \operatorname{div}(\mathbf{p}_i^2\mathbf{p}_i)d\mathcal{S}_i = \frac{1}{4}\int_{\mathcal{C}_i} \mathbf{p}_i^2(\mathbf{p}_i\cdot\overline{\mathbf{n}}_i)d\mathcal{C}_i \tag{6.47b}$$

$$\mathbf{b}_i = \frac{1}{5}\int_{\mathcal{S}_i} \operatorname{div}(\mathbf{p}_i^2\mathbf{p}_i\mathbf{p}_i^T)d\mathcal{S}_i = \frac{1}{5}\int_{\mathcal{C}_i} \mathbf{p}_i^2\mathbf{p}_i(\mathbf{p}_i\cdot\overline{\mathbf{n}}_i)d\mathcal{C}_i \tag{6.47c}$$

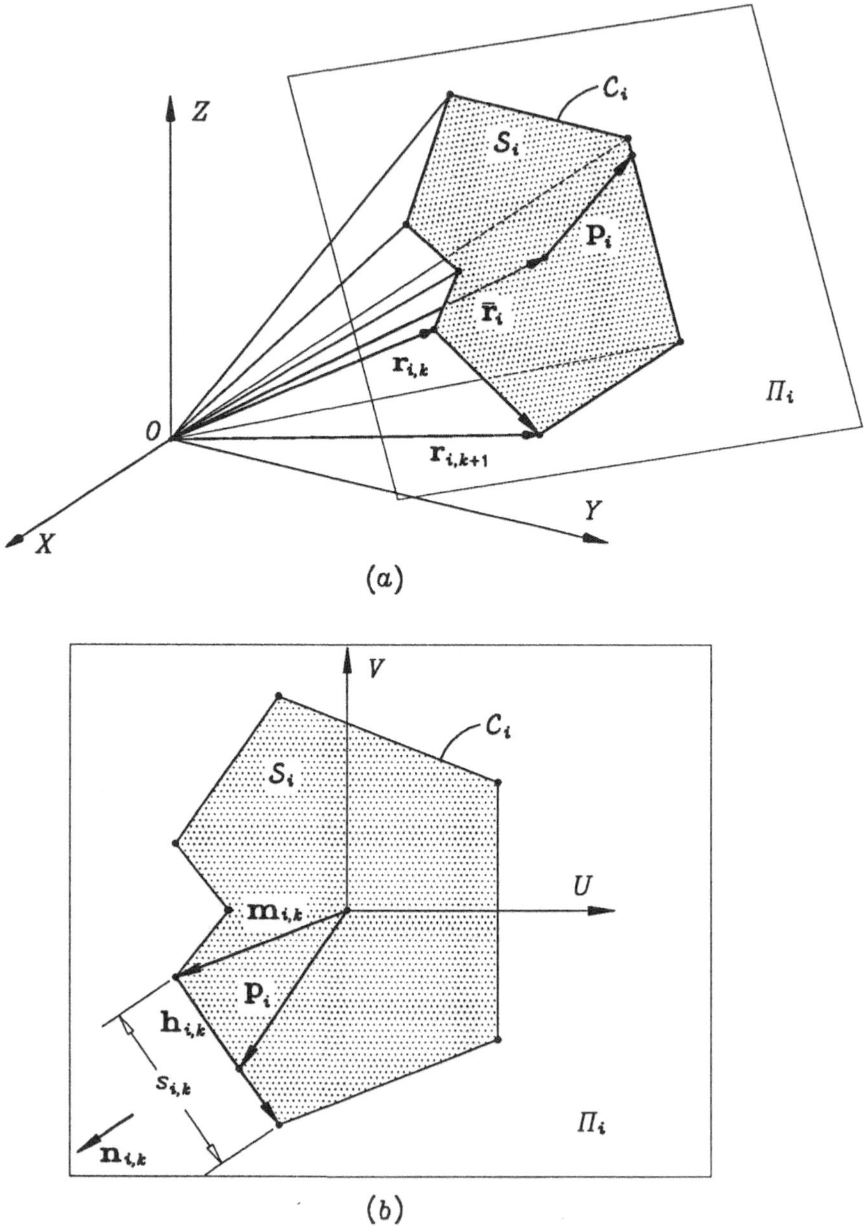

Figure 6.3 *a*) Polygon representing one face of a polyhedron approximating a closed surface; *b*) planar view of Π_i.

Here $\mathbf{P}_i \equiv \mathbf{p}_i\mathbf{p}_i\mathbf{p}_i$ is a *third-rank* tensor that is cubic and homogeneous in $\mathbf{p}_i$, whose divergence is given as

$$\mathrm{div}(\mathbf{P}_i) \equiv \mathrm{div}(\mathbf{p}_i\mathbf{p}_i\mathbf{p}_i) = (2+\nu)\mathbf{p}_i\mathbf{p}_i^T \tag{6.48}$$

Moreover, $\mathcal{C}_i$ denotes the polygonal boundary of $\mathcal{S}_i$ and $\overline{\mathbf{n}}_i$ the outward unit normal vector of $\mathcal{C}_i$ contained in the plane Π_i.

Now, let $\mathcal{C}_{i,k}$ denote the kth side of the m_i-sided polygon $\mathcal{S}_i$ joining the kth and the $(k+1)$st vertices, which are numbered counterclockwise when the face of interest is viewed from outside of the polyhedron. Moreover, a sum over subscript k is to be understood, henceforth, as modulo m_i. Furthermore, the position vector $\mathbf{p}_i$, shown in Fig. 6.3*b*, of any point of $\mathcal{C}_{i,k}$ is defined in plane Π_i as

$$\mathbf{p}_i = \mathbf{m}_{i,k} + \mathbf{h}_{i,k}s, \qquad 0 \le s \le 1, \quad \mathbf{p}_i \in \mathcal{C}_{i,k} \tag{6.49}$$

where $\mathbf{m}_{i,k}$ and $\mathbf{h}_{i,k}$ are defined as

$$\mathbf{m}_{i,k} \equiv \mathbf{r}_{i,k} - \overline{\mathbf{r}}_i, \qquad \mathbf{h}_{i,k} \equiv \mathbf{r}_{i,k+1} - \mathbf{r}_{i,k} \tag{6.50}$$

$\mathbf{r}_{i,k}$ being the position vector of the kth vertex of polygon $\mathcal{S}_i$. Similar to vector $\mathbf{p}_i$, vectors $\mathbf{m}_{i,k}$ and $\mathbf{h}_{i,k}$ lie entirely in the plane Π_i. Consequently, their representation as 2-D vectors in that plane is used in order to apply the GDT to reduce the surface integrals defined over $\mathcal{S}_i$ to line integrals defined over $\mathcal{C}_{i,k}$.

Let $\mathbf{n}_{i,k}$ be the unit normal vector to $\mathcal{C}_{i,k}$, and $s_{i,k}$ be the length of the kth side of $\mathcal{C}_i$, as shown in Fig. 6.3*b*. Thus, quantities $\mathbf{A}_i$, a_i, and $\mathbf{b}_i$ can be evaluated as indicated below, keeping in mind that $s_{i,k} = ||\mathbf{h}_{i,k}||$ and $\mathbf{p}_i \cdot \mathbf{n}_{i,k} = \mathbf{m}_{i,k} \cdot \mathbf{n}_{i,k}$, since $\mathbf{h}_{i,k} \cdot \mathbf{n}_{i,k} = 0$. Then,

$$\mathbf{A}_i = \frac{1}{4}\sum_{k=1}^{m} s_{i,k}(\mathbf{m}_{i,k} \cdot \mathbf{n}_{i,k}) \int_0^1 \mathbf{p}_i\mathbf{p}_i^T ds \tag{6.51a}$$

where

$$\mathbf{p}_i\mathbf{p}_i^T = (\mathbf{m}_{i,k}\mathbf{m}_{i,k}^T) + (\mathbf{m}_{i,k}\mathbf{h}_{i,k}^T + \mathbf{h}_{i,k}\mathbf{m}_{i,k}^T)s + (\mathbf{h}_{i,k}\mathbf{h}_{i,k}^T)s^2 \tag{6.51b}$$

and hence,

$$\int_0^1 \mathbf{p}_i\mathbf{p}_i^T ds = (\mathbf{m}_{i,k}\mathbf{m}_{i,k}^T) + \frac{1}{2}(\mathbf{m}_{i,k}\mathbf{h}_{i,k}^T + \mathbf{h}_{i,k}\mathbf{m}_{i,k}^T) + \frac{1}{3}(\mathbf{h}_{i,k}\mathbf{h}_{i,k}^T) \tag{6.51c}$$

Also,

$$a_i = \frac{1}{4}\sum_{k=1}^{m} s_{i,k}(\mathbf{m}_{i,k} \cdot \mathbf{n}_{i,k}) \int_0^1 \mathbf{p}_i^2 ds \tag{6.52a}$$

where

$$\mathbf{p}_i^2 = \mathbf{m}_{i,k}^2 + 2\mathbf{m}_{i,k} \cdot \mathbf{h}_{i,k}s + \mathbf{h}_{i,k}^2 s^2 \tag{6.52b}$$

and hence,

$$\int_0^1 \mathbf{p}_i^2 ds = \mathbf{m}_{i,k}^2 + \mathbf{m}_{i,k} \cdot \mathbf{h}_{i,k} + \frac{1}{3}\mathbf{h}_{i,k}^2 \tag{6.52c}$$

Furthermore,

$$\mathbf{b}_i = \frac{1}{5}\sum_{k=1}^{m} s_{i,k}(\mathbf{m}_{i,k} \cdot \mathbf{n}_{i,k}) \int_0^1 \mathbf{p}_i^2 \mathbf{p}_i ds \tag{6.53a}$$

where

$$\mathbf{p}_i^2 \mathbf{p}_i = \mathbf{p}_i^2 \mathbf{m}_{i,k} + \mathbf{p}_i^2 s \mathbf{h}_{i,k} \tag{6.53b}$$

whence,

$$\int_0^1 \mathbf{p}_i^2 \mathbf{p}_i ds = \mathbf{m}_{i,k}(\mathbf{m}_{i,k}^2 + \mathbf{m}_{i,k} \cdot \mathbf{h}_{i,k} + \frac{1}{3}\mathbf{h}_{i,k}^2) + \mathbf{h}_{i,k}(\frac{1}{2}\mathbf{m}_{i,k}^2 + \frac{2}{3}\mathbf{m}_{i,k} \cdot \mathbf{h}_{i,k} + \frac{1}{4}\mathbf{h}_{i,k}^2) \tag{6.53c}$$

Now that the three surface integrals (6.46a, b, c) have been reduced to line integrals and evaluated in plane Π_i, the results obtained in this 2-D subspace should be mapped to produce the results necessary for calculating $\mathbf{w}_i$ in the 3-D space, eq.(6.45). The scalar quantity a_i poses no problems and is readily multiplied by the 3×3 identity matrix $\mathbf{1}$ in the expression for $\mathbf{w}_i$. The second-rank tensor $\mathbf{A}_i$ and the vector $\mathbf{b}_i$ in the 2-D subspace are transformed to their counterpart second-rank tensor and vector, respectively, in the 3-D space, before being substituted in the expression for $\mathbf{w}_i$.

This completes the calculation of $\mathbf{w}_i$, in terms of which the second moment matrix of the piecewise-linear approximation of $\mathcal{R}$ is determined—see eqs.(6.43a & b). The above formulas have been implemented in a C code called MOMENT (Al-Daccak, 1989) for computing the global volumetric properties of arbitrary solids approximated by polyhedra. In Section 6.5 we explain how to prepare the database needed to run MOMENT.

6.5 An Outline of MOMENT

In previous sections we derived the formulation for the linear approximation of the volumetric properties of complex regions bounded by a closed surface. Thus, the problem is reduced to decomposing the bounding surface of a volume into a polyhedron, and hence, it is only necessary to know the position vectors of the vertices in a given coordinate frame.

As mentioned above, MOMENT was written for computing the global volumetric properties of arbitrary solids. The database needed to run MOMENT, called MOMENTDB, should be prepared in a systematic form. We can proceed to create MOMENTDB in three steps, namely, *i*) Limit values:

The number of vertices a and faces f of the solid must be given. Although this part is the simplest one, this information is important and should be consistent with the rest of the file, i.e., the program will take the values of a and f as limits of the two *loops* that will read the values provided by items *ii*) and *iii*).

ii) Vertex coordinates:

The list of vertices should be provided in this part. They are read and stored in an array of $a \times 3$ elements. Moreover, the kth element, for $k = 1, 2, \cdots, a$, is defined as vert_k.

iii) Vertex connectivity:

Because the boundary of a solid is approximated by f polygons, it is expected that one vertex will be common to several polygons, while the polygons may have different numbers of sides. Therefore, we have two sets of values to provide at the ℓth entry, where $\ell = 1, 2, \cdots, f$. The first set is composed of a single integer s denoting the number of sides of the ℓth polygon. The second is composed of s integers, which, based on the order of the data provided in item *ii*), will describe the order of connection of the vertices. As mentioned before, the vertices must be numbered counterclockwise when the polyhedron is viewed from outside. Furthermore, each entry being composed by $s_k + 1$ elements, we denote it by poly_k.

Example 6.5.1: *Create the data base needed to compute the volumetric properties of the parallelepiped of sides* $2u$, $2v$ *and* $2w$ *in the coordinate frame shown in Fig. 6.4, where* O *is located at the centroid and the coordinate axes are parallel to the edges.*

Solution:

For the first part of the database we know that $a = 8$ and $f = 6$. For the second part, we write the coordinates of the vertices in the same order as they are numbered in Fig. 6.4, and for the third part, it is obvious that $n = 4$ for all polygons. Hence, we must give 4 integers describing the order of connection at each entry. The required database is shown in Table 6.1. ∎

We can imagine that the boundary surface of the solid representing a cam or a follower is composed of three regions, namely, $\mathcal{R}_C$, $\mathcal{R}_B$ and $\mathcal{R}_T$, which denote the contact surface, the bottom surface and the top surface, respectively. Furthermore, the way we have defined $\mathcal{R}_C$, $\mathcal{R}_B$ and $\mathcal{R}_T$ in Chapters 3 and 4 allows us to approximate those regions as polygonal faces of three and four sides, as we will show below. Thus, $\mathcal{R}_C$ can be expressed in parametric form as

$$\mathcal{R}_C : \mathbf{r}(\psi, \lambda) = \mathbf{p}(\psi) + \lambda \mathbf{e}(\psi) \tag{6.54a}$$

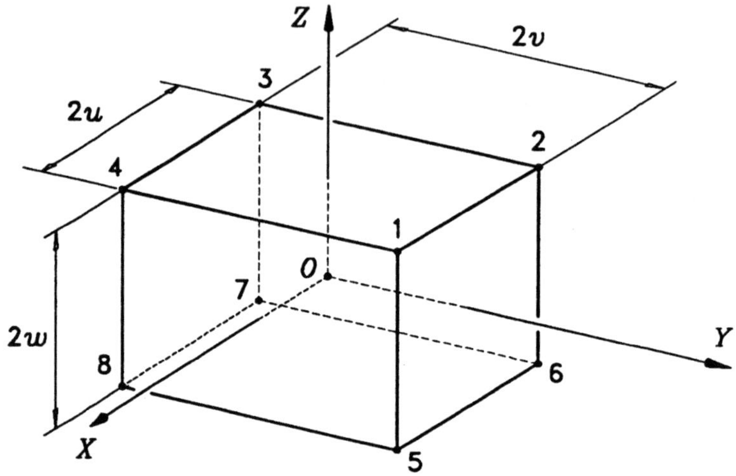

Figure 6.4 Connectivity of the vertices of a parallelepiped.

Table 6.1 Database of Example 6.5.1.

8	6			
u	v	w		
$-u$	v	w		
$-u$	$-v$	w		
u	$-v$	w		
u	v	$-w$		
$-u$	v	$-w$		
$-u$	$-v$	$-w$		
u	$-v$	$-w$		
4	1	2	3	4
4	1	5	6	2
4	2	6	7	3
4	3	7	8	4
4	4	8	5	1
4	8	7	6	5

If now we discretize the values of ψ and λ, we can define the *mesh* shown in Fig. 6.5*a*, which is mapped into $\mathcal{R}_C$ as illustrated in Fig. 6.5*b* with the aid of eq.(6.54a). Moreover, a set of polygons for which i is constant while j varies from 1 to n is called a *strip*. We can say that the mesh of Fig. 6.5*a* is composed of m strips of n polygons per strip. Of course, a strip can be defined the other way around, i.e., as a set of polygons for which j is constant while i varies from 1 to m.

Similarly, we can create the meshes of $\mathcal{R}_T$ and $\mathcal{R}_B$, which, in view of eqs.(4.37), are defined as

$$\mathcal{R}_T : \mathbf{u}(\psi, \nu) = \mathbf{S}^T(\psi)[\mathbf{p}(\nu) + \lambda_{max}\mathbf{e}(\nu)] \tag{6.54b}$$

$$\mathcal{R}_B : \mathbf{b}(\psi, \nu) = \mathbf{S}^T(\psi)[\mathbf{p}(\nu) + \lambda_{min}\mathbf{e}(\nu)] \tag{6.54c}$$

where $\mathbf{u}$ and $\mathbf{b}$ denote the position vectors defining the upper and bottom surfaces, respectively, while $\mathbf{S}(\psi)$ denotes a rotation about the Z-axis by the angle ψ, namely,

$$\mathbf{S}(\psi) = \begin{bmatrix} \cos\psi & -\sin\psi & 0 \\ \sin\psi & \cos\psi & 0 \\ 0 & 0 & 1 \end{bmatrix} \tag{6.55}$$

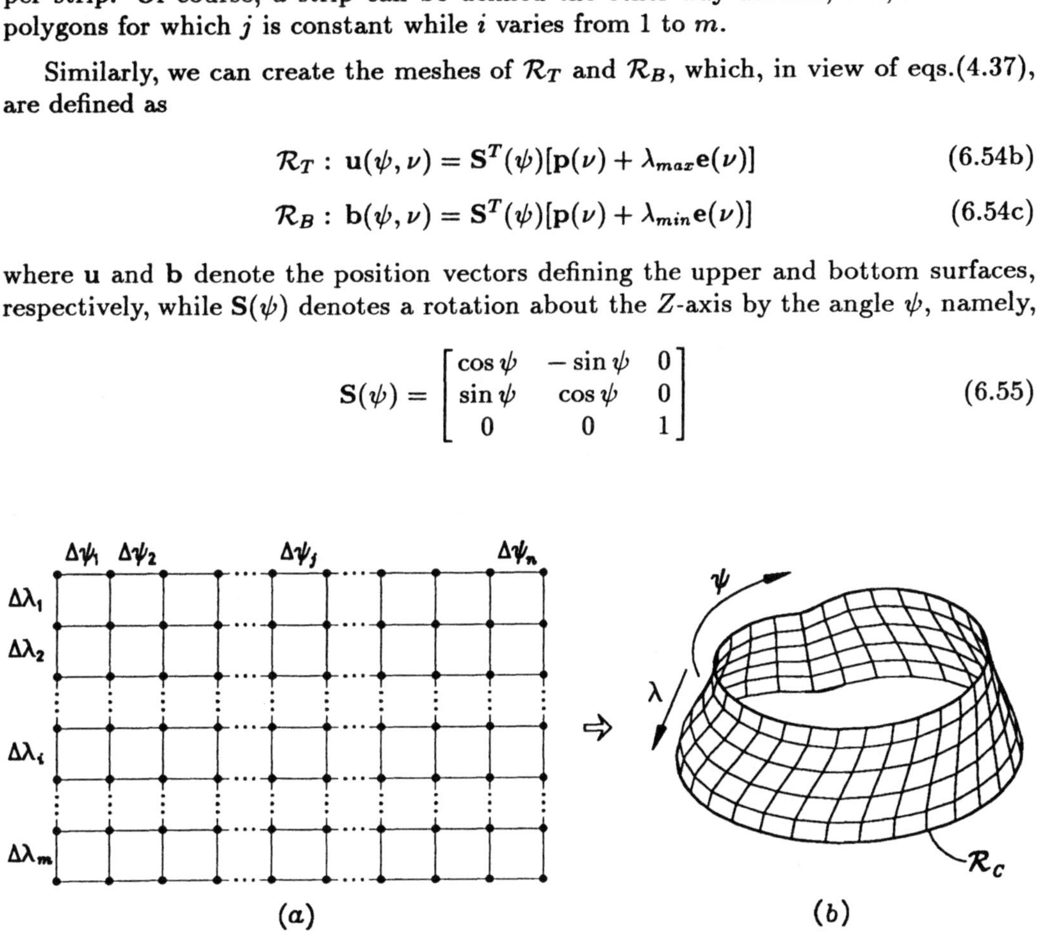

Figure 6.5 The contact surface of a cam defined by the region $\mathcal{R}_C$.

In the applications we present here, the number of strips q of $\mathcal{R}_T$ is equal to the number of strips of $\mathcal{R}_B$. Since $\mathcal{R}_C$ is a ruled surface, we can define it with only one strip. Thus, if we join the three regions, we will obtain the total number of strips of the solid, namely,

$$m = 2q + 1 \tag{6.56}$$

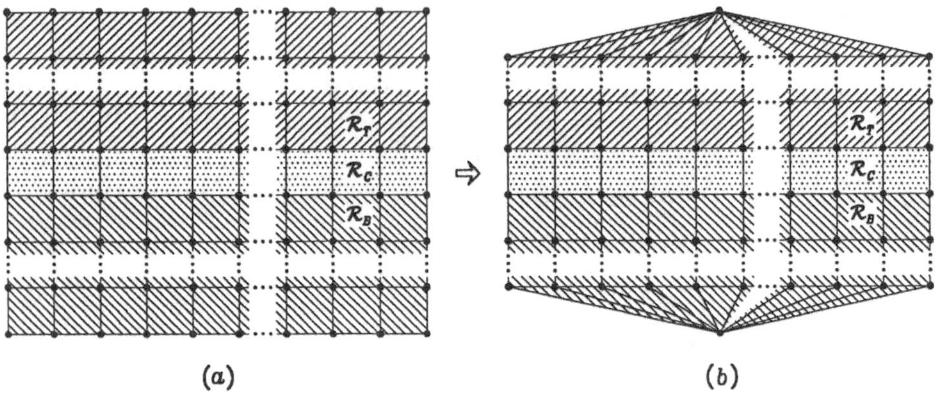

Figure 6.6 One row of vertices of the first and last strips of the whole region of a solid cam collapsing into one single vertex.

Let us refer now to Fig. 6.6*a*, in which we have joined the three regions. In all the cases we present here, one set of points of the first and last strips coincides in one vertex, giving the mesh shown in Fig. 6.6*b*. The *nodes* of the mesh of Fig. 6.6*b* are precisely the vertices of the polyhedron approximating the solid. With the mesh of Fig. 6.6*b*, we can obtain two types of solids, that we call here *first-* and *second-class* solids.

a) First-class solids.

In this case, the polyhedron is closed by connecting the last column of the vertices of Fig. 6.6*b* with the first column. An example of this case is a solid of revolution. Now, we are ready to write the algorithm needed to prepare MOMENTDB:

a.i) For the first part, the number of vertices a and polygons f are readily obtained in terms of m and n, namely,

$$a = (m-1)n + 2 \tag{6.57a}$$
$$f = mn \tag{6.57b}$$

a.ii) The algorithm applied to write the second part of MOMENTDB is described below:

```
vert_1 ← r_11
k = 2
For i = 2 till m − 1 do
    For j = 1 till n do
        vert_k ← r_ij
        k = k + 1
    enddo
enddo
vert_a ← r_m1
```

a.iii) The third part of MOMENTDB, representing the *connectivity* of the nodes, can be written with the aid of the algorithm below, that shows three loops. The first loop is the composition of the strip for $m = 1$, the second is the composition of the strips from $m = 2$ to $m = a - 1$, and the third, the strip for $m = a$, namely,

```
k = 2, ℓ = 1
For j = 1 till n do
    if j = n then
        poly_ℓ ← 3, 1, k, k + 1 − n
    else
        poly_ℓ ← 3, 1, k, k + 1
    endif
    k = k + 1, ℓ = ℓ + 1
enddo
```

```
k = 2
For i = 2 till m − 1 do
    For j = 1 till n do
        if j = n then
            poly_ℓ ← 4, k, k + n, k + 1, k + 1 − n
        else
            poly_ℓ ← 4, k, k + n, k + n + 1, k + 1
        endif
        k = k + 1, ℓ = ℓ + 1
    enddo
enddo
```

```
For j = 1 till n do
    if j = n then
        poly_ℓ ← 3, k, a, k + 1 − n
    else
        poly_ℓ ← 3, k, a, k + 1
    endif
    k = k + 1, ℓ = ℓ + 1
enddo
```

b) Second-class solids.

In this case, the polyhedron is closed by forming extra polygons with the first and last columns of the vertices of $\mathcal{R}_T$, with the corresponding opposite vertices of $\mathcal{R}_C$ of Fig. 6.6*b*. In order to illustrate this case, we have prepared an example having a mesh with $q = 3$ and $n = 5$, as shown in Fig. 6.7*a*, where the shaded area represents the extra polygons. We include below the algorithm to obtain MOMENTDB.

b.i) The numbers of vertices a and polygons f are readily obtained in terms of m and n, namely,

$$a = (m-1)(n+1) + 2 \tag{6.58a}$$

$$f = m(n+1) - 1 \tag{6.58b}$$

b.ii) The algorithm applied to write the second part of MOMENTDB is described below:

```
vert_1 ← r_11
k = 2
For i = 2 till m − 1 do
    For j = 1 till n + 1 do
        vert_k ← r_ij
        k = k + 1
    enddo
enddo
vert_a ← r_m1
```

b.iii) The third part of MOMENTDB, representing the connectivity of the nodes, can be written with the aid of the algorithm below, that shows four loops. The first loop is the composition of the strip for $m = 1$, while the second is

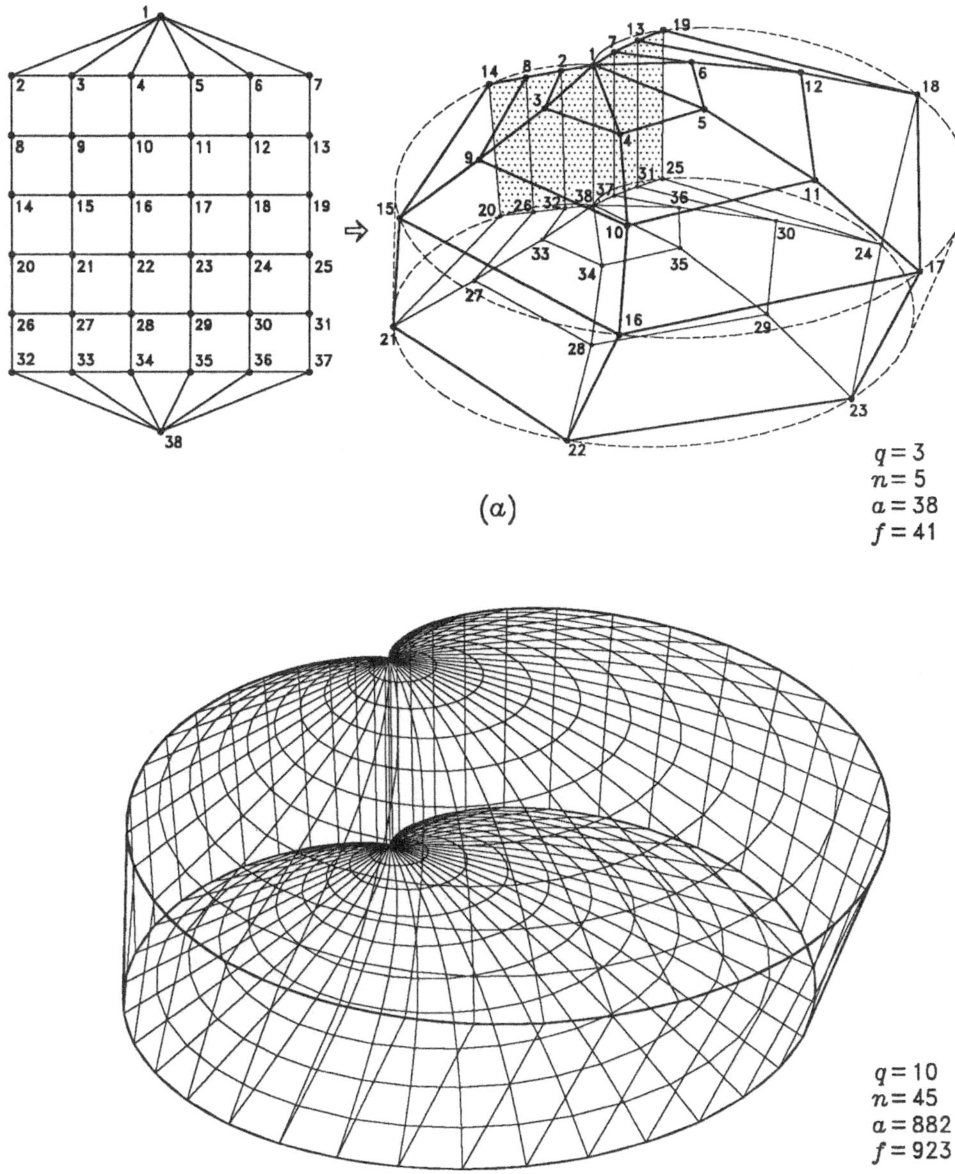

Figure 6.7 *a*) Rough approximation of a solid of the second class to illustrate the interconnection of vertices; *b*) more realistic approximation of the same solid.

the composition of the strips from $m = 2$ to $m = a-1$, the third loop being the strip for $m = a$, and the fourth, the strip with the extra polygons:

$k = 2$, $\ell = 1$
For $j = 1$ **till** n **do**
  $\text{poly}_\ell \leftarrow 3,\ 1,\ k,\ k+1$
  $k = k+1$, $\ell = \ell + 1$
enddo

$k = 2$
For $i = 2$ **till** $m-1$ **do**
  For $j = 1$ **till** n **do**
    $\text{poly}_\ell \leftarrow 4,\ k,\ k+n,\ k+n+1,\ k+1$
    $k = k+1$, $\ell = \ell + 1$
  enddo
  $k = k+1$
enddo

For $j = 1$ **till** n **do**
  $\text{poly}_\ell \leftarrow 3,\ k,\ a,\ k+1$
  $k = k+1$, $\ell = \ell + 1$
enddo

For $k = 1$ **till** $q-1$ **do**
  $\text{poly}_\ell \leftarrow 4,\ 2+(n+1)(k-1),\ 2+(n+1)(m-k-1),\ 2+(n+1)(m-k-2),\ (n+1)k$
  $\text{poly}_{\ell+1} \leftarrow 4,\ 1+(n+1)(m-k),\ 1+(n+1)k,\ 1+(n+1)(k+1),\ 1+(n+1)(m-k-1)$
  $\ell = \ell + 2$
enddo
$\text{poly}_{f-1} \leftarrow 4,\ a,\ a-(n+1),\ 2$
$\text{poly}_f \leftarrow 4,\ a,\ 1,\ 1+(n+1), a-1$

6.6 Volumetric Properties of Cams in Indexing Mechanisms

In this section we apply the algorithms presented above to produce MOMENTDB and thus run MOMENT to compute the volumetric properties of cams and followers. The first two subsections below are devoted to the elements of the prototypes of PRICAM, while, in the third subsection, we compute the volumetric properties of the elements of a spatial three-link indexing cam mechanism.

In order to compare the results of MOMENT, we compute the volumetric properties of planar PRICAM with AutoCAD^R Advanced Modeling Extension^{TM}, that uses a ray classification technique (AutoCAD^R AME^{TM}, 1992). In all cases below, the subdivision level is 6 and the direction of the rays is parallel to the Y-axis.

To our knowledge, AutoCAD AME provides six commands that create simple primitive solids, such as boxes or cubes, wedges, cylinders, spheres and tori. It is also possible to create a solid by extruding or revolving a 2-D region. By combining those primitives, one can create *composite solids* with the application of CSG. Thus, one can calculate the volumetric properties of any solid that is composed of these primitives. However, the solids we present below, except for those of the planar PRICAM, are not composite solids; in other words, they cannot be created with the aid of the tools provided by AutoCAD AME, and hence, we compare MOMENT with AutoCAD AME only for the aforementioned planar case.

6.6.1 Planar PRICAM

In this section we compute the volumetric properties of the planar PRICAM with the original dimensions of the prototype shown in Figs. 5.14 and 5.15, whose design parameters are shown in Table 6.2. Since $\mathcal{R}_T$ and $\mathcal{R}_B$ are flat surfaces, it is sufficient to define $q \equiv 1$; consequently, $m = 3$ for all these cases, i.e., for the primary and secondary systems.

Table 6.2 Planar PRICAM design parameters

a_1 mm	α_1 deg.	a_3 mm	α_3 deg.	a_4 mm	α_4 deg.	$\Delta\psi$ deg.	N
101.6	180	37.59	180	9.5	0	120	4

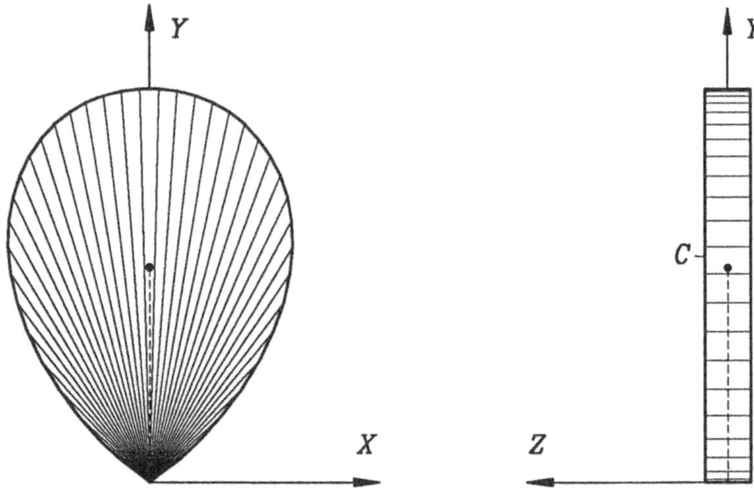

Figure 6.8 Cam of the primary system of the planar PRICAM.

Primary System

From previous chapters we know that the primary system is an RHR mechanism, with both cam and follower in direct contact. We compute below the volumetric properties of both.

i) Cam:

This element is classified as a second-class solid. The position vectors of the vertices have been computed with respect to the frame shown in Fig. 6.8, where $\lambda_{min} = -3.5$ mm and $\lambda_{max} = 3.5$ mm. We define the number of polygons per strip as $n = 197$. Equations (6.58) give

$$a = (3-1)(197+1)+2 = 398$$
$$f = 3(197+1)-1 = 593$$

The computed values of V, $\bar{\mathbf{r}}$ and $\mathbf{I}$ are:

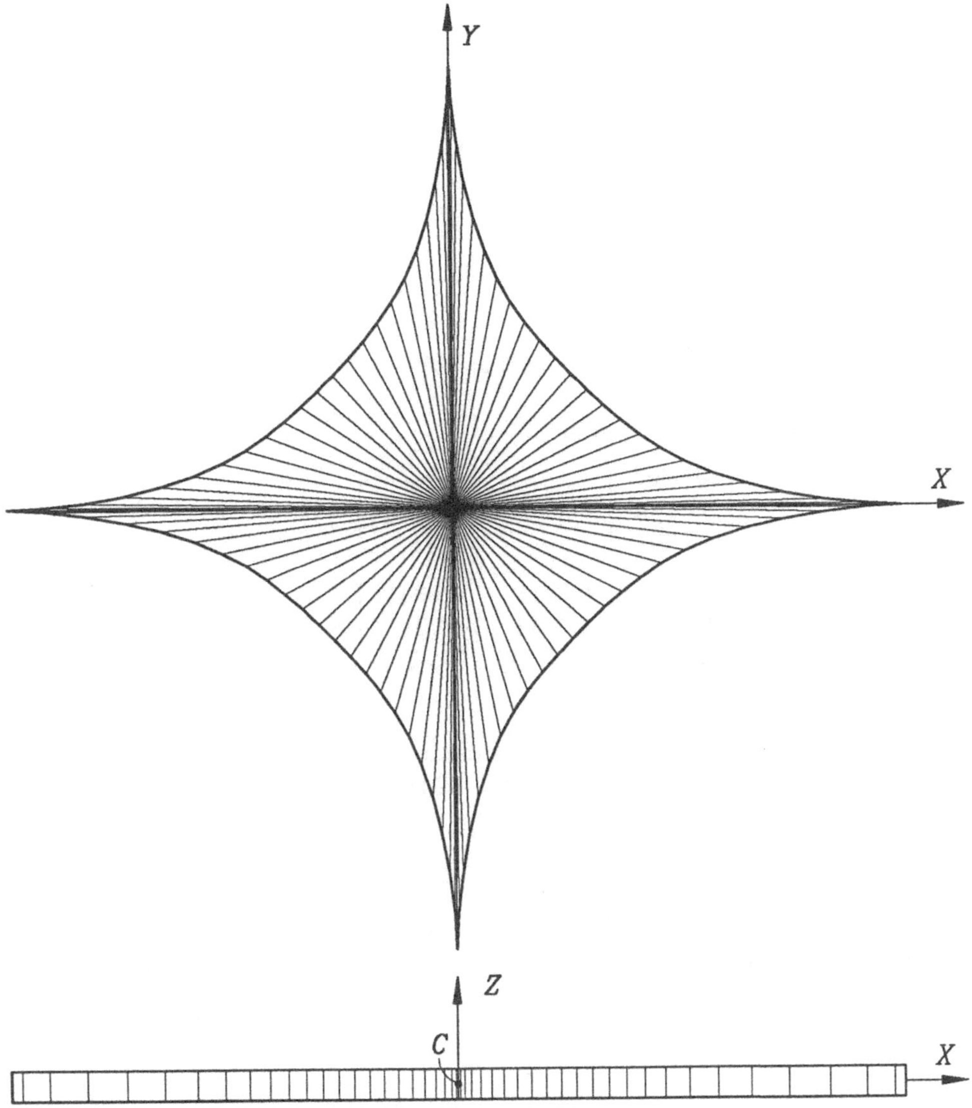

Figure 6.9 Follower of the primary system of the planar PRICAM.

$$V = 13452.29 \quad \text{mm}^3$$

$$\bar{\mathbf{r}} = \begin{bmatrix} 0.0000 \\ 33.4606 \\ 0.0000 \end{bmatrix} \quad \text{mm}$$

$$\mathbf{I} = \begin{bmatrix} 17.9502 & 0.0000 & 0.0000 \\ 0.0000 & 1.5799 & 0.0000 \\ 0.0000 & 0.0000 & 19.4203 \end{bmatrix} \times 10^6 \quad \text{mm}^5$$

Notice that $\bar{\mathbf{r}} = \mathbf{q}/V$.

On the other hand, the results obtained by AutoCAD are

$$V = 13408.39 \quad \text{mm}^3$$

$$\bar{\mathbf{r}} = \begin{bmatrix} 0.0206 \\ 33.4608 \\ -0.0090 \end{bmatrix} \quad \text{mm}$$

$$\mathbf{I} = \begin{bmatrix} 17.8915 & 0.0716 & 0.0041 \\ 0.0716 & 1.5748 & -0.0147 \\ 0.0041 & -0.0147 & 19.3567 \end{bmatrix} \times 10^6 \quad \text{mm}^5$$

Note the error in the x- and z-components of $\bar{\mathbf{r}}$ and in the off-diagonal entries of $\mathbf{I}$ that are produced with AutoCAD AME.

ii) Follower:

This element is considered as a first-class solid. The frame used as reference to obtain the position vectors of the vertices is shown in Fig. 6.9, where $\lambda_{min} = -3.5$ mm and $\lambda_{max} = 3.5$ mm. For this case, $n = 796$ and, from eqs.(6.57), one obtains

$$a = (3-1)796 + 2 = 1594$$
$$f = 3 \times 796 = 2388$$

with the resulting values of V, $\bar{\mathbf{r}}$ and $\mathbf{I}$ given below:

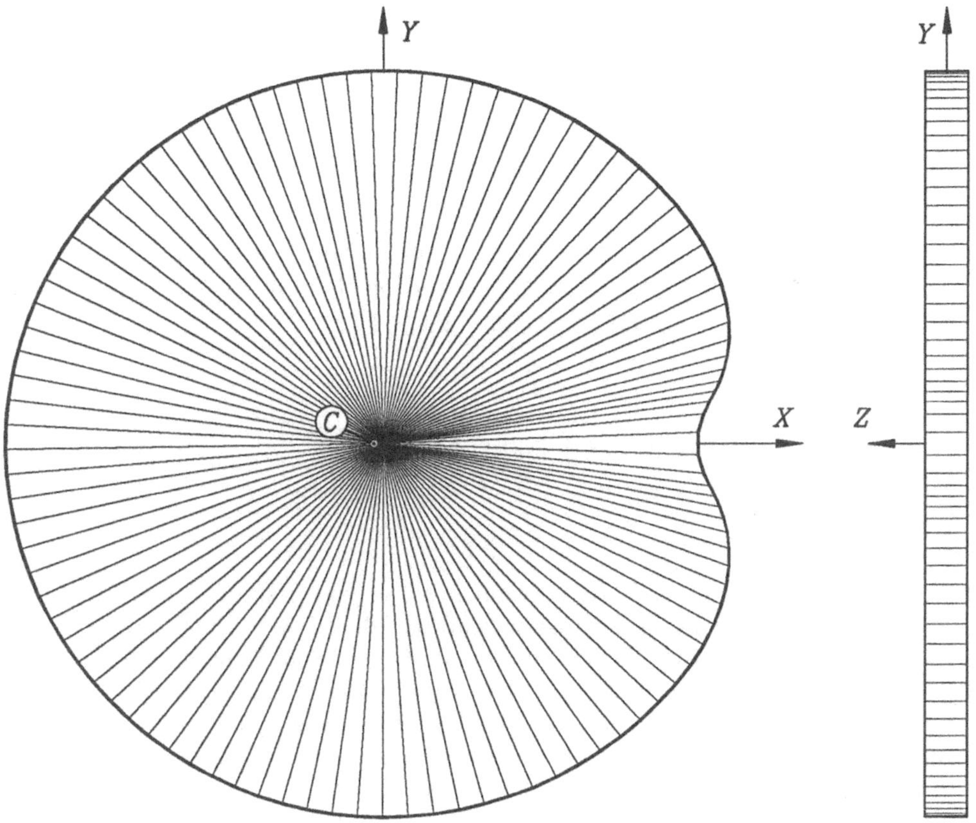

Figure 6.10 Cam of the secondary system of the planar PRICAM.

$$V = 57431.01 \quad \text{mm}^3$$

$$\bar{\mathbf{r}} = \mathbf{0}$$

$$\mathbf{I} = \begin{bmatrix} 48.4064 & 0.0000 & 0.0000 \\ 0.0000 & 48.4064 & 0.0000 \\ 0.0000 & 0.0000 & 96.3438 \end{bmatrix} \times 10^6 \quad \text{mm}^5$$

The results obtained by AutoCAD are, in turn,

$$V = 57488.65 \quad \text{mm}^3$$

$$\bar{\mathbf{r}} = \begin{bmatrix} -0.0063 \\ 0.0000 \\ 0.0045 \end{bmatrix} \quad \text{mm}$$

$$\mathbf{I} = \begin{bmatrix} 48.6170 & 0.0000 & -0.0468 \\ 0.0000 & 48.3567 & 0.0000 \\ -0.0468 & 0.0000 & 96.5050 \end{bmatrix} \times 10^6 \quad \text{mm}^5$$

Note again the errors obtained with AutoCAD AME.

Secondary System

The secondary system is an RHRR mechanism, in which the roller-follower is composed of primitive elements like cylinders, and parallelepipeds whose volumetric properties can be computed without difficulty by applying direct formulas, thus dispensing with linear approximations. However, the cam is rather a complex solid and, for this case, we apply a linear approximation by considering it as a first-class solid. The coordinates of the vertices are computed with respect to the frame shown in Fig. 6.10. In this case, $\lambda_{min} = -4$ mm, $\lambda_{max} = 4$ mm and $n = 601$. Thus, from eqs.(6.57) one obtains,

$$a = (3 - 1)601 + 2 = 1204$$
$$f = 3 \times 601 = 1803$$

The values of V, $\bar{\mathbf{r}}$ and $\mathbf{I}$ reported by MOMENT are given below:

$$V = 120094.91 \quad \text{mm}^3$$

$$\bar{\mathbf{r}} = \begin{bmatrix} -1.7883 \\ 0.0000 \\ 0.0000 \end{bmatrix} \quad \text{mm}$$

$$\mathbf{I} = \begin{bmatrix} 151.6848 & 0.00002 & 0.0000 \\ 0.00002 & 138.5061 & 0.0000 \\ 0.0000 & 0.0000 & 289.0099 \end{bmatrix} \quad \text{mm}^5$$

while the results obtained by AutoCAD are

$$V = 119702.70 \quad \text{mm}^3$$

$$\bar{\mathbf{r}} = \begin{bmatrix} -1.8493 \\ 0.0000 \\ -0.0087 \end{bmatrix} \quad \text{mm}$$

$$\mathbf{I} = \begin{bmatrix} 151.1397 & 0.00002 & -0.1369 \\ 0.00002 & 138.4635 & 0.0000 \\ -0.1369 & 0.0000 & 288.3321 \end{bmatrix} \quad \text{mm}^5$$

6.6.2 Spherical PRICAM

In this subsection we present the volumetric properties of the complex elements of the spherical PRICAM by using the design parameters of the prototype shown in Figs. 5.16 and 5.17, which are summarized in Table 6.3.

Table 6.3 Spherical PRICAM design parameters

a_1 mm	α_1 deg.	a_3 mm	α_3 deg.	a_4 mm	α_4 deg.	$\Delta\psi$ deg.	N
0	90	0	149.7	0	-7.87	120	4

Primary System

The primary system of the spherical PRICAM is an RHR mechanism. The volumetric properties of cam and follower are given below:

i) Cam:

This element is a solid of the second class. The position vectors of the vertices have been computed with respect to the frame shown in Fig. 6.11, where $\lambda_{min} =$ 62 mm and $\lambda_{max} = 72$ mm. Moreover, $m = 33$ and $n = 197$. Equations (6.58) give

$$a = (33-1)(197+1)+2 = 6338$$
$$f = 33(197+1)-1 = 6533$$

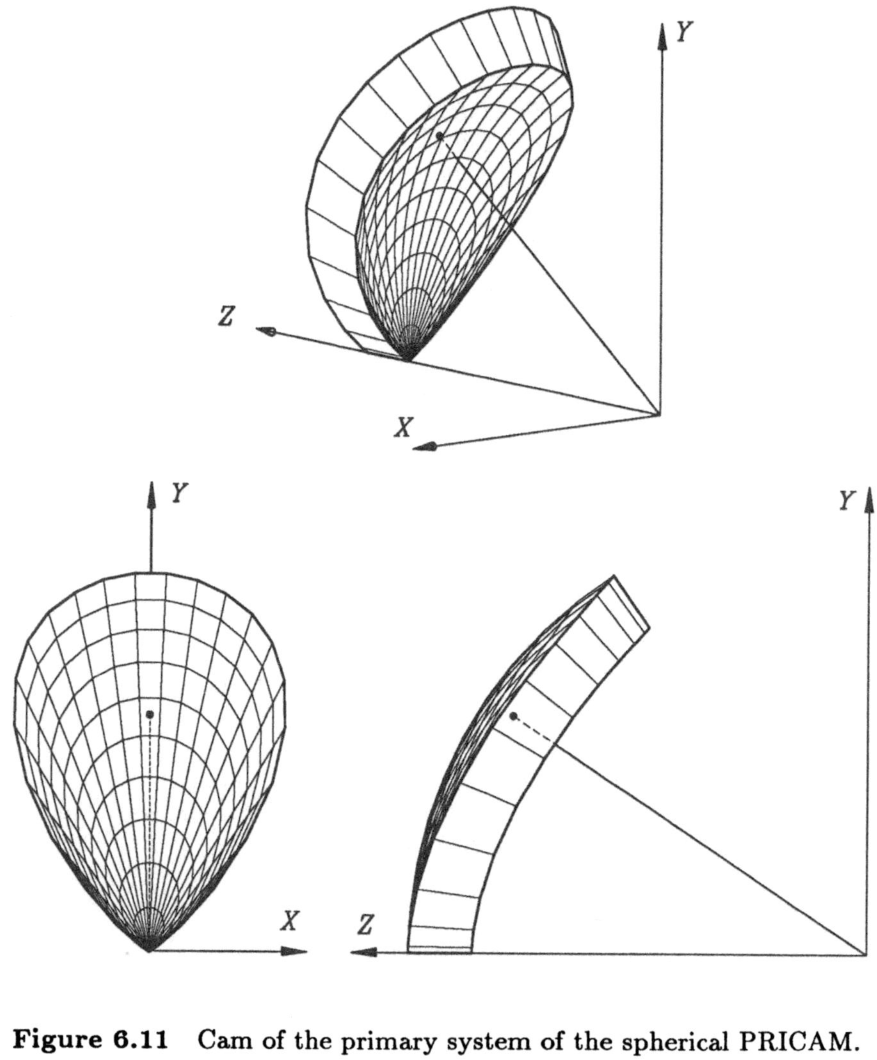

Figure 6.11 Cam of the primary system of the spherical PRICAM.

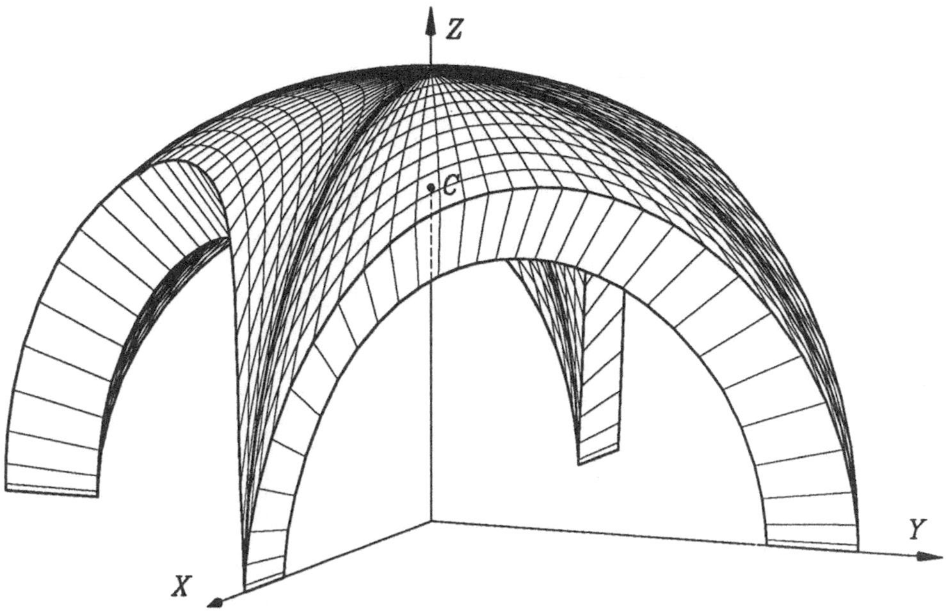

Figure 6.12 Follower of the primary system of the spherical PRICAM.

The computed values of V, $\bar{\mathbf{r}}$ and $\mathbf{I}$ are:

$$V = 19515.40 \quad \text{mm}^3$$

$$\bar{\mathbf{r}} = \begin{bmatrix} -0.7202 \\ 37.2333 \\ 55.6596 \end{bmatrix} \quad \text{mm}$$

$$\mathbf{I} = \begin{bmatrix} 84.2808 & 0.0181 & 0.0548 \\ -0.4418 & 65.6913 & -33.824 \\ -1.0247 & -28.8412 & 26.7892 \end{bmatrix} \times 10^6 \quad \text{mm}^5$$

ii) Follower:

The follower of the primary mechanism is a first-class solid. The frame used as a reference to obtain the position vectors of the vertices is shown in Fig. 6.12, where $\lambda_{min} = 55$ mm and $\lambda_{max} = 70$ mm. For this case, $m = 33$ and $n = 284$. Hence, from eqs.(6.57), one obtains

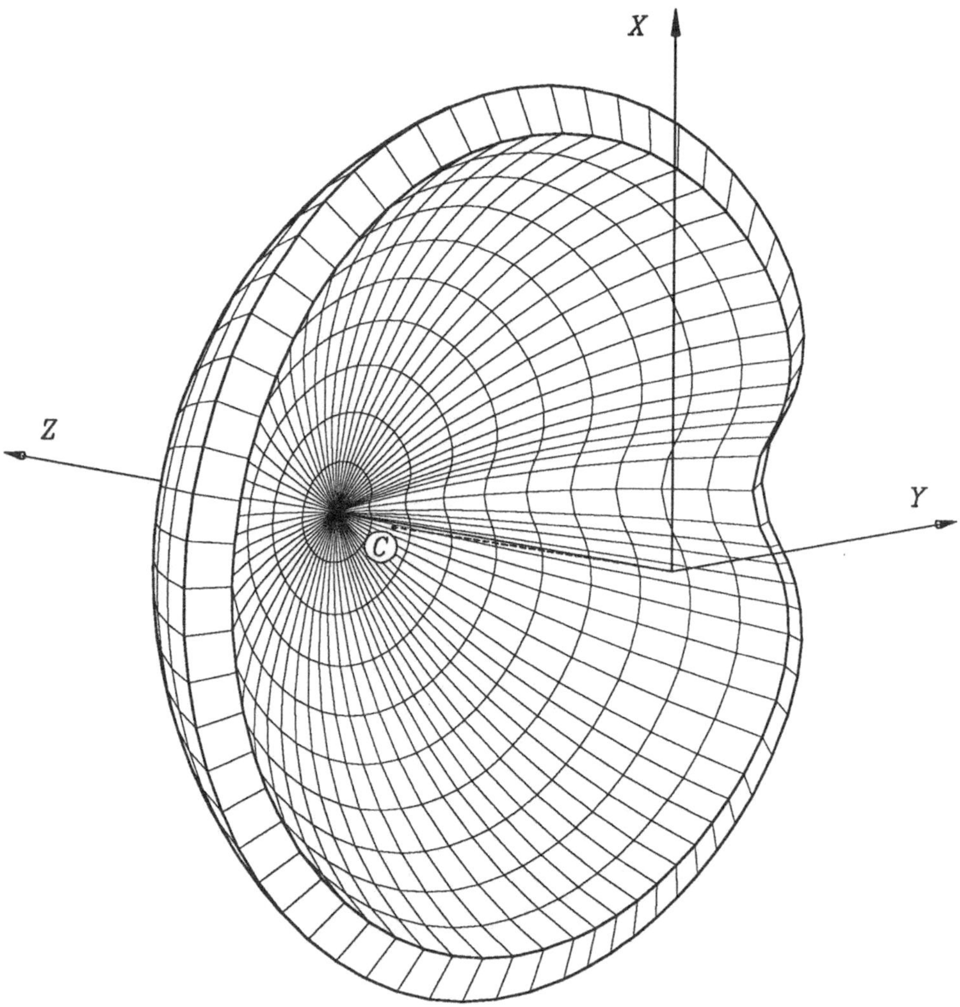

Figure 6.13 Cam of the secondary system of the spherical PRICAM.

$$a = (33 - 1)284 + 2 = 9090$$
$$f = 33 \times 284 = 9372$$

while the resulting values of V, $\bar{\mathbf{r}}$ and $\mathbf{I}$ are

$$V = 104947.64 \quad \text{mm}^3$$

$$\bar{\mathbf{r}} = \begin{bmatrix} 0.0000 \\ 0.0000 \\ 52.0738 \end{bmatrix} \quad \text{mm}$$

$$\mathbf{I} = \begin{bmatrix} 353.8211 & 10.1114 & 0.0000 \\ -10.1114 & 353.8211 & 0.0000 \\ 0.0000 & 0.0000 & 131.8547 \end{bmatrix} \times 10^6 \quad \text{mm}^5$$

Secondary System

Here we analize the cam of the secondary system as a solid of the first class. The coordinates of the position vectors of the vertices are computed with respect to the frame shown in Fig. 6.13. Here, $\lambda_{min} = 72$ mm and $\lambda_{max} = 80$ mm, while $m = 21$ and $n = 399$. Equations (6.57) give $a = 7982$ and $f = 8379$. With these values we can complete MOMENTDB and run MOMENT to obtain V, $\bar{\mathbf{r}}$ and $\mathbf{I}$, namely,

$$V = 147808.57 \quad \text{mm}^3$$

$$\bar{\mathbf{r}} = \begin{bmatrix} 0.00646 \\ -0.70032 \\ 56.61330 \end{bmatrix} \quad \text{mm}$$

$$\mathbf{I} = \begin{bmatrix} 667.2946 & 0.0224 & 0.06977 \\ 0.1087 & 681.7080 & 6.3141 \\ 0.1131 & 9.3424 & 363.0885 \end{bmatrix} \quad \text{mm}^5$$

6.6.3 Spatial Mechanisms

In this subsection we analize a spatial RHR indexing mechanism, i.e., a mechanism whose input and output axes are non-parallel and non-intersecting. The design parameters are shown in Table 6.4, followed by the results obtained for the two elements.

i) Cam:

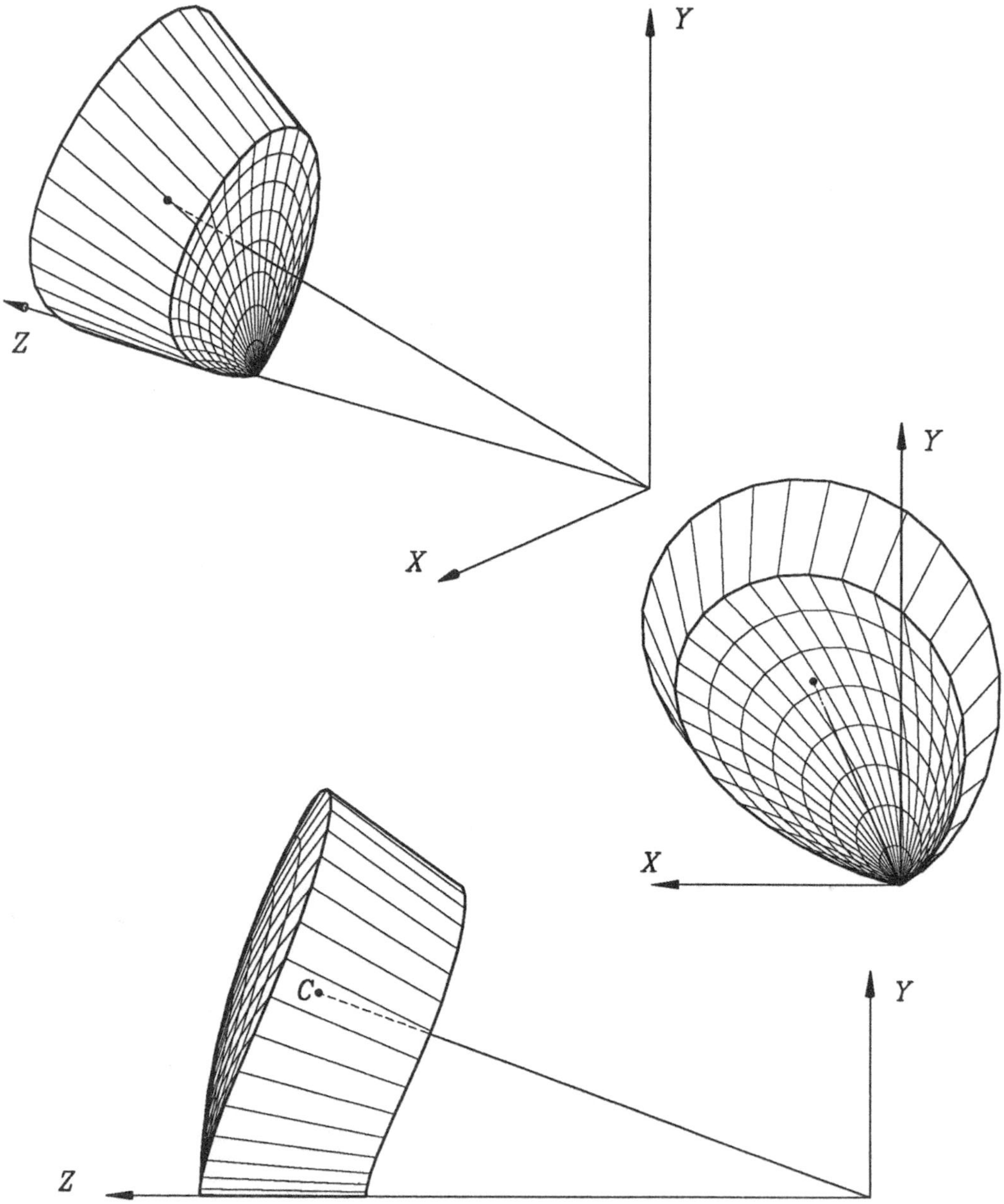

Figure 6.14 Cam of a spatial ICM.

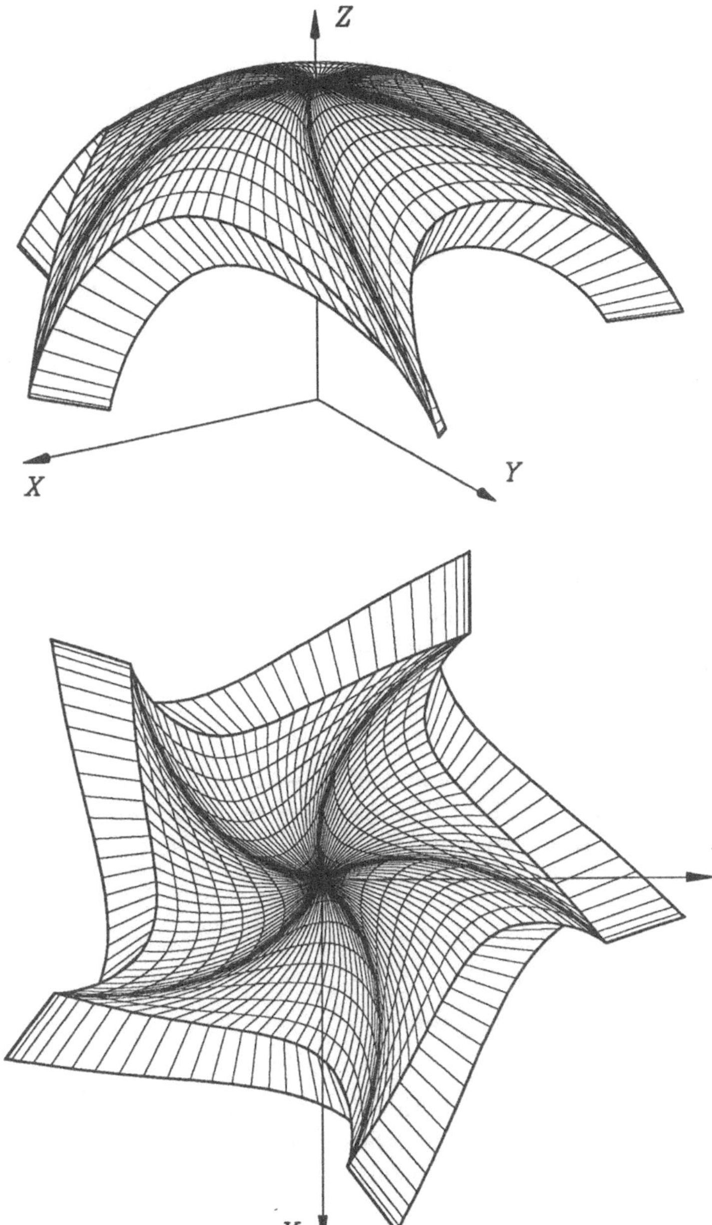

Figure 6.15 Follower of a spatial ICM.

Table 6.4 RHR Indexing mechanism design parameters

a_1 mm	α_1 deg.	λ_{min} mm	λ_{max} mm	$\Delta\psi$ deg.	N
50	75	75	100	150	5

The cam is a second-class solid. With $m = 21$ and $n = 107$, we obtain, from eqs.(6.58), the number of vertices and polygons approximating the boundary of the cam, namely, $a = 3962$ and $f = 4157$. The coordinates of the vertices are computed with respect to the frame shown in Fig. 6.14. The volumetric properties are:

$$V = 49815.20 \quad \text{mm}^3$$

$$\bar{\mathbf{r}} = \begin{bmatrix} 13.3146 \\ 30.7274 \\ 82.4824 \end{bmatrix} \quad \text{mm}$$

$$\mathbf{I} = \begin{bmatrix} 387.4937 & -24.6628 & -40.5894 \\ -7.6414 & 351.2679 & -117.8718 \\ -36.3799 & -110.7701 & 67.1700 \end{bmatrix} \times 10^6 \quad \text{mm}^5$$

ii) Follower:

The follower is a solid of the first class with $m = 21$ and $n = 475$; eqs.(6.57) give $a = 9502$ and $f = 9975$. The coordinates of the vertices are obtained with respect to the frame shown in Fig. 6.15. Thus, the volumetric properties are given as

$$V = 381106.08 \quad \text{mm}^3$$

$$\bar{\mathbf{r}} = \begin{bmatrix} 0.0000 \\ 0.0000 \\ 77.9070 \end{bmatrix} \quad \text{mm}$$

$$\mathbf{I} = \begin{bmatrix} 2568.468 & 112.046 & 0.0000 \\ -112.046 & 2568.468 & 0.0000 \\ 0.0000 & 0.0000 & 1049.2261 \end{bmatrix} \times 10^6 \quad \text{mm}^5$$

For clarity in the presentation, the number of polygons shown in Figs. 6.8–6.15 is smaller than the number used to compute the volumetric properties.

Chapter 7

AN OUTLINE OF USYCAMS

7.1 Introduction

In the last twenty years or so, *computer graphics* has become a fundamental tool for the study of mechanisms, as made apparent by the software available in this area. We can cite IMP (Integrated Mechanisms Program) as one of the earliest software packages developed for the analysis of arbitrary linkages (Sheth and Uicker, 1972). A few years later KINSIN III, a package meant for the synthesis of mechanisms, was introduced (Rubel and Kaufman, 1977). The graphical part of these early packages consisted essentially of 2-D lines. More mechanism software with improvements in their graphics has been produced, such as LINCAGES (Erdman and Gustafson, 1977), MINN-DWELL (Kota et al., 1987) and SIXPAQ (González-Palacios and Angeles, 1991*b*).

Software for the CAD/CAM of planar cams was developed by De Fraine (1982), while Koloc and Václavíc (1988) produced software for the analysis and synthesis of 2-D and 3-D cam mechanisms. In this chapter we outline USYCAMS (Unified Synthesis of Cam Mechanisms), our platform for the *interactive synthesis* of cams, with strong graphics support.

Given that modern workstations provide a highly integrated environment for computation and interaction, our goal in the development of USYCAMS is that it should be easy to use and as general as possible. Following these criteria, the user of USYCAMS need not be a CAD expert, although a certain familiarity with the theory of cam mechanisms is expected, in order to exploit the capabilities of USYCAMS. Solving complex problems like undercutting with visualization aids reduces to moving the mouse around the appropriate menus.

Although this chapter is based on the theory presented in the previous chapters, it is self-contained and can be studied once the basic terminology and concepts of the first four chapters have been grasped. We intend that this chapter should be understandable for readers with only fundamental knowledge of *cam kinematics*. However, the reader unfamiliar with this discipline is advised to read Chapter 1 first.

USYCAMS has many applications. It can be used to give an introduction to cam mechanisms in an undergraduate course, or to solve complex design problems involving, e.g., undercutting of spatial cam mechanisms, in a graduate design course. Furthermore, USYCAMS can be regarded as a sophisticated design tool for industrial applications involving dimensioning, balancing, dynamic analysis, simulation and *finite-element analysis* (FEA) for stress, strain and thermal calculations. Note that USYCAMS provides a data base for useful mesh generation in FEA, but is limited to the handling and production of geometric and kinematic information.

7.2 On-Line vs. Off-Line Synthesis

The *off-line synthesis* of mechanical systems in general, and of cams, in particular, is defined here as the process whereby the user precomputes, with a set of geometric parameters, the information needed as input data for a graphics program that produces both numerical and graphical renderings of the system under synthesis. For example, the off-line synthesis of a four-bar linkage consists of inputting precomputed link lengths and one value of the input angle, while producing *i*) numerical values of the Cartesian coordinates of the joint centers involved and all remaining joint angles, and *ii*) a graphical rendering of the foregoing numerical results. In *on-line* synthesis the user supplies the geometric information directly to the graphics program with the aid of an input device like a mouse or a joystick; the user *interacts* with the program producing graphical results.

We have developed USYCAMS to supply the mechanical designer with a universal tool for the synthesis of planar, spherical and spatial cam mechanisms; it is a platform for the interactive synthesis of such mechanisms.

7.3 General Description of USYCAMS

We give an outline of USYCAMS for a *Silicon Graphics Inc. IRIS* workstation. It can run on other *UNIX* workstations with suitable graphics software and hardware, but then, obviously, all device-dependent features must be modified.

At the outset, we divide the window into five *viewports*, VP1, VP2, VP3, VP4 and VP5, as shown in Fig. 7.1. Each of these viewports serves a specific function:

VP1: This part of the screen is devoted to rendering the solid models and animating the mechanism under design.

VP2: This viewport displays the main menu in two modes, namely, *passive* and *active*. In passive mode, VP2 tells the user what kind of mechanism is in

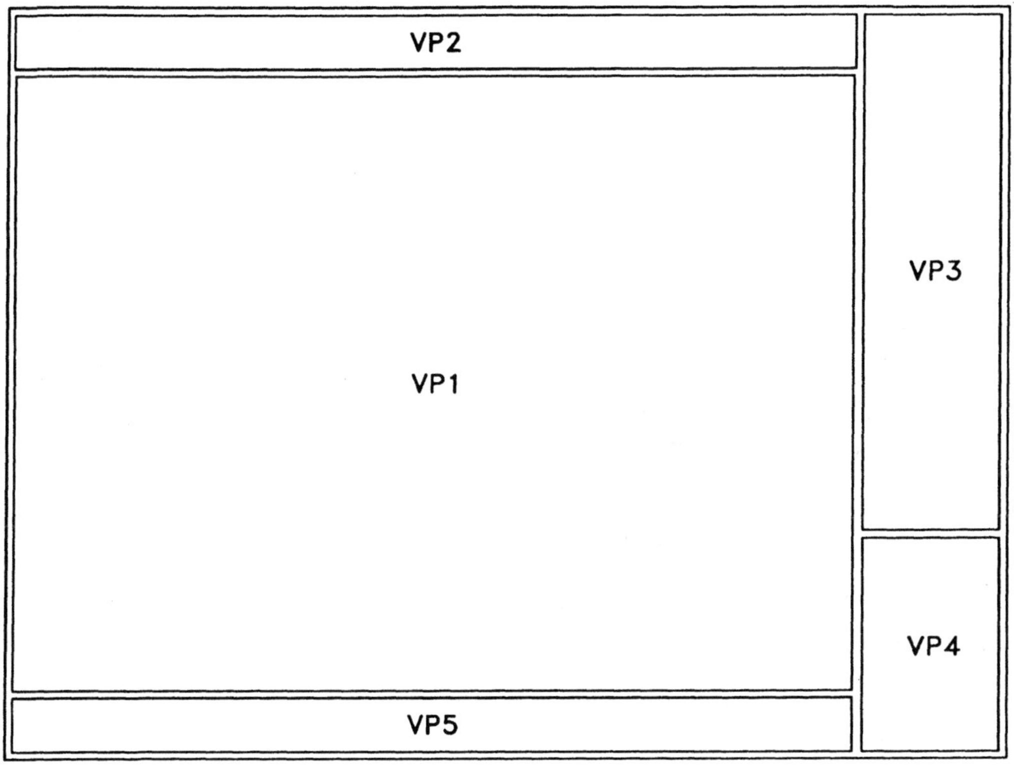

Figure 7.1 Distribution of view ports.

the process of synthesis; in active mode, the user can interact with the program and choose the type of mechanism desired.

VP3: This viewport shows the design parameters such as $\hat{\alpha}_1$ $\hat{\alpha}_3$, $\hat{\alpha}_4$ and λ, introduced in Chapters 3 and 4. VP3 interacts simultaneously with VP1, so that a change in any of the parameters is reflected in the solid models of the mechanism.

VP4: Like VP3, VP4 interacts with VP1; this viewport shows the parameters pertaining to the input-output function.

VP5: If at least one of the parameters of VP3 or VP4 is active, a sliding bar appears in VP5, so that the user can modify the parameter values by moving the cursor, with the aid of the mouse, along this bar.

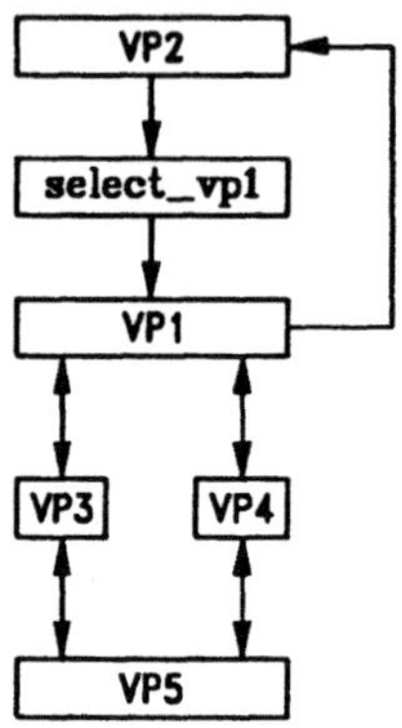

Figure 7.2 Interactivity of the five viewports.

Figure 7.2 shows the global flowchart of interaction of the five viewports. `select_vp1` is the procedure that decides which type of solid model will be displayed on VP1.

7.4 Main Menu

The main menu is iconized and displayed in VP2; the user can choose the type of mechanism to be synthesized by clicking these icons with the mouse. Thus, VP2 is divided into seven sections called POSi, for $i = 0, 1, 2, \cdots, 6$, as described below:

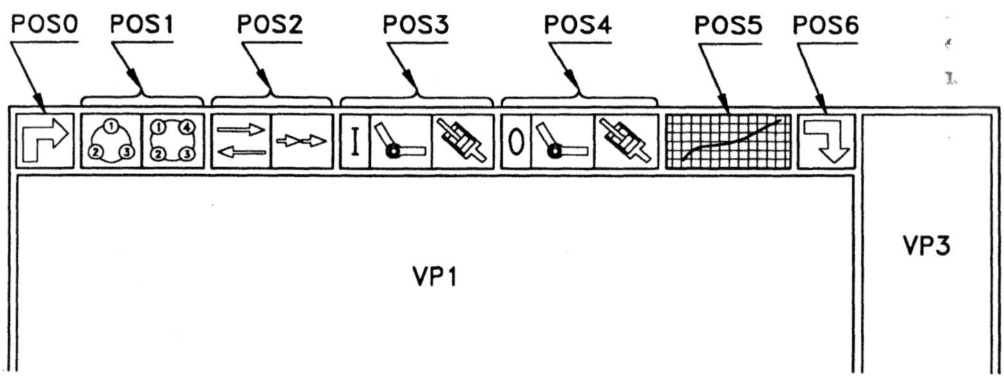

Figure 7.3 Icons of the main menu.

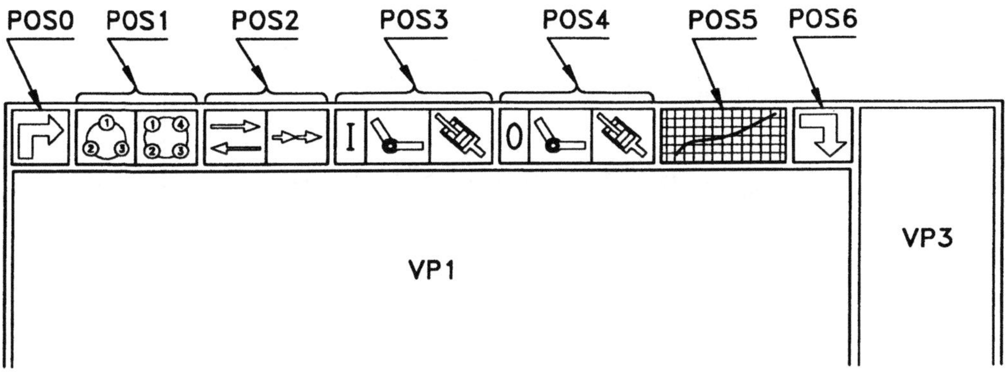

Figure 7.3 Icons of the main menu.

POS6: The user can exit VP2 any time by selecting this icon, VP2 thus switching into passive mode.

The icons selected will stay highlighted in order to remind the user which type of mechanism has been chosen. Both the icons and the flowchart of the main menu are shown in Figs. 7.3 and 7.4, respectively.

7.4.1 Main Loop

As mentioned above, viewport VP1 is used to display the solid model representation of the mechanism selected in VP2. If we look at Fig. 7.5, we notice that there are up to sixteen different types of mechanisms with independent *synthesis procedures*, which are identified with the labels shown at the right-hand side of Fig. 7.5; for example, if the user wants to synthesize a four-link cam mechanism with oscillating motion in the follower, with a revolute pair as input and a prismatic pair as output, the procedure chosen for this selection is defined as **vp1_4_osc_rp**. The flowchart sample of these procedures is shown in Fig. 7.6. In the synthesis of one of the sixteen mechanisms the user will be working most of the time in its corresponding procedure, which is the reason why we call it the main loop.

All sixteen procedures have similar structures; what changes in each case is only the synthesis algorithm, **construct_surfaces**, the procedures to generate the solid models of cam and follower, **draw_solids**, and the procedures to draw the icons of VP3 and VP4, namely, **draw_icons**.

At the main-loop level, USYCAMS performs the algorithm of the flowchart of Fig. 7.6, namely,

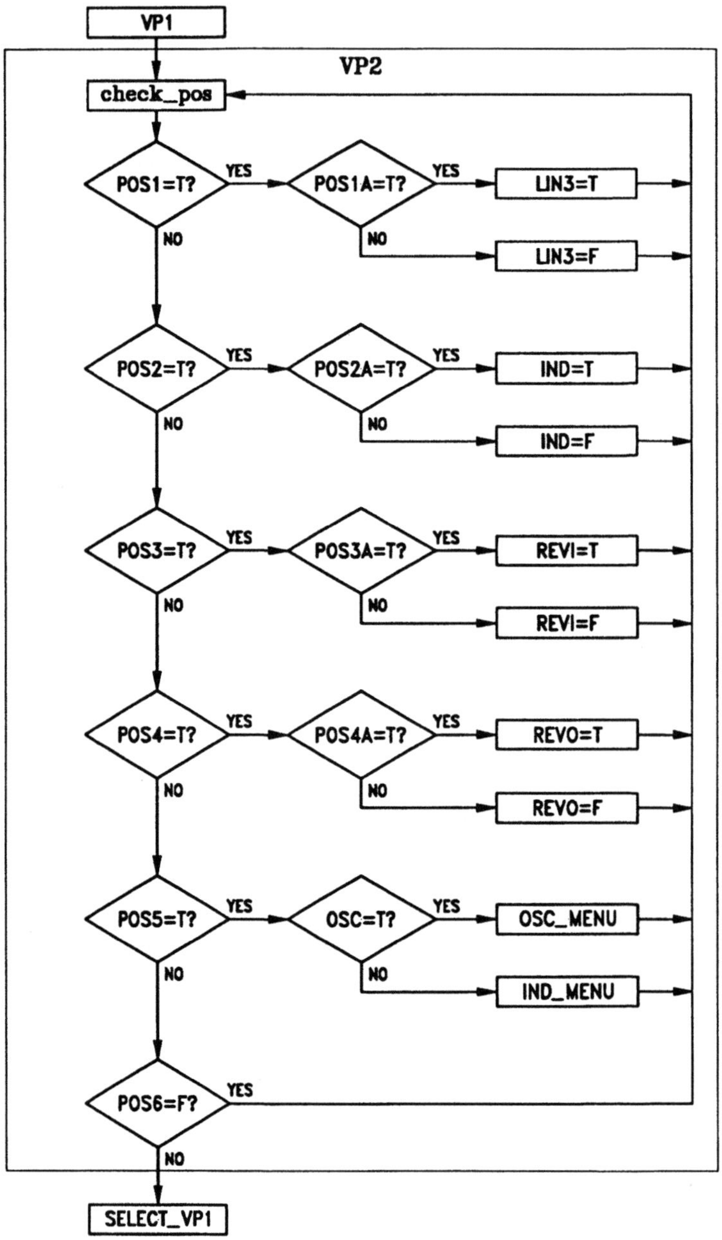

Figure 7.4 Flowchart of the main menu.

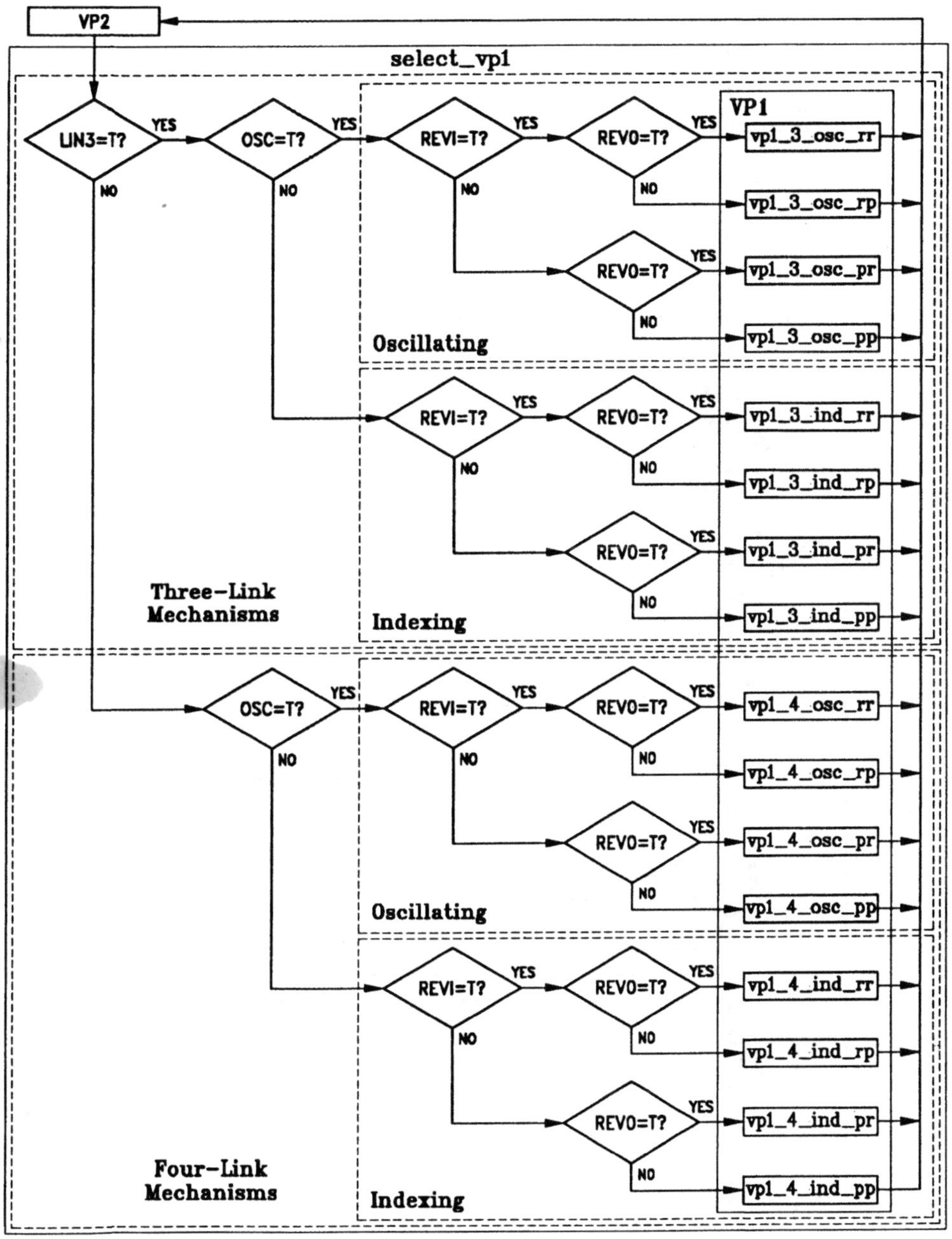

Figure 7.5 Flowchart of the `select_vp1` procedure.

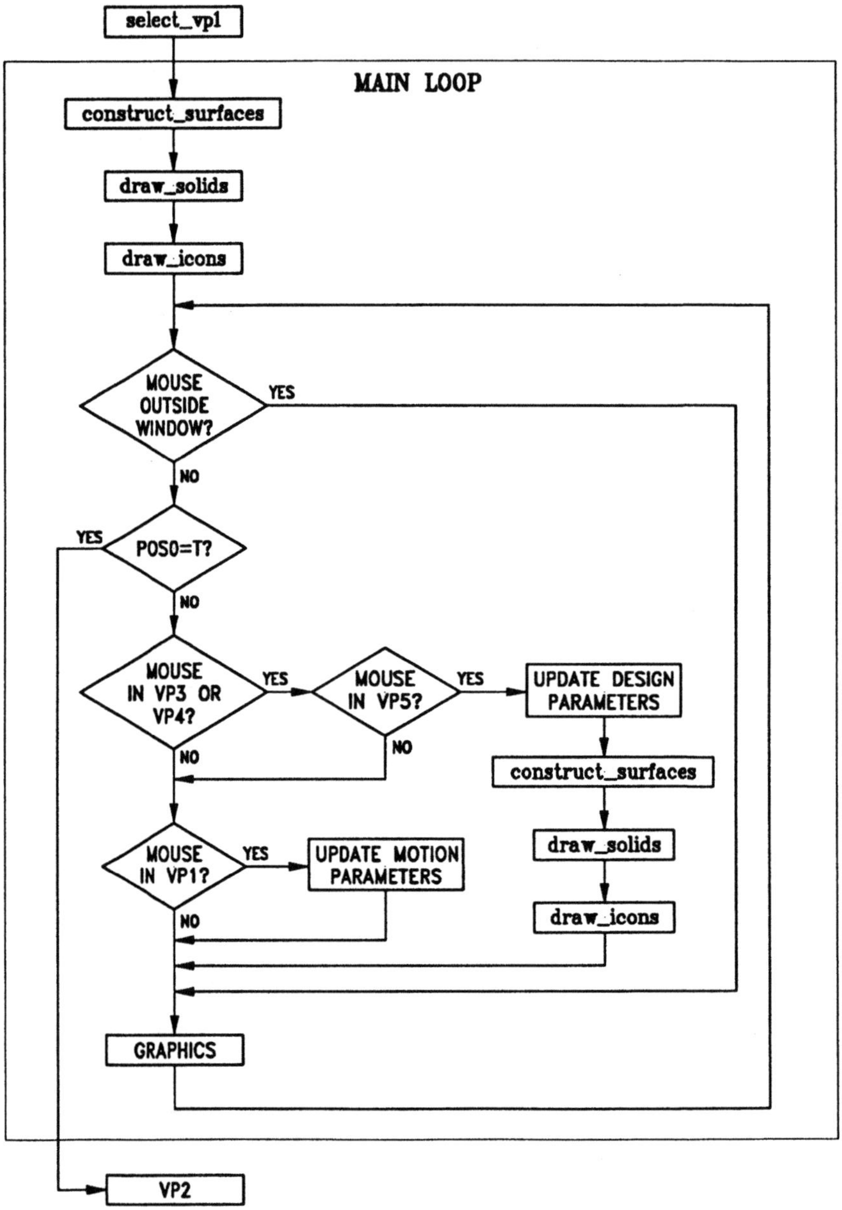

Figure 7.6 Flowchart of the procedures of VP1.

1. The procedure `construct_surfaces` is called. Here, the cam and the follower surfaces are synthesized, based upon results obtained in previous chapters.

2. With the aid of the graphics library supported by the *Silicon Graphics Power Series Workstation (IRIS 4D/420VGX)* or compatible, the procedure `draw_solids` is called to generate the contact surfaces. The information needed in this procedure is the set of position vectors and the corresponding normals of the surfaces, which are generated by `construct_surfaces`.

3. The icons of VP3 and VP4 are updated according to the design-parameter selection given in the main menu.

4. The position of the mouse is verified, so that, if it is outside the five viewports, the program goes to step 14; otherwise, it continues with the next step.

5. The position of the mouse is verified, so that, if it is in POS0 and the left button is pressed, the user exits the loop and returns to VP2. If the latter condition is not reached, the program goes to step 6.

6. The user can choose any design parameter at this time by locating the mouse and pressing the left button in any of the icons shown in the viewports VP3 and VP4. If at least one of these is chosen, USYCAMS goes to the next step, leaving highlighted the icons chosen; otherwise, USYCAMS goes to step 12. This step works in the same way to deactivate any highlighted icon.

7. The user can vary the value of the parameters highlighted in VP3 and VP4 by locating the mouse in VP5 and moving the sliding bar as desired with the left button pressed. If the mouse is not in VP5, USYCAMS goes to step 12.

8. The design parameters highlighted in VP3 and VP4 are updated according to the motion of the sliding bar of VP5.

9. Same as step 1.

10. Same as step 2.

11. Similar to step 3. The difference here is that only the highlighted icons are updated according to the numerical variations given by the sliding bar.

12. By moving the mouse around VP1, while pressing the left or middle buttons, or both simultaneously, we can store the difference of two consecutive events for the subsequent step. If we do not press any button, USYCAMS goes to step 14.

13. Rotations, translations and scalings are updated here according to the button combination given in step 12. With the left button pressed, the solid models rotate in order to have different points of view; with the middle button pressed,

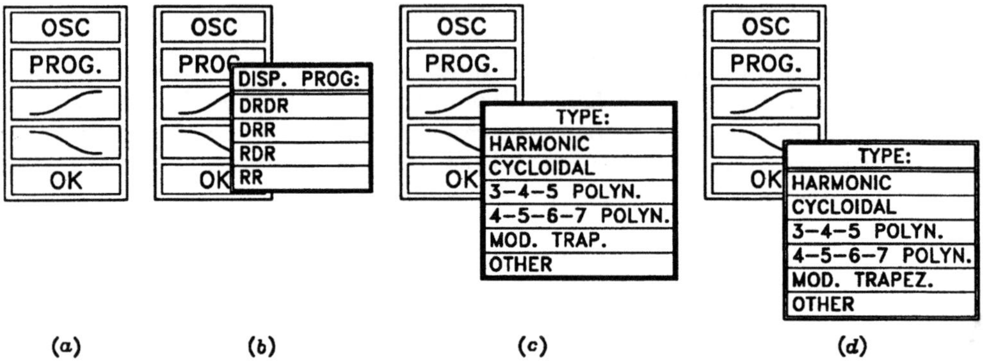

Figure 7.7 `sm_io_selection` menu for oscillating mechanisms.

the models are translated along any direction parallel to the plane of the screen; with both pressed simultaneously, the models are scaled so that they can be zoomed in.

14. The solid model of cam and follower are regenerated according to the information given in the previous steps, while the program continues with step 4.

The names of the sixteen subroutines labeled generically as `construct_surfaces`, `draw_solids` and `draw_icons` of the corresponding procedures of VP1 are listed in Table 7.1.

7.5 Submenus

The icons displayed in the viewports VP3 and VP4 are the submenus of USYCAMS. First we outline the submenu that appears when VP2 is active and POS5 is selected; we call this submenu the input-output function selection `sm_io_selection`. Then we describe the three subroutines `draw_icons` shown in Table 7.1.

7.5.1 `sm_io_selection`

When we discussed POS5 in Section 7.4, we mentioned that this submenu is displayed with the aim of providing more information about the input-output function. As shown in Fig. 7.4, this submenu has two possibilities, namely,

1. `osc_menu`: When an oscillating mechanism is chosen, the submenu shown in Fig. 7.7*a* is displayed in VP4. By selecting icon PROG., the user can choose one of the displacement programs listed below:

Table 7.1 Names of VP1 subroutines

VP1	construct_surfaces	draw_solids	draw_icons
vp1_3_osc_rr	cs_3_osc_rr	ds_3_osc_rr	di_3_osc
vp1_3_osc_rp	cs_3_osc_rp	ds_3_osc_rp	di_3_osc
vp1_3_osc_pr	cs_3_osc_pr	ds_3_osc_pr	di_3_osc
vp1_3_osc_pp	cs_3_osc_pp	ds_3_osc_pp	di_3_osc
vp1_3_ind_rr	cs_3_ind_rr	ds_3_ind_rr	di_3_ind
vp1_3_ind_rp	cs_3_ind_rp	ds_3_ind_rp	di_3_ind
vp1_3_ind_pr	cs_3_ind_pr	ds_3_ind_pr	di_3_ind
vp1_3_ind_pp	cs_3_ind_pp	ds_3_ind_pp	di_3_ind
vp1_4_osc_rr	cs_4_osc_rr	ds_4_osc_rr	di_4
vp1_4_osc_rp	cs_4_osc_rp	ds_4_osc_rp	di_4
vp1_4_osc_pr	cs_4_osc_pr	ds_4_osc_pr	di_4
vp1_4_osc_pp	cs_4_osc_pp	ds_4_osc_pp	di_4
vp1_4_ind_rr	cs_4_ind_rr	ds_4_ind_rr	di_4
vp1_4_ind_rp	cs_4_ind_rp	ds_4_ind_rp	di_4
vp1_4_ind_pr	cs_4_ind_pr	ds_4_ind_pr	di_4
vp1_4_ind_pp	cs_4_ind_pp	ds_4_ind_pp	di_4

i) Dwell-rise-dwell-return (DRDR).

ii) Dwell-rise-return (DRR).

iii) Rise-dwell-return (RDR).

iv) Rise-return (RR).

As in item 1, the user can choose the type of function for the rise and return phases, as shown in Figs. 7.7*c* and *d*, respectively. The user exits this submenu by pointing the OK icon. Figure 7.8 shows the flowchart of this procedure.

2. `ind_menu`: This submenu appears when the user chooses an indexing mechanism. In this case the type of function desired for the rise phase is selected by pressing the left button on the icon with the curve as shown in Fig. 7.9*a*; then, another submenu will be displayed showing a list of typical input-output functions, as shown in Fig. 7.9*b*. Once the function is selected, the user exits this submenu by pointing the OK icon. This procedure is illustrated in the flowchart of Fig. 7.10.

7.5.2 `di_3_osc`

Three-link cam mechanisms with constant pressure angle can be synthesized with the aid of this submenu. The possible four combinations of kinematic pairs used at the input and the output share the same icons displayed in VP3 and VP4. The procedures that draw these icons are called `vp3_3_osc` and `vp4_3_osc`.

The icons displayed in VP3 are described below:

1. With this icon, the user can animate the mechanism. There are three options to select with this icon: *a*) counterclockwise rotation of the cam for the case of revolute pair and translation for the case of prismatic pairs; *b*) no motion; and *c*) clockwise rotation of the cam for the case of revolute pair and translation for the case of prismatic pairs. Moreover, the numerical value of the relative position of the cam with respect to the frame is printed.

   Icons 2–7 below are used to print the numerical values of the various parameters, namely,

2. The distance between input and output axes, a_1, in the selected units;

3. The angle between input and output axes, α_1, in degrees;

4. The radius of the auxiliary roller, a_4, in the selected units;

5. The angle of the auxiliary roller, α_4, in degrees;

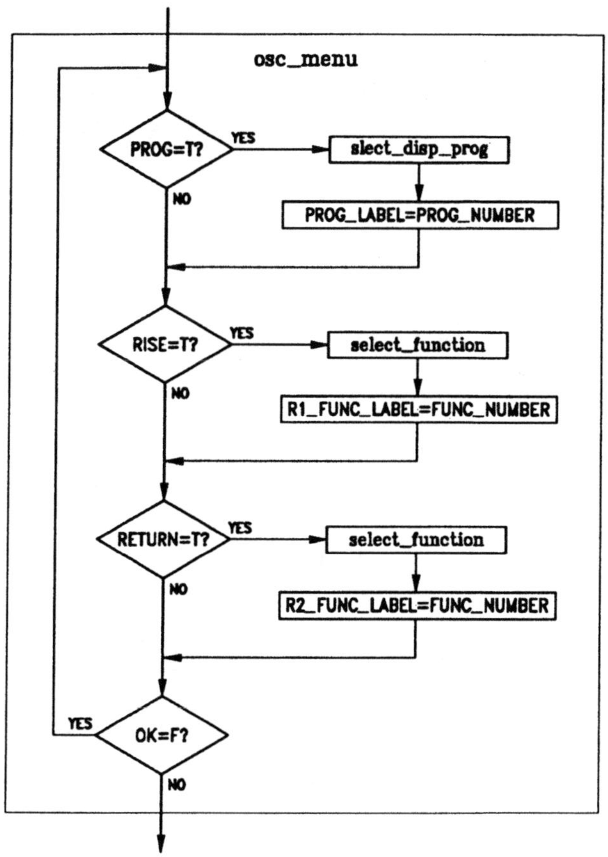

Figure 7.8 Flowchart of osc_menu.

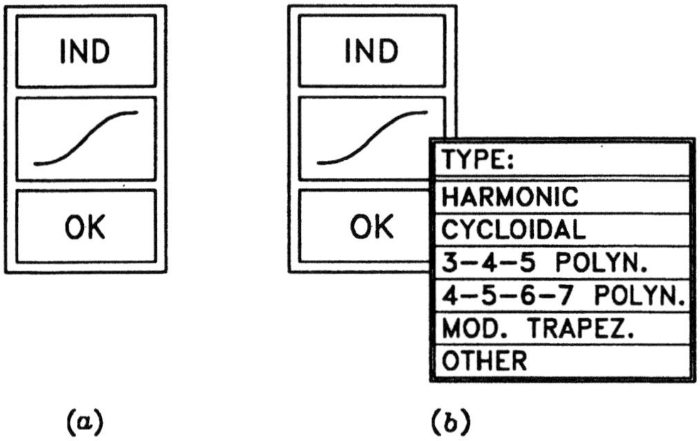

Figure 7.9 `sm_io_selection` menu for indexing mechanisms.

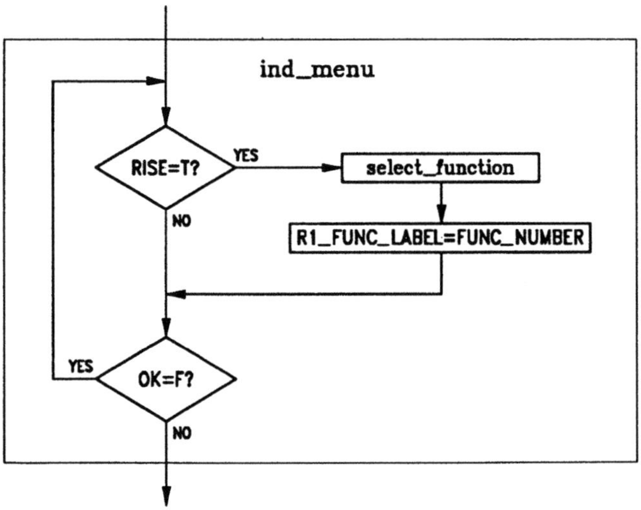

Figure 7.10 Flowchart of `ind_menu`.

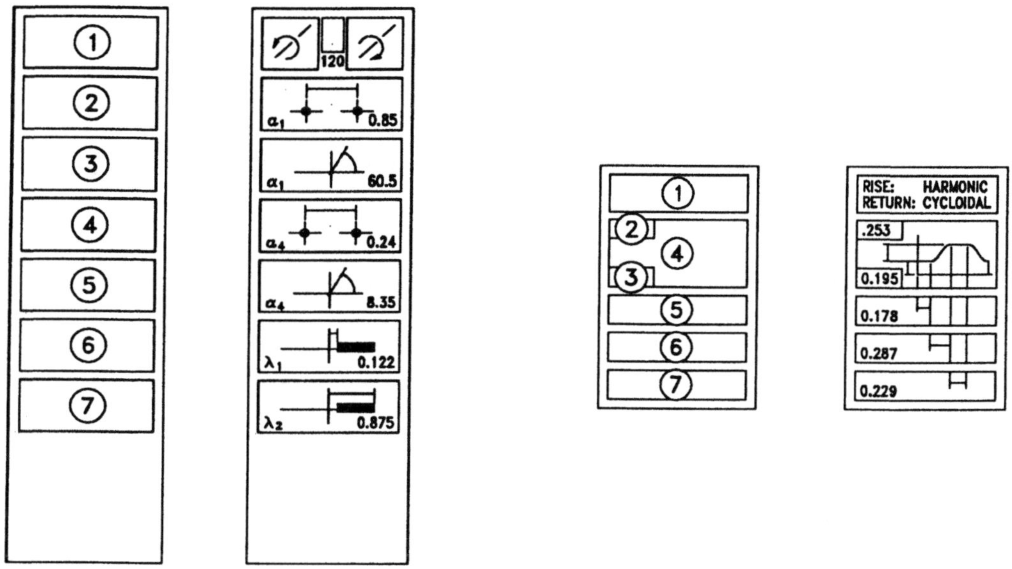

(a) (b)

Figure 7.11 Icons of the `di_3_osc` submenu; *a*) VP3 viewport and *b*) VP4 viewport.

6. The initial point along the contact line of cam and follower, determined by λ_1;

7. The end point along the contact line of cam and follower, determined by λ_2.

Now we describe the icons displayed in VP4. Note that there are four types of displacement programs. However, all of them can be generated from the DRDR type by defining either one of the two dwells or both equal to zero. Therefore, the icon displayed in VP4 will contain the form DRDR.

1. The type of input-output function is printed in this area.

2. The interval of oscillation of the follower is printed in this icon. When the output pair is a revolute, the value is given in degrees; if the output pair is prismatic, it is given in the chosen units of length.

3. The starting value of the input-output function is printed in this icon. When the output pair is a revolute, the value is given in degrees; if the output pair is prismatic, it is given in the chosen units of length.

4. The displacement program for one period of the cam is drawn in this icon.
5. The first interval of the function is printed as a percentage of one period of the cam motion. When the displacement program is either RDR or RR, this value is initialized as zero.
6. The second interval of the function is printed as a percentage of one period of the cam motion.
7. The third interval of the function is printed as a percentage of one period of the cam motion. When the displacement program is either DRR or RR, this value is initialized as zero.

The icons of VP3 and VP4 of the submenu `di_3_osc` are shown in Fig. 7.11.

7.5.3 di_3_ind

When three-link indexing cam mechanisms are chosen from the main menu, the possible four combinations of kinematic pairs used at the input and the output share the same icons displayed in VP3 and VP4. The procedures that draw these icons are called `vp3_3_ind` and `vp4_3_ind`.

The icons displayed in VP3 are described below:

1. With this icon, the user can animate the mechanism. There are three options to select: *a*) counterclockwise rotation of the cam for the case of revolute pair and translation for the case of prismatic pairs; *b*) no motion; and *c*) clockwise rotation of the cam for the case of revolute pair and translation for the case of prismatic pairs. Moreover, the numerical value of the relative position of the cam with respect to the frame is printed.

   Icons 2–7 below are used to print the numerical values of the various parameters, namely,

2. The distance between input and output axes, a_1, in the selected units;
3. The angle between input and output axes, α_1, in degrees;
4. The initial point along the contact line of cam and follower, determined by λ_1;
5. The end point along the contact line of cam and follower, determined by λ_2.

The icons displayed in VP4 are described below:

1. The type of input-output function is printed in this area.

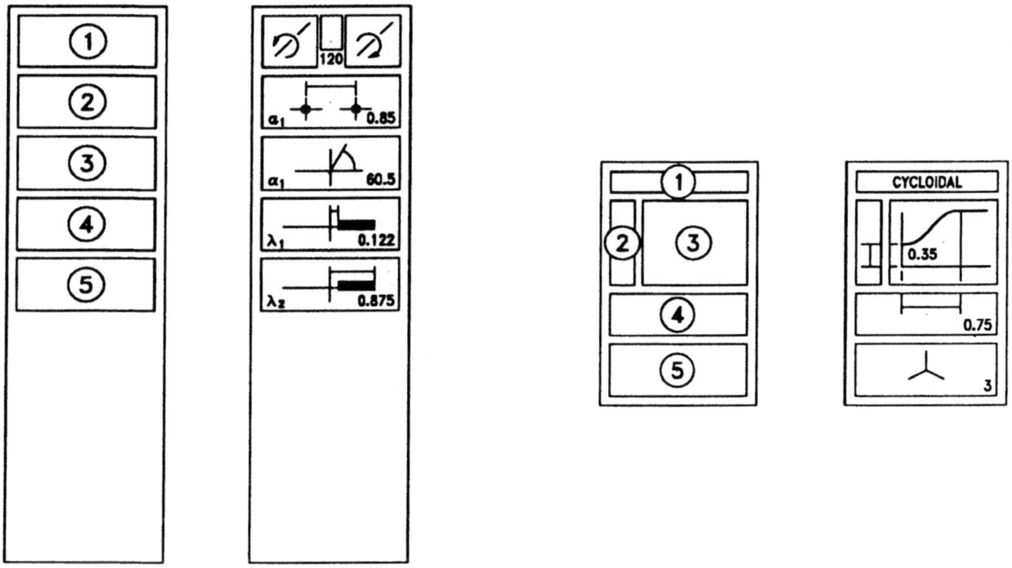

(*a*) (*b*)

Figure 7.12 Icons of the `di_3_ind` submenu;
a) VP3 viewport and *b*) VP4 viewport.

2. The starting value of the input-output function is printed in this icon. When the output pair is a revolute, the value is given in degrees; if the output pair is prismatic, in the chosen units of length.

3. The displacement program for one indexing step is drawn in this icon.

4. The interval of the function is printed as a percentage of one period of the cam motion.

5. The number of periods of the cam for one period of the follower is shown in this icon.

The icons of VP3 and VP4 of the submenu `di_3_ind` are shown in Fig. 7.12.

7.5.4 di_4

When four-link cam mechanisms are chosen from the main menu, both oscillating and indexing mechanisms share the same icons displayed in VP3, whose procedure to draw the icons is called **vp3_4**. If the user selects an oscillating mechanism, USYCAMS will display in VP4 the icons shown in Fig. 7.11*b*. On the other hand, if the user selects an indexing mechanism, USYCAMS will display in VP4 the icons shown in Fig. 7.12*b*.

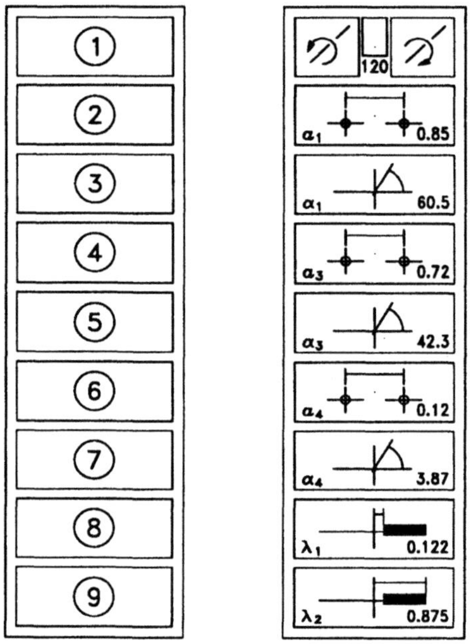

Figure 7.13 Icons of VP3 of the DI_4 submenu.

The two possible types of the icons of VP4 are already described in the last two subsections. We describe the icons of VP3 below:

1. With this icon, the user can animate the mechanism. Three options are available, namely, *a*) counterclockwise rotation of the cam for the case of revolute pair and translation for the case of prismatic pairs; *b*) no motion; and *c*) clockwise rotation of the cam for the case of revolute pair and translation for the case of prismatic pairs. The numerical value of the relative position of the cam with respect to the frame is printed.

   Icons 2–9 below are used to print the numerical values of the various parameters, namely,

2. The distance between input and output axes, a_1, in the selected units;

3. The angle between input and output axes, α_1, in degrees;

4. The distance between the output axis and the axis of the roller, a_3, in the selected units;

5. The angle between the output and the axis of the roller, α_3, in degrees;

6. The radius of the roller, a_4, in the selected units;

7. The angle of the roller, α_4, in degrees;

8. The initial point along the contact line of cam and follower, determined by λ_1;

9. The end point along the contact line of cam and follower, determined by λ_2.

The icons of VP3 of the submenu `di_4` are shown in Fig. 7.13.

7.6 Synthesis Procedures

In this section we outline the sixteen `construct_surfaces` synthesis procedures listed in the second column of Table 7.1. We specify the synthesis parameters needed to execute each procedure and describe the most important underlying algorithm, namely, the generation of the database containing the nodes of cam and follower surfaces, called `cam_surf` and `fol_surf`, respectively.

`cs_3_osc_rr`:

Procedure `cs_3_osc_rr` is called for the synthesis of three-link oscillating cam mechanisms with constant pressure angle. There are two possible configurations, namely, planar and spherical. Both input and output pairs are revolutes, and hence, the input-output function is given as

$$\phi = \phi(\psi)$$

which is defined in the procedure `osc_menu` and updated in the procedure `vp4_3_osc`, where ψ_j, ϕ_j and ϕ'_j, for $j = 1, 2, \cdots, n$, are generated.

i) Planar mechanism.
For reference, this type of mechanism is best known as *cam mechanism with oscillating flat-face follower.* When the user points the cursor in the icon a_1, USYCAMS uses this part of the procedure. Other than ψ_j, ϕ_j and ϕ'_j, this procedure expects values of a_1, a_4, λ_1 and λ_2.

The algorithm that generates **cam_surf** is displayed below:

For $i = 1$ till 2 do
    For $j = 1$ till n do

$$a_3 \leftarrow \frac{a_1 \cos \phi_j}{\phi_j' - 1}$$

$$\texttt{cam\_surf}_{ij} \leftarrow \begin{bmatrix} a_1 \cos \psi_j + a_3 \cos(\psi_j - \phi_j) + a_4 \sin(\psi_j - \phi_j) \\ -a_1 \sin \psi_j - a_3 \sin(\psi_j - \phi_j) + a_4 \cos(\psi_j - \phi_j) \\ 0 \end{bmatrix} + \lambda_i \begin{bmatrix} 0 \\ 0 \\ 1 \end{bmatrix}$$

    enddo
enddo

ii) Spherical mechanism. When the user points the cursor in the icon α_1, USYCAMS switches to spherical mechanisms. Besides ψ_j, ϕ_j and ϕ_j', this procedure expects values of α_1, α_4, λ_1 and λ_2.
The algorithm that generates **cam_surf** is displayed below:

```
For i = 1 till 2 do
   For j = 1 till n do
```

$$\alpha_3 \leftarrow \arctan\left(\frac{\sin\alpha_1 \cos\phi_j}{\phi'_j - \cos\alpha_1}\right)$$

$$k_1 \leftarrow \cos\alpha_1 \cos\alpha_3 - \sin\alpha_1 \sin\alpha_3 \cos\phi_j$$

$$k_3 \leftarrow \sin\alpha_1 \cos\alpha_3 + \cos\alpha_1 \sin\alpha_3 \cos\phi_j$$

$$k_5 \leftarrow \cos\psi_j \cos\phi_j + \sin\psi_j \sin\phi_j \cos\alpha_1$$

$$k_6 \leftarrow \sin\psi_j \cos\phi_j - \cos\psi_j \sin\phi_j \cos\alpha_1$$

$$h_1 \leftarrow k_3 \sin\psi_j - \sin\alpha_3 \sin\phi_j \cos\psi_j$$

$$h_2 \leftarrow k_3 \cos\psi_j + \sin\alpha_3 \sin\phi_j \sin\psi_j$$

$$\texttt{cam\_surf}_{ij} \leftarrow \lambda_i \begin{bmatrix} k_5 \sin\alpha_4 - h_1 \cos\alpha_4 \\ -k_6 \sin\alpha_4 - h_2 \cos\alpha_4 \\ \sin\alpha_1 \sin\phi_j \sin\alpha_4 + k_1 \cos\alpha_4 \end{bmatrix}$$

```
   enddo
enddo
```

cs_3_osc_rp:

Procedure `cs_3_osc_rp` is called for the synthesis of three-link oscillating cam mechanisms with constant pressure angle, revolute coupling as input and prismatic coupling as output, known as *cam mechanism with translating flat-face follower.*

The input-output function is given as

$$z_3 = z_3(\psi)$$

which is defined in the `osc_menu` procedure and updated in the `vp4_3_osc` procedure, where ψ_j, z_{3j} and z'_{3j}, for $j = 1, 2, \cdots, n$, are generated.

This procedure expects values of ψ_j, z_{3j} and z'_{3j}, which are generated internally, and α_1, a_4, λ_1 and λ_2, whose values can be changed by using the VP3 viewport.

The algorithm that generates `cam_surf` is described below:

For $i = 1$ till 2 do
  For $j = 1$ till n do

$$\texttt{cam\_surf}_{ij} \leftarrow \begin{bmatrix} -z'_{3j}\sin\alpha_1\cos\psi_j + (a_4 - z_{3j}\sin\alpha_1)\sin\psi_j \\ z'_{3j}\sin\alpha_1\sin\psi_j + (a_4 - z_{3j}\sin\alpha_1)\cos\psi_j \\ 0 \end{bmatrix} + \lambda_i \begin{bmatrix} 0 \\ 0 \\ 1 \end{bmatrix}$$

  enddo
enddo

cs_3_osc_pr:

The **cs_3_osc_pr** procedure is called for the synthesis of three-link oscillating cam mechanisms with constant pressure angle, prismatic coupling as input and revolute coupling as output.

The input-output function is given as

$$\phi = \phi(z_2)$$

which is defined in the **osc_menu** procedure and updated in the **vp4_3_osc** procedure, where z_{2j}, ϕ_j, and ϕ'_j, for $j = 1, 2, \cdots, n$, are generated.

This procedure expects values of z_{2j}, ϕ_j, and ϕ'_j, which are generated internally, and the values of a_1, α_1, a_4, λ_1 and λ_2, which the user provides with the aid of the VP3 viewport.

Now we include the algorithm that generates **cam_surf**:

For $i = 1$ **till** 2 **do**
  For $j = 1$ **till** n **do**

$$a_3 \leftarrow \frac{\sin\alpha_1 \cos\phi_j}{\phi_j'}$$

$$\texttt{cam\_surf}_{ij} \leftarrow \begin{bmatrix} a_1 + a_3\cos\phi_j - a_4\sin\phi_j \\ a_3\cos\alpha_1\sin\phi_j + a_4\cos\alpha_1\cos\phi_j - z_{2j}\cos\alpha_1\sin\alpha_1 \\ a_3\sin\alpha_1\sin\phi_j + a_4\cos\phi_j\sin\alpha_1 - z_{2j}\sin^2\alpha_1 \end{bmatrix} + \lambda_i \begin{bmatrix} 0 \\ -\sin\alpha_1 \\ \cos\alpha_1 \end{bmatrix}$$

  enddo
enddo

It is apparent that this type of mechanism has the restriction that $\phi' \neq 0$; otherwise a_3 becomes unbounded.

cs_3_osc_pp:

Procedure **cs_3_osc_pp** is called for the synthesis of three-link oscillating cam mechanisms with constant pressure angle and prismatic pair in both input and output couplings.

The input-output function is now given as

$$z_3 = z_3(z_2)$$

which is defined in the **osc_menu** procedure and updated in the **vp4_3_osc** procedure, where z_{2j}, z_{3j}, and z_{3j}', for $j = 1, 2, \cdots, n$, are generated.

This procedure expects values of z_{2j}, z_{3j}, and z_{3j}', which are produced by USYCAMS, and the values of α_1, a_4, λ_1 and λ_2, which the user provides with the aid of the VP3 viewport.

The algorithm that generates **cam_surf** is now described:

For $i = 1$ till 2 do
  For $j = 1$ till n do

$$\mu_0 \leftarrow \arctan\left(\frac{z'_{3j} - \cos\alpha_1}{\sin\alpha_1}\right)$$

$$\texttt{cam\_surf}_{ij} \leftarrow \begin{bmatrix} 0 \\ -a_4\sin(\alpha_1 - \mu_0) - z_{3j}\sin\alpha_1 \\ a_4\cos(\alpha_1 - \mu_0) + z_{3j}\cos\alpha_1 - z_{2j} \end{bmatrix} + \lambda_i \begin{bmatrix} 1 \\ 0 \\ 0 \end{bmatrix}$$

  enddo
enddo

`cs_3_ind_rr`:

Three-link indexing cam mechanisms with minimum sliding along the line of contact can be synthesized with this procedure. In this case, both input and output pairs are revolutes.

The input-output function is given as

$$\phi = \phi(\psi)$$

which is defined in the **ind_menu** procedure and updated in the **vp4_3_ind** procedure, where ψ_j, ϕ_j and ϕ'_j, for $j = 1, 2, \cdots, n$, are generated.

This procedure expects values of ψ_j, ϕ_j and ϕ'_j, which are produced internally, and the values of a_1, α_1, λ_1 and λ_2, which the user can choose from the icons displayed in VP3.

The algorithm that generates **cam_surf** and **fol_surf** is displayed below:

For $i = 1$ **till** 2 **do**
  For $j = 1$ **till** n **do**

$$\theta_2 \leftarrow \arctan\left(\frac{\phi'_j \sin\alpha_1}{\phi'_j \cos\alpha_1 - 1}\right)$$

$$c \leftarrow \sqrt{\phi'^2_j - 2\phi'_j \cos\alpha_1 + 1}$$

$$b_2 \leftarrow \frac{\phi'^2_j - \phi'_j \cos\alpha_1}{c^2} a_1$$

$$\texttt{cam\_surf}_{ij} \leftarrow b_2 \begin{bmatrix} \cos\psi_j \\ -\sin\psi_j \\ 0 \end{bmatrix} + \lambda_i \begin{bmatrix} -\sin\psi_j \sin\theta_2 \\ -\cos\psi_j \sin\theta_2 \\ \cos\theta_2 \end{bmatrix}$$

$$\texttt{fol\_surf}_{ij} \leftarrow (b_2 - a_1) \begin{bmatrix} \cos\phi_j \\ -\sin\phi_j \\ 0 \end{bmatrix} + \lambda_i \begin{bmatrix} -\sin\phi_j \sin(\theta_2 - \alpha_1) \\ -\cos\phi_j \sin(\theta_2 - \alpha_1) \\ \cos(\theta_2 - \alpha_1) \end{bmatrix}$$

  enddo
enddo

cs_3_ind_rp:

Procedure **cs_3_ind_rp** is called for the synthesis of three-link indexing cam mechanisms with minimum sliding at the higher pair contact, revolute coupling as input and prismatic coupling as output.

The input-output function is given as

$$z_3 = z_3(\psi)$$

which is defined in the **osc_menu** procedure and updated in the **vp4_3_osc** procedure, where ψ_j, z_{3j} and z'_{3j}, for $j = 1, 2, \cdots, n$, are generated.

This procedure receives internally the values of ψ_j, z_{3j} and z'_{3j}, and, from the user, the values of a_1, α_1, λ_1 and λ_2, which are chosen from the icons of VP3.

Below we describe the algorithm that generates **cam_surf** and **fol_surf**:

For $i = 1$ till 2 do
  For $j = 1$ till n do

$$\texttt{cam\_surf}_{ij} \leftarrow z'_{3j} \sin\alpha_1 \begin{bmatrix} -\cos\psi_j \\ \sin\psi_j \\ 0 \end{bmatrix} + \lambda_i \begin{bmatrix} 0 \\ 0 \\ 1 \end{bmatrix}$$

$$\texttt{fol\_surf}_{ij} \leftarrow \begin{bmatrix} -(z'_{3j} \sin\alpha_1 + a_1) \\ z_{3j} \sin\alpha_1 \cos\alpha_1 \\ -z_{3j} \sin\alpha_1 \end{bmatrix} + \lambda_i \begin{bmatrix} 0 \\ \sin\alpha_1 \\ \cos\alpha_1 \end{bmatrix}$$

  enddo
enddo

cs_3_ind_pr:

Procedure **cs_3_ind_pr** is called for the synthesis of three-link indexing cam mechanisms with minimum sliding at the higher pair, prismatic coupling as input and revolute coupling as output.

The input-output function is given as

$$\phi = \phi(z_2)$$

which is defined in the **osc_menu** procedure and updated in the **vp4_3_osc** procedure, where z_{2j}, ϕ_j, and ϕ'_j, for $j = 1, 2, \cdots, n$, are generated.

This procedure receives internally the values of z_{2j}, ϕ_j, and ϕ'_j, and, from the user, the values of a_1, α_1, λ_1 and λ_2, which can be chosen from the icons of VP3.

The algorithm that generates **cam_surf** and **fol_surf** is included below:

```
For i = 1 till 2 do
   For j = 1 till n do
```

$$\texttt{cam\_surf}_{ij} \leftarrow \begin{bmatrix} a_1 + \sin\alpha_1/\phi'_j \\ -z_{2j}sa_1\cos\alpha_1 \\ -z_{2j}\sin^2\alpha_1 \end{bmatrix} + \lambda_i \begin{bmatrix} 0 \\ -\sin\alpha_1 \\ \cos\alpha_1 \end{bmatrix}$$

$$\texttt{fol\_surf}_{ij} \leftarrow \frac{\sin\alpha_1}{\phi'_j} \begin{bmatrix} \cos\phi_j \\ -\sin\phi_j \\ 0 \end{bmatrix} + \lambda_i \begin{bmatrix} 0 \\ 0 \\ 1 \end{bmatrix}$$

```
   enddo
enddo
```

It is apparent that this type of mechanism has the restriction that $\phi' \neq 0$; otherwise, a_3 becomes unbounded.

cs_3_ind_pp:

Although theoretically three-link indexing cam mechanisms with minimum sliding and prismatic pair in both input and output couplings can be synthesized, according to the results obtained in Section 3.2.4, this configuration is not feasible. If the user choses this option, USYCAMS will display a message saying that it is not possible to synthesize this type of mechanism.

cs_4_osc_rr and cs_4_ind_rr:

The synthesis procedures of four-link oscillating and indexing cam mechanisms with revolute pairs at the input and output couplings are similar. What makes the difference between these procedures is the displacement program.

The input-output function is given as

$$\phi = \phi(\psi)$$

which is defined in the **osc_menu** or **ind_menu** procedures and updated in the **vp4_3_osc** and **vp4_3_ind** procedures, respectively. In either case, ψ_j, ϕ_j and ϕ'_j, for $j = 1, 2, \cdots, n$, are generated.

Procedures **cs_4_osc_rr** and **cs_4_ind_rr** involve all planar, spherical and spatial cases. Here the user can visualize the smooth transformation from one configuration to another. The algorithm that generates **cam_surf** is relatively long, and for this reason it is not included here. Nevertheless, we outline the algorithms that generate the database for planar and spherical cam surfaces. The reader interested in the general algorithm is referred to Section 4.2.1, where the corresponding formulation is introduced.

Besides the values of ψ_j, ϕ_j and ϕ'_j, which are generated by USYCAMS, this procedure expects values of a_1, α_1, a_3, α_3, a_4, α_4, λ_1 and λ_2, which are given by the user with the aid of the VP3 viewport.

i) Planar mechanisms. This mechanism is best known as ***planar cam mechanism with oscillating roller-follower.***

For $i = 1$ till 2 do
  For $j = 1$ till n do

$$b_2 \leftarrow \frac{\phi_j'^2 - \phi_j' \cos\alpha_1}{\phi_j'^2 - 2\phi_j' \cos\alpha_1 + 1} a_1$$

$$b_3 \leftarrow \sqrt{(a_3 c\phi_j + a_1 - b_2)^2 + a_3{}^2 s^2 \phi_j}$$

$$\delta \leftarrow \arctan\left(\frac{a_3 s\phi_j}{a_3 c\phi_j + a_1 - b_2}\right)$$

$$\texttt{cam\_surf}_{ij} \leftarrow \begin{bmatrix} \cos\psi_j b_2 + \cos(\psi_j - \delta)(b_3 - a_4) \\ -\sin\psi_j b_2 - \sin(\psi_j - \delta)(b_3 - a_4) \\ 0 \end{bmatrix} + \lambda_i [0, \quad 0, \quad 1]$$

  enddo
enddo

ii) Spherical mechanisms. This mechanism is best known as ***spherical cam mechanism with oscillating roller-follower.***

$$
\begin{array}{l}
\texttt{For } i = 1 \texttt{ till } 2 \texttt{ do} \\
\quad \texttt{For } j = 1 \texttt{ till } n \texttt{ do} \\
\qquad \theta_2 \leftarrow \arctan\left(\dfrac{\phi'_j \sin\alpha_1}{\phi'_j \cos\alpha_1 - 1}\right) \\
\qquad \theta_3 \leftarrow \arctan\left(\dfrac{\sqrt{[c(\alpha_1-\theta_2)c\phi_j s\alpha_3 + c\alpha_3 s(\alpha_1-\theta_2)]^2 + s^2\alpha_3 s^2\phi_j}}{c\alpha_3 c(\alpha_1-\theta_2) - c\phi_j s\alpha_3 s(\alpha_1-\theta_2)}\right) \\
\qquad \delta \leftarrow \left(\dfrac{s\alpha_3 s\phi_j}{s\alpha_3 c(\alpha_1-\theta_2)c\phi_j + c\alpha_3 s(\alpha_1-\theta_2)}\right) \\
\qquad \tilde{k}_3 \leftarrow \sin\theta_2\cos(\theta_3-\alpha_4) + \cos\theta_2\sin(\theta_3-\alpha_4)\cos\delta \\
\qquad \texttt{cam\_surf}_{ij} \leftarrow \lambda_i \begin{bmatrix} -\tilde{k}_3\sin\psi_j + \sin(\theta_3-\alpha_4)\sin\delta\cos\psi_j \\ -\tilde{k}_3\cos\psi_j - \sin(\theta_3-\alpha_4)\sin\delta\sin\psi_j \\ \cos\theta_2\cos(\theta_3-\alpha_4) - \sin\theta_2\sin(\theta_3-\alpha_4)\cos\delta \end{bmatrix} \\
\quad \texttt{enddo} \\
\texttt{enddo}
\end{array}
$$

cs_4_osc_rp and cs_4_ind_rp:

Procedures **cs_3_osc_rp** and **cs_3_ind_rp** are called for the synthesis of four-link oscillating and indexing cam mechanisms with revolute coupling as input and prismatic coupling as output, the difference between these two procedures being the displacement program. For oscillating mechanisms, the mechanism is best known as *cam mechanism with translating roller-follower.*

The input-output function is given as

$$z_3 = z_3(\psi)$$

which is defined either in the **osc_menu** or **ind_menu** procedures and updated in the **vp4_3_osc** or **vp4_3_ind** procedures, depending on the user selection. In these procedures ψ_j, z_{3j} and z'_{3j}, for $j = 1, 2, \cdots, n$, are generated.

Besides ψ_j, z_{3j} and z'_{3j}, which are generated by USYCAMS, these procedures expect values of a_1, α_1, a_3, a_4, λ_1 and λ_2, which are given by the user with the aid of the icons of the VP3 viewport.

The algorithm that generates **cam_surf** is now outlined:

```
For i = 1 till 2 do
   For j = 1 till n do
```

$$b_2 \leftarrow -z_j'3 \sin\alpha_1$$

$$b_3 \leftarrow \sqrt{(a_3 + a_1 + z_{3j}' \sin\alpha_1)^2 + z_{3j}^2 \sin^2\alpha_1}$$

$$\delta \leftarrow \arctan\left(\frac{-z_{3j} s a_1}{a_1 + a_3 + z_{3j}' \sin\alpha_1}\right)$$

$$\mathtt{cam\_surf}_{ij} \leftarrow \begin{bmatrix} b_2\cos\psi_j + (b_3 - a_4)\cos(\delta - \psi_j) \\ -b_2\sin\psi_j + (b_3 - a_4)\sin(\delta - \psi_j) \\ 0 \end{bmatrix} + \lambda_i \begin{bmatrix} 0 \\ 0 \\ 1 \end{bmatrix}$$

```
   enddo
enddo
```

cs_4_osc_pr and cs_4_ind_pr:

Procedures **cs_3_osc_pr** and **cs_3__pr** are called for the synthesis of four-link oscillating and indexing cam mechanisms, respectively, with prismatic coupling as input and revolute coupling as output.

The input-output function is given as

$$\phi = \phi(z_2)$$

which is defined either in the **osc_menu** or **vp4_3_ind** procedures and updated in the **vp4_3_osc** or **vp4_3_ind** procedures, depending on the user selection. In these procedures z_{2j}, ϕ_j, and ϕ_j', for $j = 1, 2, \cdots, n$, are generated.

Besides z_{2j}, ϕ_j, and ϕ_j', which are generated by USYCAMS, these procedures expect values of a_1, α_1, a_3, a_4, λ_1 and λ_2, which are given by the user with the aid of the icons of the VP3 viewport.

We now outline the algorithm that generates **cam_surf**:

For $i = 1$ till 2 do
  For $j = 1$ till n do

$$b_3 \leftarrow \frac{\sqrt{(a_3\phi'_j \cos\phi_j - \sin\alpha_1)^2 + (a_3\phi'_j \sin\phi_j)^2}}{\phi'_j}$$

$$\delta \leftarrow \arctan\left(\frac{a_3\phi'_j \sin\phi_j}{a_3\phi'_j \cos\phi_j - \sin\alpha_1}\right)$$

$$\texttt{cam\_surf}_{ij} \leftarrow \begin{bmatrix} a_1 + \sin\alpha_1/\phi'_j + (b_3 - a_4)\cos\delta \\ [(b_3 - a_4)\sin\delta - z_{2j} s a_1]\cos\alpha_1 \\ [(b_3 - a_4)\sin\delta - z_{2j}\sin\alpha_1]\sin\alpha_1 \end{bmatrix} + \lambda_i \begin{bmatrix} 0 \\ -\sin\alpha_1 \\ \cos\alpha_1 \end{bmatrix}$$

  enddo
enddo

cs_4_osc_pp and cs_4_ind_pp:

Procedures cs_3_osc_pp and cs_3_ind_pp are called for the synthesis of four-link indexing and oscillating cam mechanisms, respectively, with prismatic pairs at both the input and the output.

The input-output function is given as

$$z_3 = z_3(z_2)$$

which is defined either in the osc_menu or ind_menu procedures and updated in the vp4_3_osc or vp4_3_ind procedures, depending on the user selection. In these procedures z_{2j}, z_{3j}, and z'_{3j}, for $j = 1, 2, \cdots, n$, are generated.

Besides z_{2j}, z_{3j}, and z'_{3j}, which are generated by USYCAMS, these procedures expect values of α_1, a_3, a_4, λ_1 and λ_2, which are given by the user with the aid of the icons of VP3.

The algorithm that generates cam_surf is included below:

```
For i = 1 till 2 do
   For j = 1 till n do
```

$$\theta_2 \leftarrow \arctan\left(\frac{z'_{3j}\sin\alpha_1}{z'_{3j}\cos\alpha_1 - 1}\right)$$

$$b_3 \leftarrow [a_3\cos(\alpha_1 - \theta_2) - z_{3j}\sin(\alpha_1 - \theta_2)]$$

$$z_{32} \leftarrow [a_3\sin(\alpha_1 - \theta_2) + z_{3j}\cos(\alpha_1 - \theta_2)]$$

$$\texttt{cam\_surf}_{ij} \leftarrow \begin{bmatrix} 0 \\ (b_3 - a_4)\cos\theta_2 - z_{32}\sin\theta_2 \\ (b_3 - a_4)\sin\theta_2 + z_{32}\cos\theta_2 - z_{2j} \end{bmatrix} + \lambda_i \begin{bmatrix} 1 \\ 0 \\ 0 \end{bmatrix}$$

```
   enddo
enddo
```

thereby completing the description of the individual synthesis algorithms.

7.7 Application Example

As an example of the application of USYCAMS for other design problems than the synthesis of cams, we take the synthesis of a drive for a shaking machine. Shaking machines find many applications in industry, e.g., for mixing fluids, for screening of grains or ore, etc. Note that these operations are essentially different from those meant to be accomplished with cam mechanisms. Nevertheless, as we show here, the underlying drive mechanism can be readily synthesized with the aid of USYCAMS.

The drive is meant to produce an advancing motion of a rotor with a periodic acceleration program that accelerates in part of the cycle and decelerates in the rest of it. Moreover, we want to produce this motion with a driving shaft rotating at a uniform angular velocity, while the driven shaft rotating intermittently, with non-zero mean velocity.

The mechanism we choose for the synthesis of this problem is of the RHR type. Thus, the input-output function is of the form

$$\phi = \phi(\psi), \quad \phi' > 0$$

Given these characteristics, the function we apply here is similar to that of the cycloidal motion, namely,

$$\phi = \frac{2\pi}{N}(x - \frac{k}{2\pi}\sin 2\pi x), \quad 0 \le x \equiv \frac{\psi}{2\pi} \le 1, \quad 0 < k < 1 \tag{7.1}$$

Table 7.2 Synthesis parameters of a shaking mechanism.

Type	a_1 mm	α_1 deg.	λ_{min} mm	λ_{max} mm	k	N
Planar	100	180	-4	4	0.5	3
Spherical	0	135	40	80	0.5	3
Spatial	50	135	40	80	0.5	3

where N is the number of rotations of the driving shaft for one full rotation of the driven shaft and k is a constant that is introduced to control motion and geometric characteristics. In cycloidal motion, $k = 1$.

With the application of USYCAMS, we have provided the synthesis parameters, shown in Table 7.2, for the solution of three configurations, namely, planar, spherical and spatial. The three solutions are displayed in Figs. 7.14–7.16.

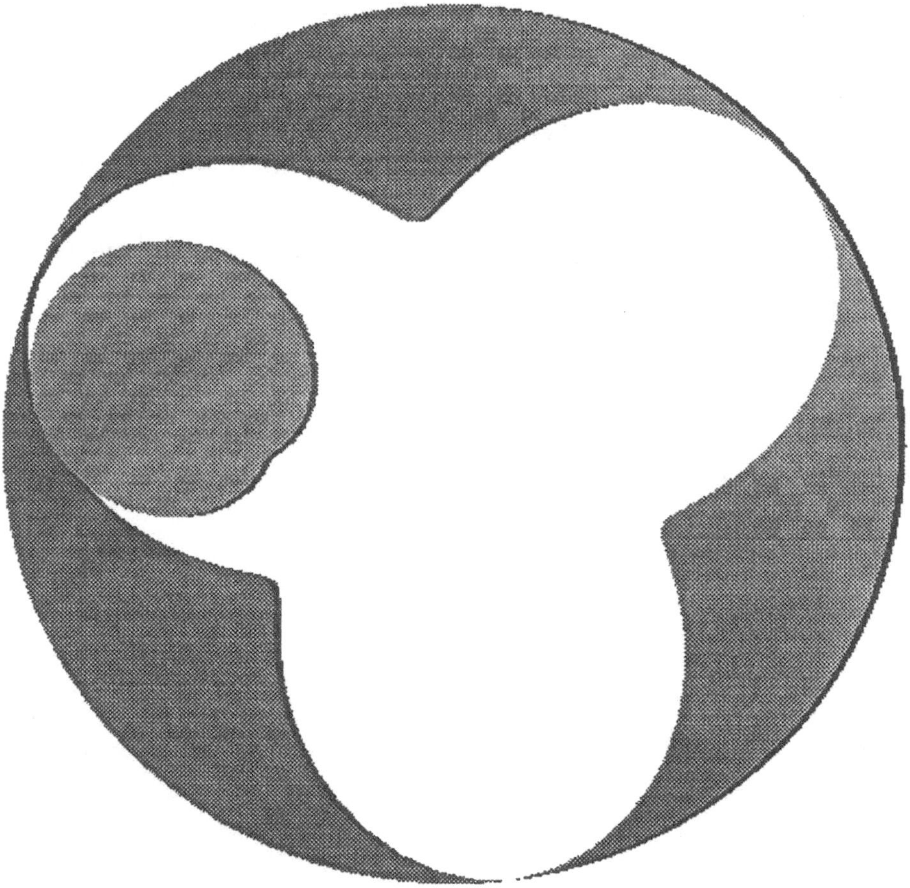

Figure 7.14 Planar shaking mechanism.

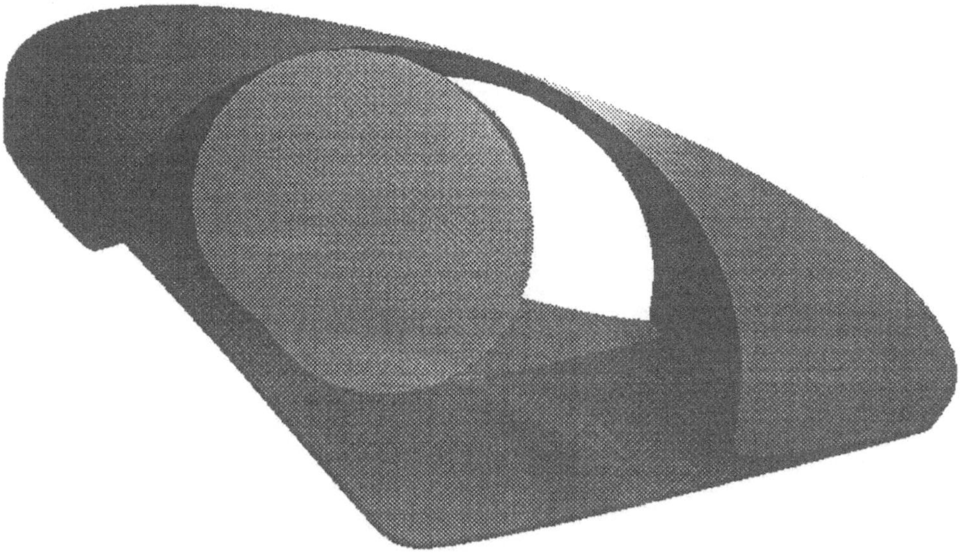

Figure 7.15 Spherical shaking mechanism.

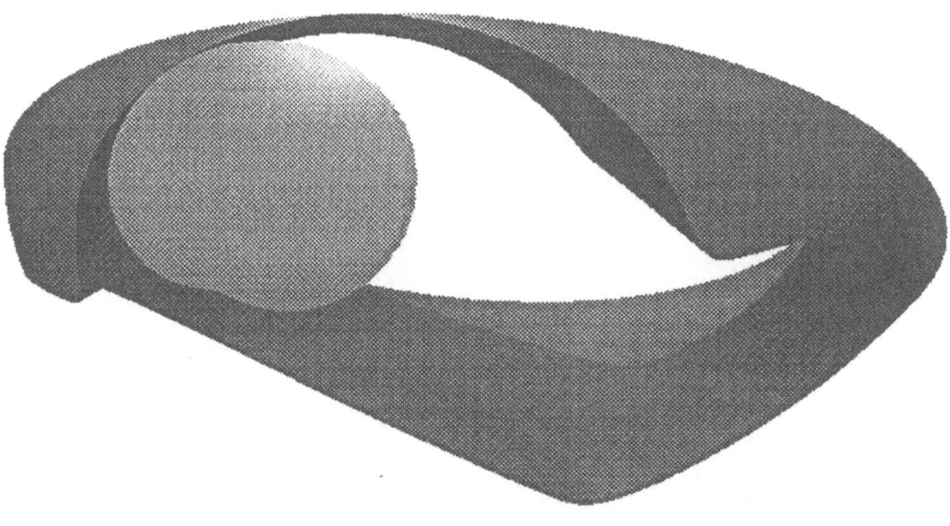

Figure 7.16 Spatial shaking mechanism.

Chapter 8

THE CAD OF CAM MECHANISMS

8.1 Introduction

This chapter concerns the use of commercial CAD systems in the synthesis and design of cam mechanisms. We divide the chapter into two parts: *a*) the synthesis of planar cam mechanisms using CAD systems; and *b*) case studies of complete mechanical designs of the complex cam mechanisms synthesized in Chapter 5, i.e., PRICAM in its two versions, planar and spherical.

The section on the CAD-based synthesis of planar mechanisms is meant to introduce a novel graphical procedure that makes extensive use of the capabilities of commercial CAD software. The features that we use here are the plotting of functions, the rotation and translation of planar rigid figures, and the determination of intersections of lines and curves. Here, we assume that the displacement function of the follower is given as a smooth explicit function of the input variable, so that its derivative is available. In some instances, such a displacement function may not be available, but only a discrete set of input-output pairs. In this case, a smooth interpolation of this set will be necessary before the derivative can be determined.

If only a discrete set of points is available for the functions needed in the synthesis procedures, interpolation does not represent any difficulty for AutoCAD^{TM}. This software has two different types of interpolation, namely, generation of a smooth curve passing through the given points, and generation of a spline curve supported by the points.

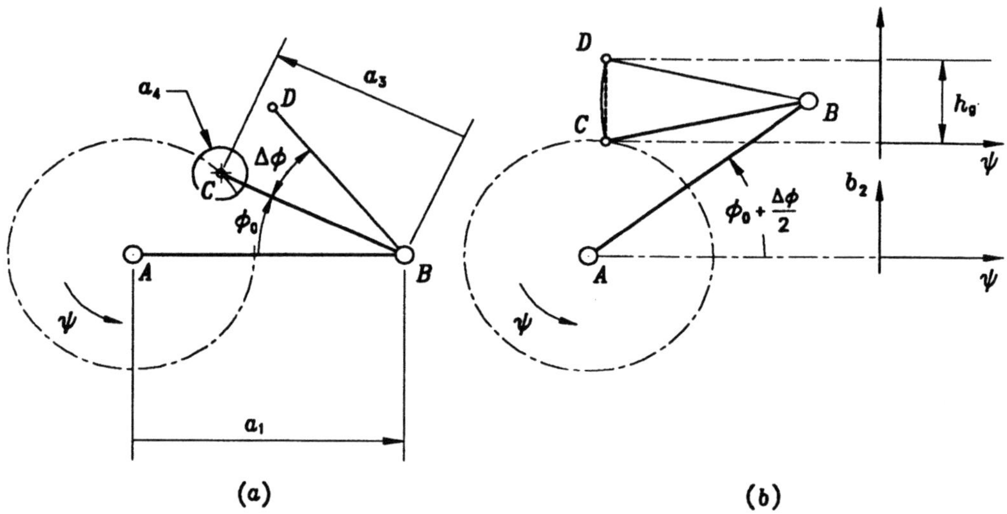

Figure 8.1 RHRR mechanism layout

8.2 CAD-Based Methods. Planar Applications

Graphical methods for the synthesis of cam mechanisms could be considered obsolete with present-day computer technology. However, the ever increasing availability of CAD systems makes graphical methods of cam synthesis worth revisiting.

In this section we revisit graphical methods of planar cam synthesis and propose a novel *CAD-based method.* In the realm of traditional graphical methods (Rothbart, 1956; Jensen, 1965; Chen 1982), the cam profile is obtained as the tangent curve to a sequence of placements of the geometric entity—circle, line or arbitrary curve—representing the follower. The accuracy of the profile thus obtained is totally dependent upon the drafting skill of the user. With CAD systems there is no need for drafting skill; now the problem is the digital approximation of the tangent curve. This problem was overcome in the early days of CAD/CAM technology with the introduction of *envelopes* (Struik, 1961; Wilson, Sadler and Michels, 1983; Backhouse and Jones, 1990), that yield the cam profile coordinates directly and eliminate the inaccuracies of tangent-curve tracing. While the envelope method is very reliable and has found extensive acceptance, it is still essentially a numerical method that does not exploit the interactive capabilities of CAD systems. This state of affairs prompted us to propose an innovation in the rendering of the profile. This innovation consists of finding the contact points between cam and follower, in a totally graphical and interactive fashion that exploits features available in commercial CAD systems.

The theory presented in Chapters 3 and 4 is the basis of the method introduced here. The profile of a three-link mechanism is used as a reference for the construction of a four-link mechanism; in other words, RHR and RHP mechanisms are the two basic groups of the four types of mechanisms discussed here. These four types pertain to two families, the {RHRR, $\mathrm{R\overline{HH}R}$} family based on RHR mechanisms, and the {RHRP, $\mathrm{R\overline{HH}P}$} family, based on RHP mechanisms. Although not included here, the synthesis of those groups pertaining to PHR and PPP mechanisms can follow the same methodology. We assume henceforth that all geometric parameters of the cam have been either prescribed by an experienced cam designer, or determined using an optimization procedure, as described in detail in (Angeles and López-Cajún, 1991). Moreover, contrary to conventional graphical methods, we assume that an accurate expression for the derivative of the output variable with respect to the input variable, $\phi'(\psi)$, is available. This assumption is plausible; even if the input-output function is given as a discrete set of input-output pairs, we can always form a smooth interpolation using spline functions.

All procedures discussed below can be readily automated with the aid of a CAD system. We have implemented them in AutoCAD™.

8.2.1 RHRR Mechanisms

An illustration of the cam mechanism discussed in this section is shown in Fig. 4.13. The parameters given at the outset are shown in Fig. 8.1a, where: a_1 is the distance between the centers of the fixed revolutes; a_3 the distance between the fixed and moving revolute centers of the follower; a_4 the radius of the roller; ϕ_0 the angle of the follower corresponding to the *lowest* position of the follower; and $\Delta\phi$ is the rise of the follower.

Once the parameters are fixed, we start by rotating the segment AB through an angle $\phi_0 + \Delta\phi/2$, so that the chord CD is parallel to the ordinate of the displacement coordinate frame, as shown in Fig. 8.1b. Thus, the displacement program is based on the length of the chord CD, namely, the chord of the arc with radius a_3 subtending angle DBC. Then, as the follower rotates, the projection of an arbitrary position of the center of the roller onto segment CD is represented by h, i.e.,

$$h = h_g \tau \tag{8.1}$$

where

$$h_g = 2a_3 \sin\frac{\Delta\phi}{2} \tag{8.2}$$

is the length of the chord, while τ is the normal input-output function defined as in eq.(B.1).

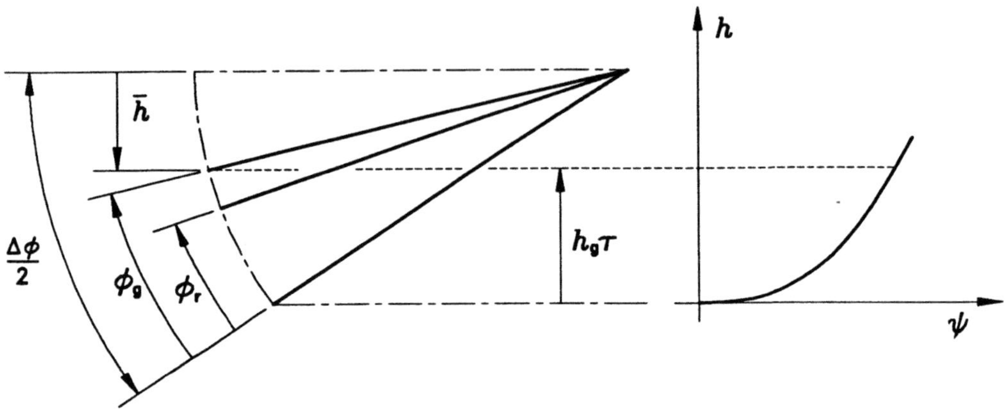

Figure 8.2 Difference between ϕ_g and ϕ_r

With the application of eq.(8.2), the actual angle of displacement of the follower, ϕ_r, and the angle of displacement obtained graphically, ϕ_g, are identical only at the ends of the follower motion and at the middle point of this motion. In any other position there is a difference, which, when normalized with respect to $\Delta\phi$, is denoted by ξ, and measures the relative error involved, i.e.,

$$\xi = \frac{\phi_g - \phi_r}{\Delta\phi} \tag{8.3}$$

Moreover,

$$\phi_r = (\Delta\phi)\tau \tag{8.4a}$$

$$\phi_g = \frac{\Delta\phi}{2} - \sin^{-1}\frac{\bar{h}}{a_3} \tag{8.4b}$$

where $\bar{h}$, as shown in Fig. 8.2, is given as

$$\bar{h} = a_3 \sin\frac{\Delta\phi}{2} - h_g\tau \tag{8.5}$$

Now, upon combining eqs.(8.2, 8.4 & 8.5) with eq.(8.3), we obtain from the latter

$$\xi(\tau) = \frac{1}{2} - \tau - \frac{1}{\Delta\phi}\sin^{-1}\left[(1-2\tau)\sin\frac{\Delta\phi}{2}\right] \tag{8.6}$$

The maximum value of ξ, for a given value of $\Delta\phi$, can be computed as $\xi_M = \xi(\tau_M)$, where τ_M is obtained from the equation $d\xi/d\tau = 0$. For example, for a cycloidal input-output motion, τ_M is given as

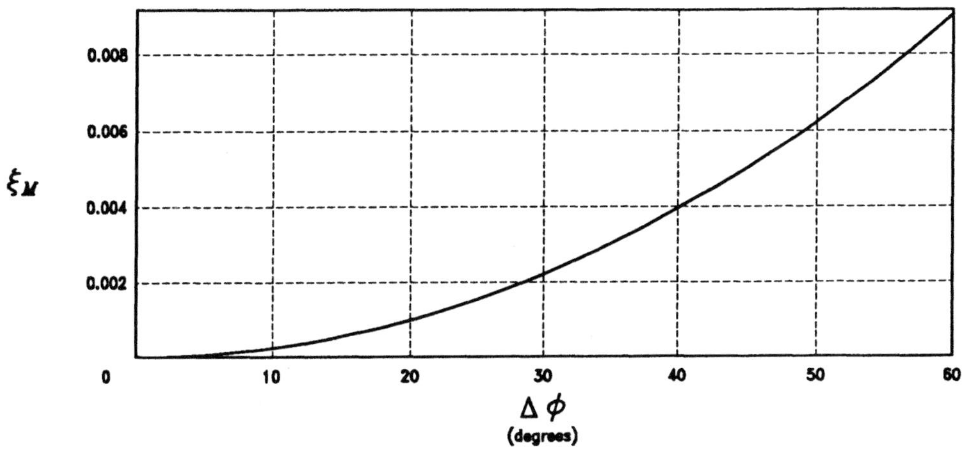

Figure 8.3 Plot of ξ_{max} vs. $\Delta\phi$

$$\tau_M = \frac{1}{2} \pm \frac{1}{2\Delta\phi}\sqrt{\frac{\Delta^2\phi}{\sin^2\frac{\Delta\phi}{2}} - 4} \tag{8.7}$$

Thus, with $\Delta\phi = 30°$ for example, $\xi_M = 0.00221$. A plot of ξ_M vs. $\Delta\phi$ is shown in Fig. 8.3. The magnitude of ξ_M is apparently negligible for the purposes of this chapter, and hence, eq.(8.2) can be considered as an acceptable approximation.

The coordinate axes b_2 and ψ are located as shown in Fig. 8.1b, where b_2 is defined in Chapter 4 as the distance from the axis of rotation of the cam to the contact line between cam and follower of the RHR mechanism, as recalled below for quick reference:

$$b_2 = \frac{\phi'}{\phi' - 1}a_1 \tag{4.49}$$

Further, the cam is considered fixed and the frame of the mechanism is rotated in the opposite direction of the desired rotation of the cam. Thus, for $\psi = \psi_5$, the corresponding points of h and b_2 are projected into the arc $\mathcal{A}$ and the line $\mathcal{L}$, respectively, as shown in Fig. 8.4. Then, the loci of C' and E for ψ_5 are readily obtained, where C' and E are nothing but I_{43} and I_{32}, respectively.

Finally, we determine F, the contact point between cam and roller, as the intersection of the circle with center C' and radius a_4 with segment EC', as indicated in Fig. 8.4. This procedure is repeated for each ψ_i, where $i = 0, 1, \cdots, n$, and n is the number of subdivisions of the full rotation of the cam. The locus of point F thus resulting, shown in Fig. 8.5, is the desired cam profile.

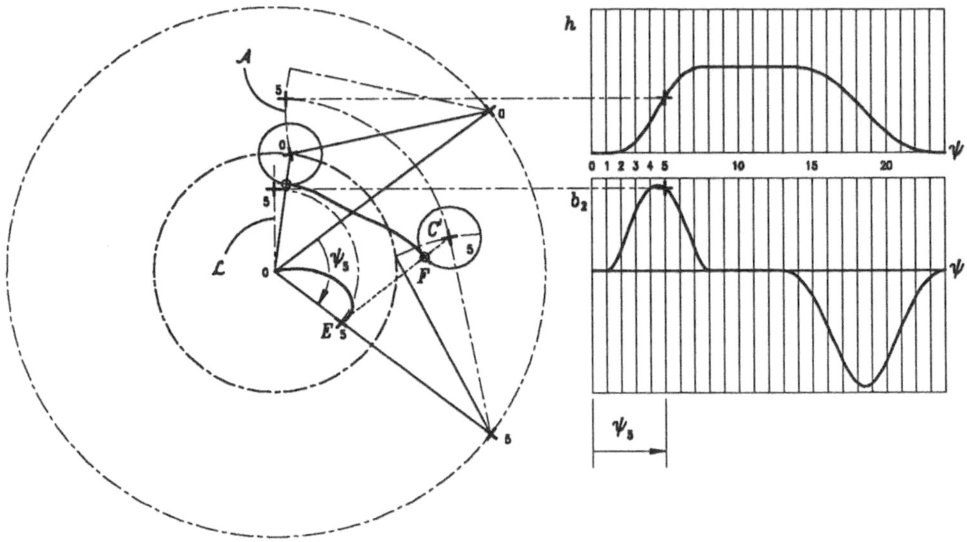

Figure 8.4 Contact point for $\psi = \psi_5$ of the RHRR mechanism

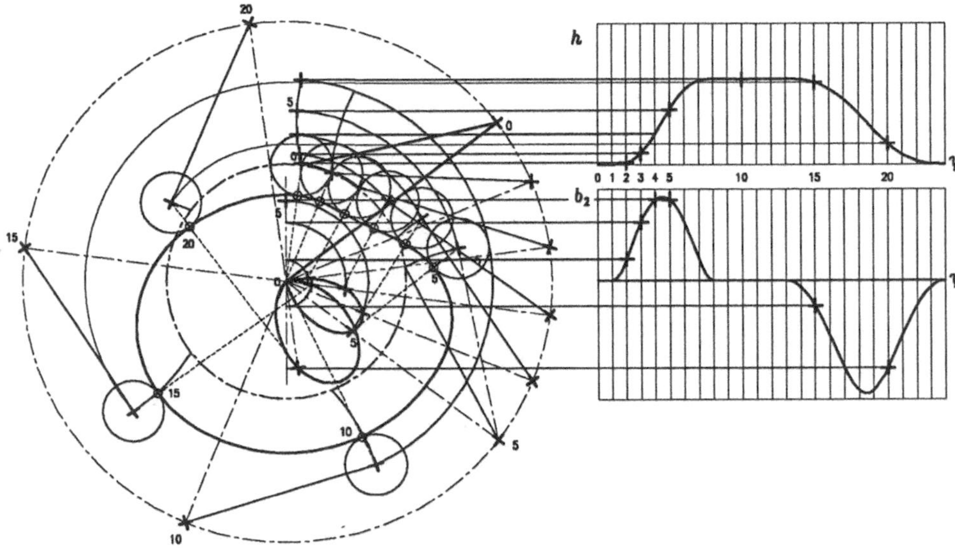

Figure 8.5 Cam profile of the RHRR mechanism

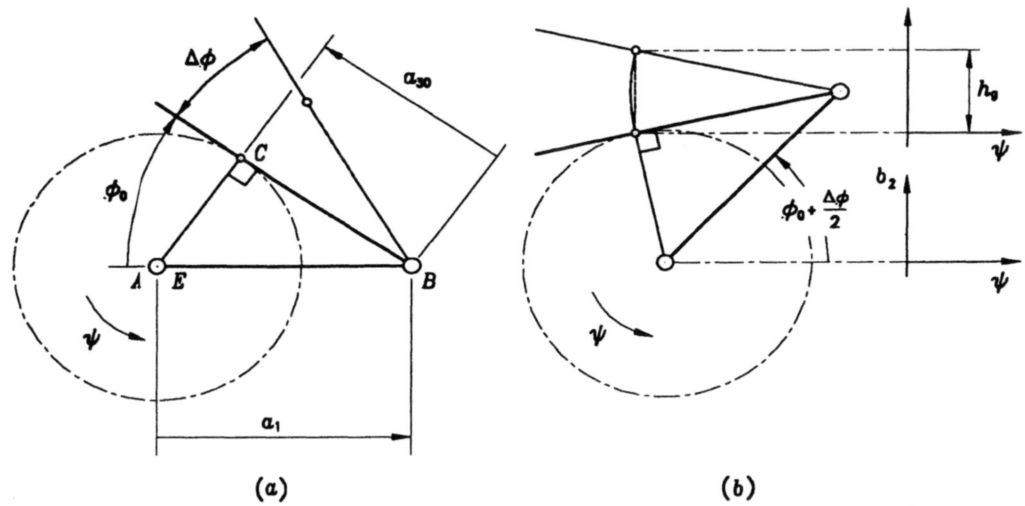

Figure 8.6 R$\overline{\text{HH}}$R mechanism layout

8.2.2 R$\overline{\text{HH}}$R Mechanisms

The type of mechanism studied here is shown in Fig. 4.20. As stated in Chapter 4, the synthesis of R$\overline{\text{HH}}$R mechanisms is similar to the synthesis of RHRR mechanisms, with only two important differences: a) the distance a_3 is variable; and b) the pressure angle is identically zero, i.e., segment EC is always perpendicular to segment BC, as shown in Fig. 8.6a.

The reference value of a_3 is $a_{30} \equiv a_1 \cos\phi_0$, and is obtained as shown in Fig. 8.6a. Thus, h_g is computed as

$$h_g = 2a_{30}\sin\frac{\Delta\phi}{2} \tag{8.8}$$

Figure 8.7 shows the procedure to find the cam-follower contact point for $\psi = \psi_5$, where a_3 is variable. This profile is designed with zero offset, i.e., with $a_4 = 0$, as shown in Fig. 8.8.

If $a_4 \neq 0$, the profile is obtained from that obtained with zero offset and auxiliary circles of radius a_4 as discussed in Section 4.4.1. The intersection of these circles with the dashed lines give the points of the desired profile, as shown in Fig. 8.9.

8.2.3 RHRP Mechanisms

A mechanism of the kind discussed in this section, with $\alpha_1 = -\pi/2$, is shown in Fig. 4.14, the given parameters being shown in Fig. 8.10. The ordinates of the two

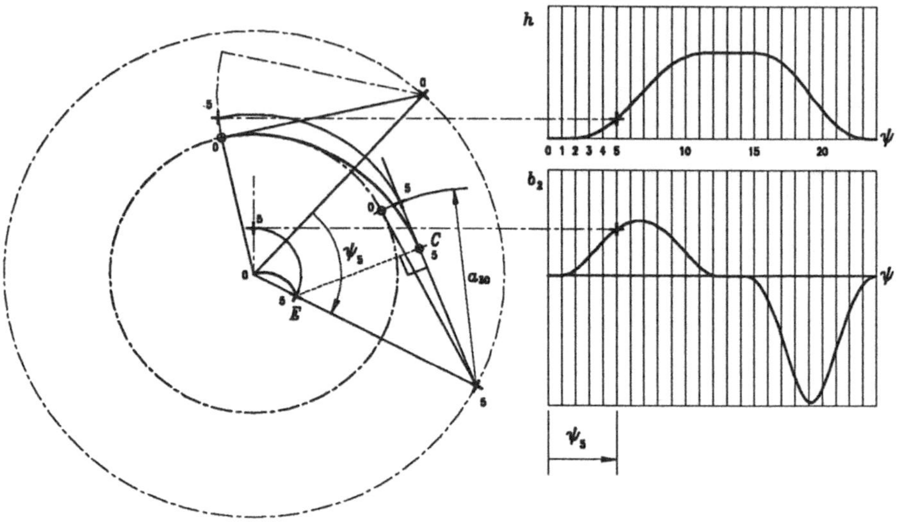

Figure 8.7 Contact point for $\psi = \psi_5$ of the $R\overline{HH}R$ mechanism

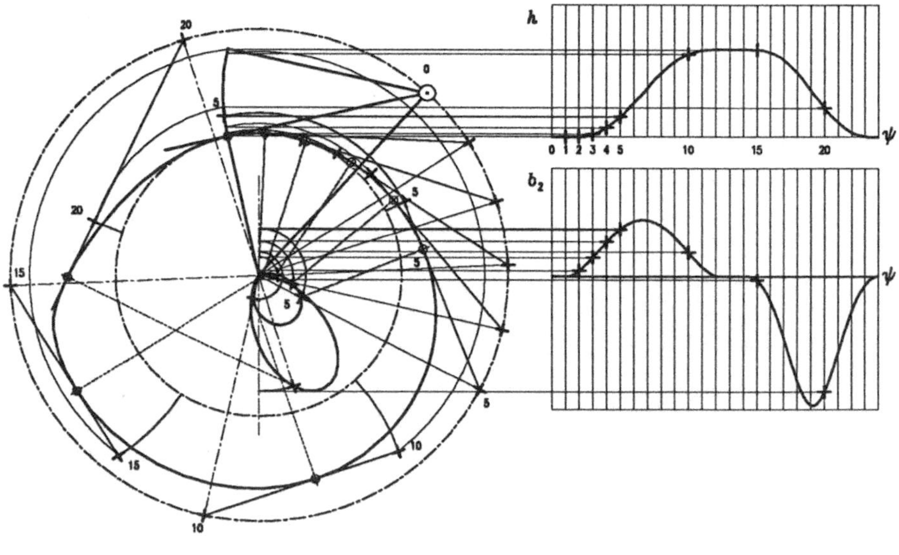

Figure 8.8 Cam profile of the $R\overline{HH}R$ mechanism with $a_4 = 0$

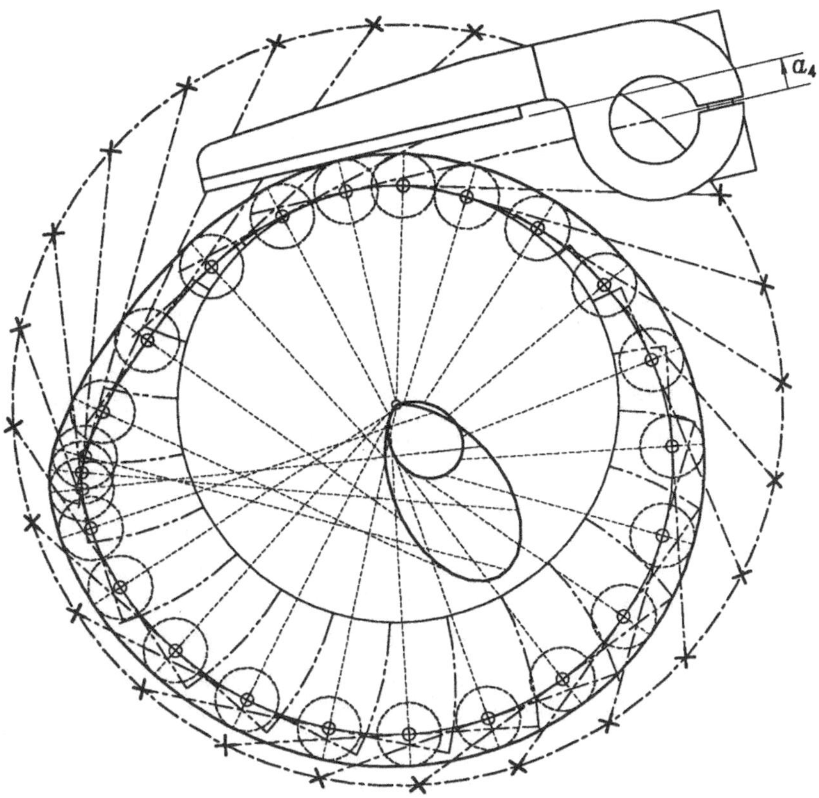

Figure 8.9 Cam profile of the R$\overline{\text{HH}}$R mechanism with $a_4 \neq 0$

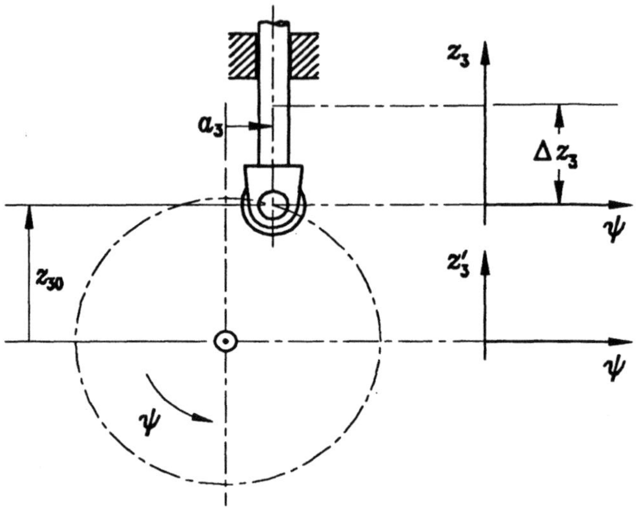

Figure 8.10 RHRP mechanism layout

coordinate frames are parallel to the direction of motion of the follower. According to eq.(3.21d), $b_2 = z_3'(\psi)$, which we abbreviated here as $b_2 = z_3'$.

Once the plots of z_3 and z_3' are obtained either graphically or numerically, their ordinates are projected onto line $\mathcal{L}$, as shown in Fig. 8.11 for the value of $\psi = \psi_5$. The i-points from z_3 are rotated through the angle ψ_i, while the i-points from z_3' are rotated through the angle $\psi_i + \pi/2$. Hence, the pitch curve of the RHRP mechanism and the cam profile of the RHP mechanism, respectively, are readily obtained.

The points of the cam profile are determined as the intersections of the circles of radius a_4 with segment EC', the profile thus obtained being shown in Fig. 8.12.

8.2.4 R$\overline{\text{HH}}$P Mechanisms

The procedure to plot the cam profile for the RHP mechanism is the same as that followed in Section 8.2.3, the points of the cam profile of the R$\overline{\text{HH}}$P mechanism being obtained likewise. The only difference here is that $a_3 = 0$ for $\psi = 0$, as shown in Fig. 8.13, and, as discussed in Chapter 4, a_3 changes so that the pressure angle is zero; in other words, segment EF is always perpendicular to the face of the follower, as shown in Fig. 8.14 for $\psi = \psi_5$. The profile obtained is shown in Fig. 8.15.

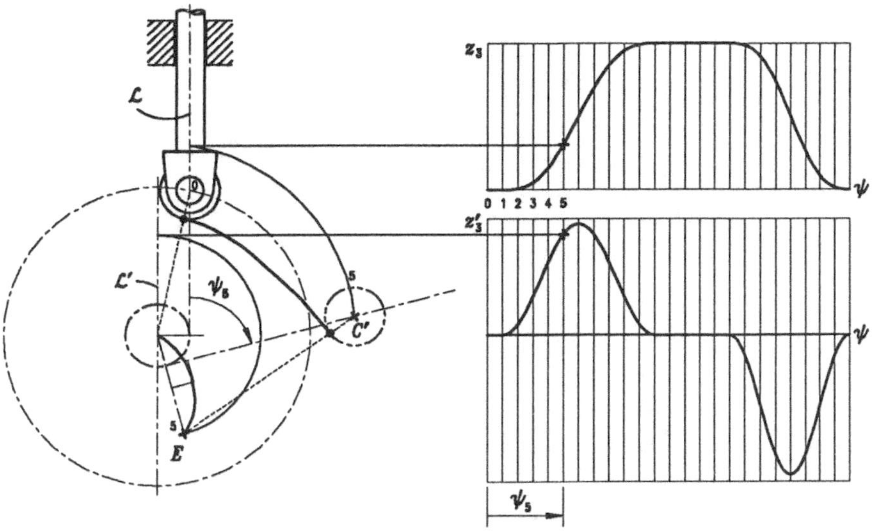

Figure 8.11 Contact point for $\psi = \psi_5$ of the RHRP mechanism

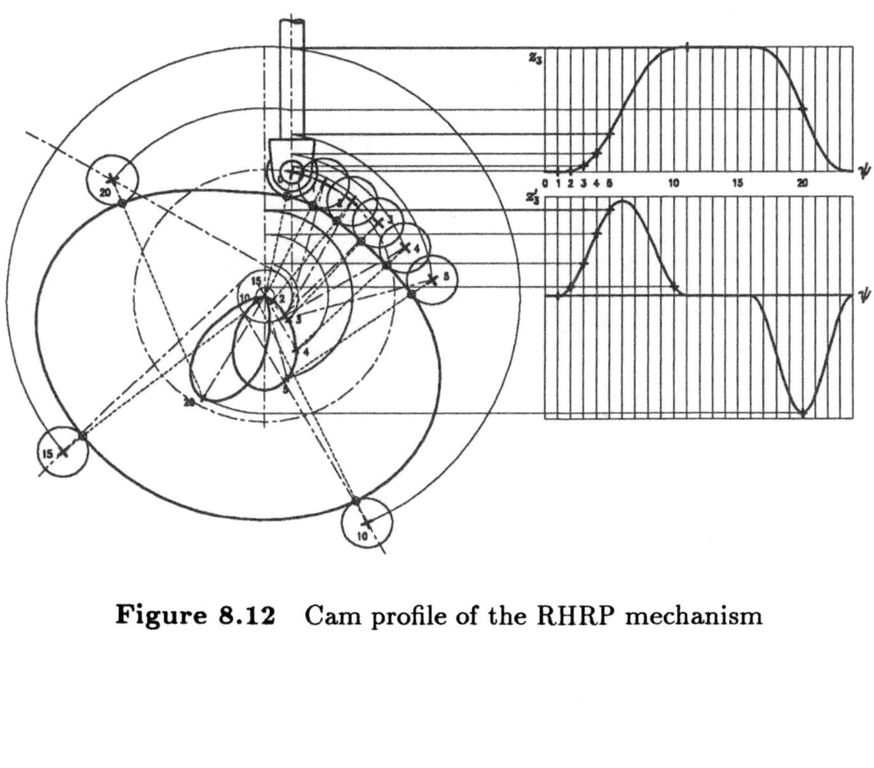

Figure 8.12 Cam profile of the RHRP mechanism

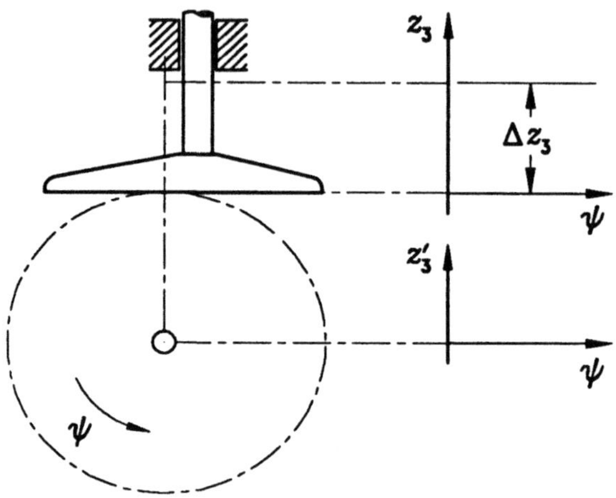

Figure 8.13 $R\overline{HH}P$ mechanism layout

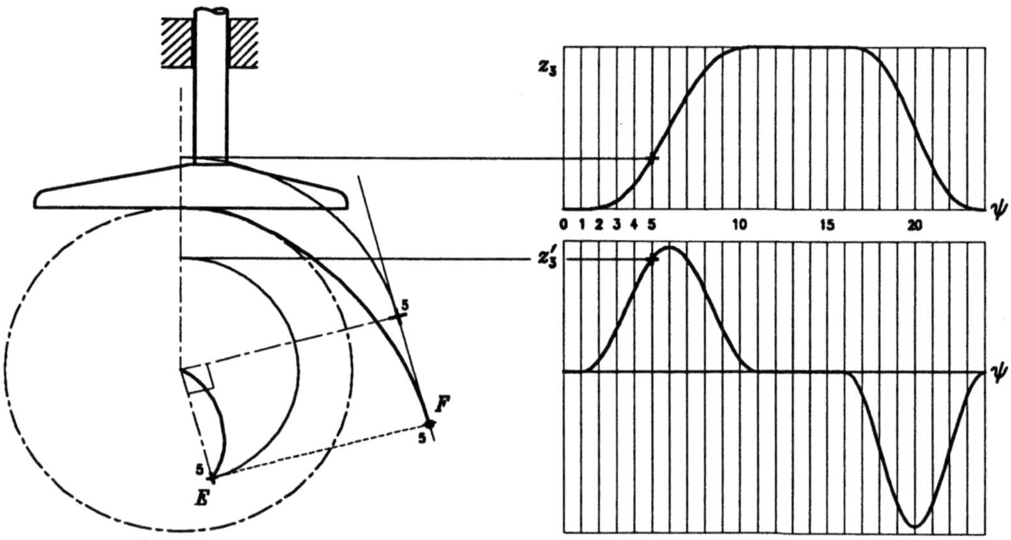

Figure 8.14 Contact point for $\psi = \psi_5$ of the $R\overline{HH}P$ mechanism

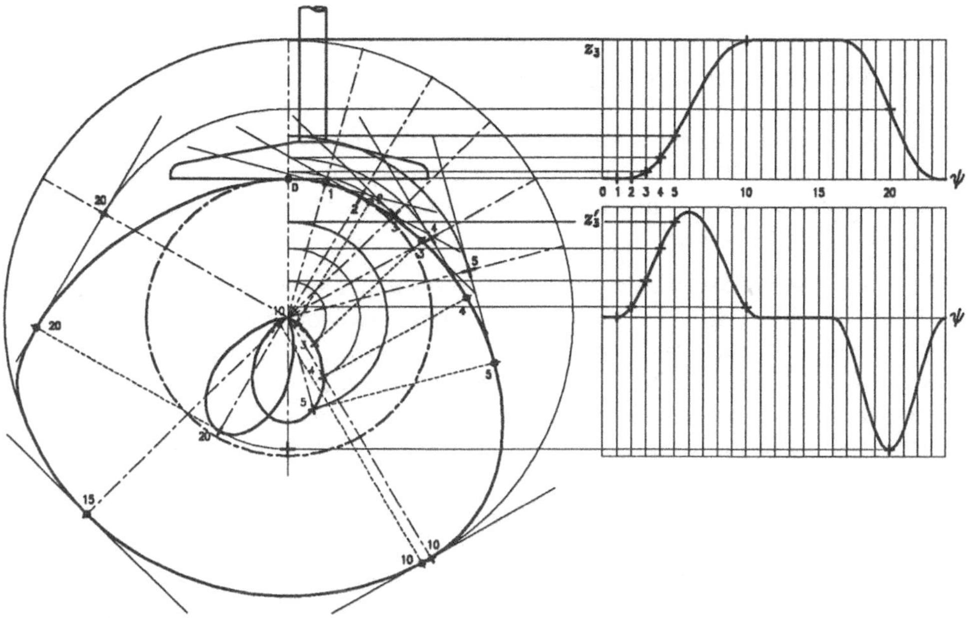

Figure 8.15 Cam profile of the R$\overline{\text{HH}}$P mechanism

8.3 Case Studies

The design of cam or other mechanisms is a much more complex process than that of mechanism synthesis. In fact, mechanism synthesis is itself a more involved task than we have discussed in this book. According to Hartenberg and Denavit (1964), the kinematic synthesis of mechanisms can be divided into *a*) number synthesis, *b*) type synthesis and *c*) dimensional synthesis. In number synthesis, the kinematician determines the number of links or moving parts in a given mechanism; in type synthesis, a decision is made about the nature of the mechanism to be used to accomplish a given function. Such a decision involves choosing among a linkage, a cam mechanism, a gear train, a belt-pulley transmission or a combination of these. Tradeoffs must be taken into account here that enable the designer to choose between the simplicity of manufacturing a linkage, and the complex shapes that must be machined to produce a cam. The ease of manufacture of linkages must be weighed against their limited versatility when it comes to meeting pointwise conditions over the whole operation cycle, as compared with the functional flexibility of a cam mechanism. Moreover, conditions permitting, it is sometimes possible to accomplish the same function with a planar, a spherical or a spatial device, whether a linkage or a cam mechanism. We have only given simple geometric conditions in Chapter 3 that allow the designer to decide on the type of cam mechanism to be used. That is, if the axes between which motion is to be transmitted are parallel, then a planar cam mechanism should be used; if the axes are intersecting, then a spherical cam mechanism is needed; if the axes are skew, then a spatial mechanism is required. We have assumed at the outset that a cam mechanism is required, thereby doing away with type synthesis. For cam mechanisms, the issue of number synthesis was simplified to deciding whether to include a roller or not. If no roller is included, then the number of links involved is three; otherwise, it is four.

The last stage of kinematic synthesis, dimensional synthesis, has been the focus of this book. In this stage, the *fundamental dimensions* of the linkage are determined, based on the kinematic and static conditions at hand. Kinematic and static conditions are those dictated only by the purely motion-transmission characteristics. These characteristics pertain to a) a given input-output function and b) force-transmission features such as pressure angle. Other conditions like functionality and those arising from a comprehensive mechanical analysis, which we shall term here *mechanical considerations*, are left aside in determining what we have called the fundamental dimensions of the mechanism. Mechanical considerations pertain to stress and strain distributions, wear, fatigue and the like. In the discussion below we focus only on some functional conditions that involve decisions based more on expert knowledge than on analysis or algorithms.

The object of this discussion is, then, the production of practical manufacturing drawings that will allow the machine-tool shop to produce the various pieces with the prescribed tolerances, and that will lead to a feasible and affordable assembly for an acceptable operation of the mechanism. In laying out the manufacturing drawings, we assume that we have a suitable piece of CAD software at our disposal. Commercial mechanical-oriented CAD software has its scope and limitations. Standard CAD software for mechanical drawings allows the user to perform a certain set of basic functions, namely,

- specify and render primitives (points, line segments, circular arcs, planes, cylinders, spheres, tori, etc.);
- solve simple geometric problems involving primitives (find intersections between lines; find centers of circles; draw perpendiculars to lines; etc.)
- distinguish between visible and hidden lines or between physical lines at the boundaries of bodies and reference lines by means of various thicknesses and types, e.g., continuous vs. dashed lines;
- obtain dimensioning and tolerancing using established norms (ANSI, ISO, etc.);
- distinguish between different moving pieces by means of patterns;

- produce standard mechanical parts like screws, chamfers, fillets, keyways, etc.

While most mechanical elements can be produced with the sole aid of primitives, cam profiles and surfaces need a special treatment, as do helical, bevel and hypoid gears. The challenge is to produce the complex shapes of cams with commercial CAD software. We do it by first producing the databases describing the complex shapes involved by means of the techniques studied in Chapters 3, 4 and 5 as well as in Section 8.2. Once these databases are available, we have to supply them to the CAD software in a suitable format, as described in the user's manuals of this software.

In the subsections below we illustrate the use of CAD software for the rendering of manufacturing drawings with two specific examples.

8.3.1 The Manufacturing Drawings of the Planar PRICAM

The success of a planar cam mechanism is measured by the accuracy obtained in the parallelism between the input and the output axes. The accuracy of this parallelism dictates that of the assembly of the remaining pieces, and the overall mechanism.

Figure 8.16*a* shows structural elements that are ancillary to the cam mechanism itself, namely, *a*) the two supporting elements with flat surfaces parallel to the surface on which the mechanism is mounted and *b*) the four-piece frame composed of plates with medium planes perpendicular to the supporting surface. The first two elements are basically plates as well, each with a vertical rib to accommodate the four-piece frame. The latter could have been fabricated in many ways, but we chose to make it in four pieces for economy of production. We coud have used a single block and removed material by machining; that would have been very costly in terms of machining operations and wasted material. The drawback of designing the frame in four parts is the loss of parallelism and perpendicularity upon assembly, which would have been very precisely controlled if the whole part had been machined out of one single block. The advantage of doing it in four parts, on the other hand, is the ease of assembly, that would have been rather challenging if a single block had been used.

Our prototype is for demonstration only, and so, the four-piece frame was cut of two transparent, colorless plexyglass 9-mm thick plates and two aluminum 15-mm thick plates.

Other design details that we took into account are the support of the input or cam shaft and of the output or follower shaft. The former is supported as a cantilever beam, while the latter as a clamped-clamped beam. For demonstration purposes the input shaft is to be driven manually, but we account in this design also for a mechanical drive, and so we assume that the input shaft will be subjected to both torsional and axial loads. We use an array of two single-row ball bearings to minimize bending. Because the contact between cam and follower takes place along cylindrical surfaces of generatrices parallel to the shaft axes, no axial forces are transmitted from the cam to the follower, and so the output shaft can be safely assumed to be subjected to no axial force. For this reason, and because the loads are moderate, we chose single-row ball bearings to support this shaft. More details of the frame-base design are visible in Fig. 8.16.

Figure 8.17*a* shows a side view and a front view of the cam. For economy of machining, we designed this element in two pieces fastened with screws. The right part of the side view is the cam of the primary mechanism, while the left part is the cam of the secondary mechanism. Both parts were machined using a three-axis NC vertical machining center with data points produced with the procedure described in Section 7.6.

The follower is shown in both side and front views in Fig. 8.17b. This element is machined of one single plate, again, in the same NC vertical machining center used for the cam profiles. The data to produce the contours were obtained as described in Section 7.6.

8.3.2 The Manufacturing Drawings of the Spherical PRI-CAM

The spherical prototype of PRICAM, as synthesized in Section 5.5, has certain similarities with its planar counterpart. It consists of a frame and the mechanism itself. The frame is composed of a parallelepiped that, at first glance, resembles a cube, i.e., the lengths of its edges are very close to each other. An essential difference between the two prototypes is that here the concern is no longer the parallelism between input and output axes; the concern here is rather twofold: intersection and perpendicularity. To accommodate both, we mount the output axis on holes that have a 1-mm clearance, in order to adjust the mounting of this shaft with screws. Here, as in the design of the planar prototype, we assume that the cam shaft is subjected to both torsion and axial force, while the follower cam is subjected only to torsion and radial force. The cam shaft is mounted as a cantilever beam on the frame, as in the planar case. Here we use a double-row ball bearing to mount the cam shaft.

Details of the cam of this mechanism are shown in Fig. 8.19a. It is apparent from this figure that the cam was cut out of a single aluminum block. The side view of the cam shows the cam of the primary mechanism on the concave side of the piece, while the convex side is the secondary mechanism. The follower was cut in one single piece as well, as shown in Fig. 8.19b. The machining of both the cam and the follower called for a 4-axis NC machine tool, given the spherical geometry involved.

The material used in the two prototypes was aluminum, for these are meant for demonstration purposes only. Aluminum is not only lighter than steel, but also much easier and cheaper to machine. For production machines, which are in operation around the clock, and subjected to heavy loads, aluminum would have been out of the question. More likely, steel would have been used and, moreover, due to the contact between surfaces, a surface treatment of the cam and follower contact surfaces would have been mandatory. We have not addressed these issues here.

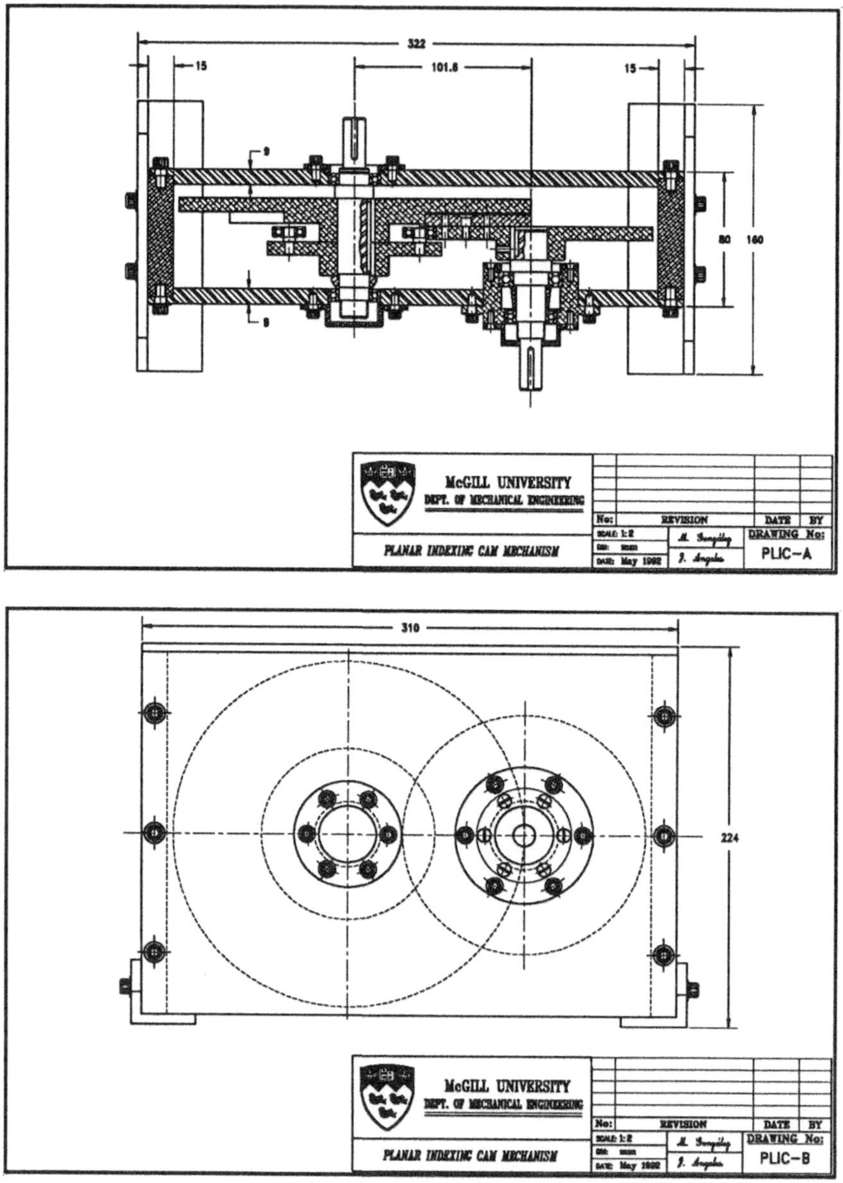

(*a*)

(*b*)

Figure 8.16 *a*) Top and *b*) front views of planar PRICAM

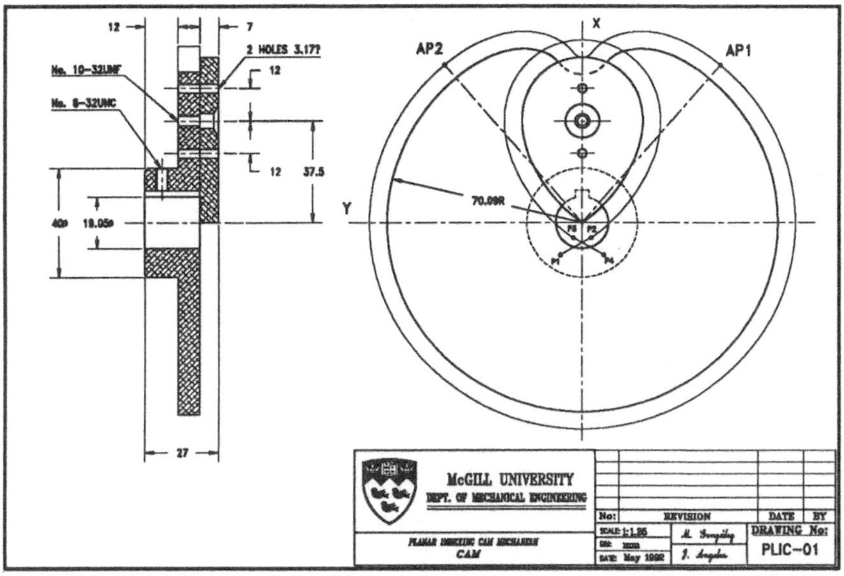

(a)

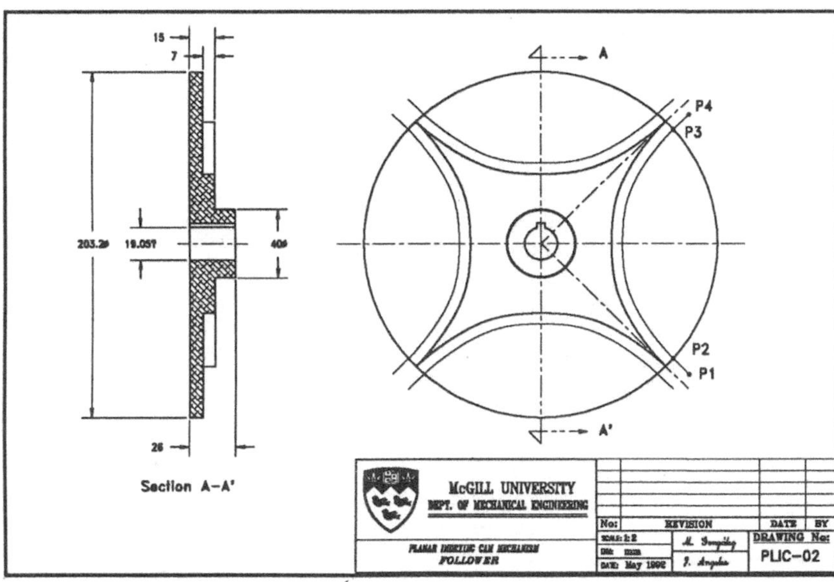

(b)

Figure 8.17 *a*) Cam and *b*) follower of planar PRICAM

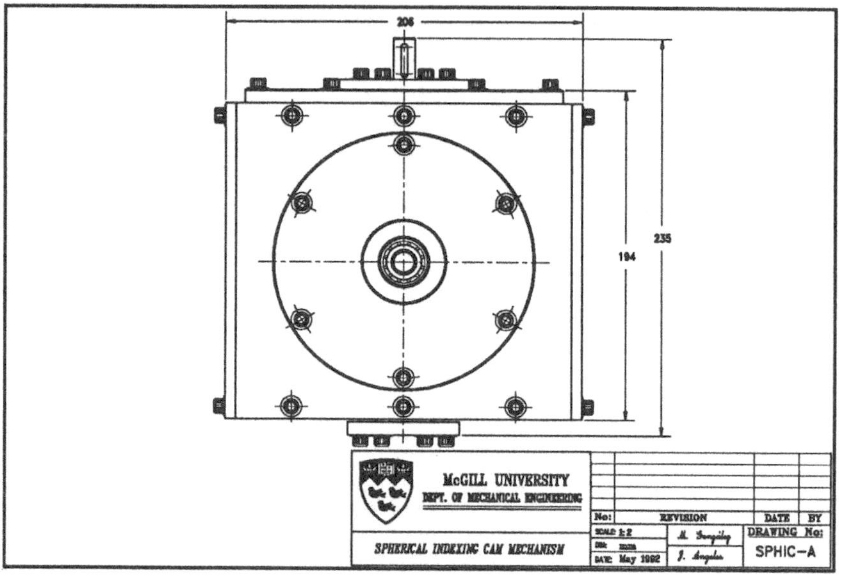

(a)

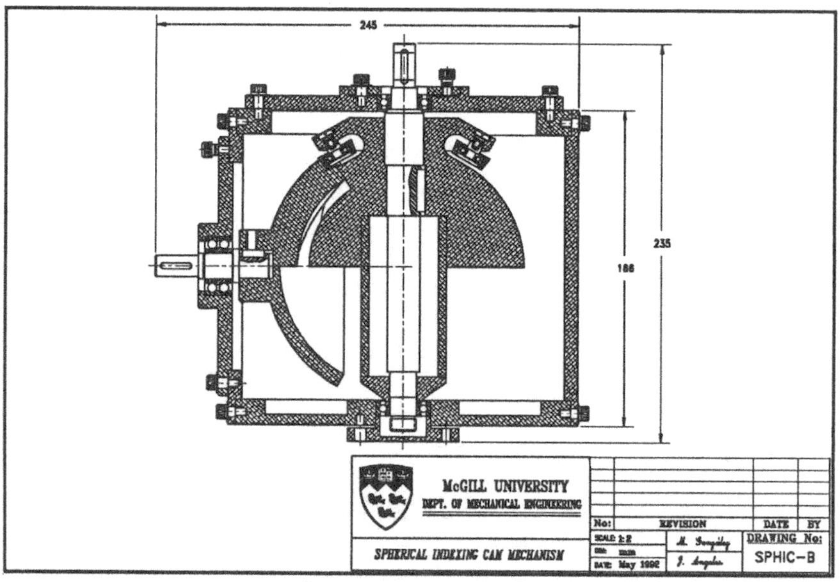

(b)

Figure 8.18 *a*) Front and *b*) lateral views of spherical PRICAM

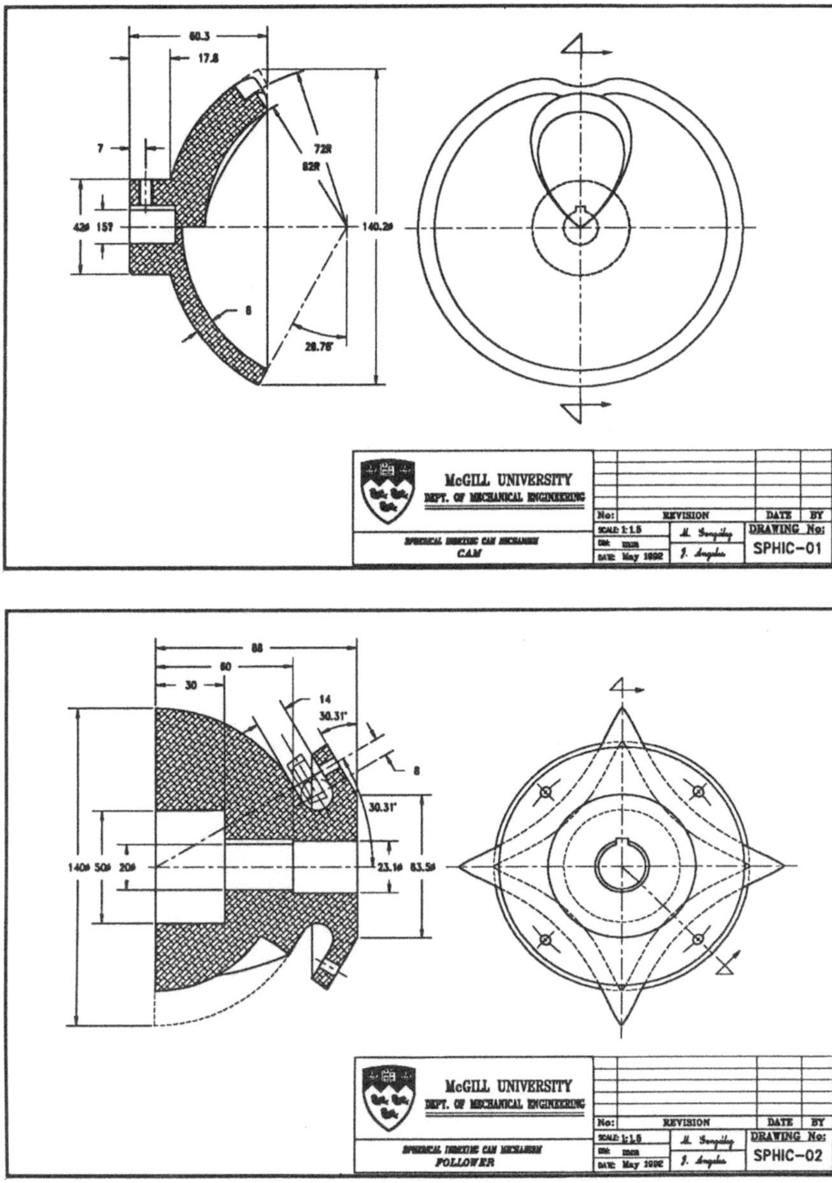

(a)

(b)

Figure 8.19 *a*) Cam and *b*) follower of spherical PRICAM

Appendix A

BIBLIOGRAPHY

Al-Daccak, M., 1989, *Contributions to the Geometric Modelling of Mechanical Elements with Complex Shapes*, M. Eng. Thesis, Department of Mechanical Engineering, McGill University, Montreal.

Al-Daccak, M., and Angeles, J., 1993, "The Calculation of the Volumetric Properties of Sweep-Generated Solids Via Line Integration", *ASME Journal of Mechanical Design*, Vol. 115, pp. 110–118

Amarnath, C., and Gupta, B. K., 1975, "On a Cam Mechanism with Multiple Dwell Capability," *Proc. 4th World Congress on Theory of Machines and Mechanisms*, Newcastle-upon-Tyne, Vol. 4, pp. 773–776.

Amarnath, C., and Gupta, B., K., 1978, "Novel Cam-Linkage Mechanisms for Multiple Dwell Generation," in J. Rees Jones, (editor), *Cams and Cam Mechanisms*, The Institution of Mechanical Engineers, London,, pp. 123–127.

Angeles, J., 1982, *Spatial Kinematic Chains*, Springer-Verlag, Berlin, pp. 149-157.

Angeles, J. and López-Cajún, C., 1991, *Optimization of Cam Mechanisms*, Kluwer Academic Publishers, Dordrecht.

Aronhold, S. H., 1872, "Outline of Kinematic Geometry", Verhandlungen des Vereins zur Beförderung des Gewerbefleißes in Preußen, Berlin, 51, pp 129–155.

AutoCAD AME, 1992, *Advanced Modeling Extension Release 2.1. Reference Manual*, Autodesk, Inc.

Backhouse, C. J. and Jones J. R., 1990, "Envelope Theory Applied to Globoidal Cam Surface Geometry", *Proc. Institution of Mechanical Engineers*, Vol. 204, pp. 409–416.

Ball, R., S., 1875, "The Theory of Screws. A Geometrical Study of the Kinematics, Equilibrium, and Small Oscillations of a Rigid Body," *Trans. R. Irish Acad.*, Vol. 25, pp. 157–217.

Ball, R., S., 1900, *A Treatise on the Theory of Screws*, Cambridge University Press, Cambridge.

Beggs, J. S., 1959, *Ein Beitrag zur Analyse Räumlicher Mechanismen*, Doctoral Thesis, Technische Hochschule Hannover, Hanover.

Berzak, N., and Freudenstein, F., 1979, "Optimization Criteria in Polydine Cam Design," *Proc. 5th World Congress on Theory of Machines and Mechanisms*, Montreal, Vol. 2, pp. 1303–1306.

Bialkowicz, B., Klimowicz, T., and Swietlik, M., 1979, "Changes of the Dynamic Properties of the Real Cam Profile During its Wear," *Proc. 5th World Congress on Theory of Machines and Mechanisms*, Montreal, Vol. 2, pp. 984–987.

Bickford J. H., 1965, "Mechanisms for Intermittent Rotary Motion", *Machine Design*, December, 23, 1965, pp. 119–131.

Bickford J. H., 1972, *Mechanisms for Intermittent Motion*, Industrial Press Inc., New York.

Bottema, O. and Roth, B., 1979, *Theoretical Kinematics*, North-Holland Publishing Co., Amsterdam.

Brand, L., 1965, *Advanced Calculus*, John Wiley & Sons, Inc., New York.

Chakraborty, J. and Dhande, S. G., 1977, *Kinematics and Geometry of Planar and Spatial Cam Mechanisms*, John Wiley and Sons, Inc., New York.

Chen, F. Y., 1982, *Mechanics and Design of Cam Mechanisms*, Pergamon Press, New York.

Chen, P., and Roth, B., 1969, "A Unified Theory for the Finitely and Infinitesimally Separated Position Problems of Kinematic Synthesis," *Journal of Engineering for Industry, Trans. ASME*, Vol. 91, pp. 208–203.

Craig, J. J., 1989, *Introduction to Robotics: Mechanics and Control*, Addison-Wesley Publishing Company, Reading, Mass.

De Fraine, J., 1982, "The Computer-Aided Design and Computer-Aided Manufacturing of Cams", Technical Report, Centrum voor Wetenschappelijk en Technisch Onderzoek der Metaalverwerkende Nijverheid, MC 75, Brussels, November.

Derek, W., 1987, *Great Engineers*, Academy Editions, London.

Dhande, S. G., Bhadoria, B., S., and Chakraborty, J., 1975, "A Unified Approach to the Analytical Design of Three-Dimensional Cam Mechanisms," *Journal of Engineering for Industry, Trans. ASME*, Vol. 97, pp. 327–333.

Dhande, S., G., and Chakraborty, J., 1975, "Mechanical Error Analysis of Cam-Follower Systems. A Stochastic Approach," *Proc. 4th World Congress on Theory of Machines and Mechanisms*, Newcastle-upon-Tyne, Vol. 4, pp. 957–962.

Di Benedetto A., 1975, "Some Methods of Kinematic Synthesis of Cam Profiles for Prescribed Jerk Pattern," *Proc. 4th World Congress on Theory of Machines and Mechanisms*, Newcastle-upon-Tyne, Vol. 4, pp. 963–968.

Dittrich, G., 1966, "Konstruktion und Herstellung sphärischer Kurvengetriebe", *Konstruktion*, Vol. 10, pp. 421–426.

Dittrich, G., and Zakel, H., 1979, "Classification and Design of Three-Dimensional Cam Mechanisms," (in German) *Proc. 5th World Congress on Theory of Machines and Mechanisms*, Montreal, Vol. 2, pp. 1086–1089.

Dizioğlu, B., 1989, "Kinematische Geometrie der Axoiden und die Bahnfläche einer Geraden," *Mechanism and Machine Theory*, Vol. 24, pp. 431–438.

DoCarmo, M., P., 1976, *Differential Geometry of Curves and Surfaces*, Prentice-Hall, Englewood Cliffs.

Dudley, W. M., 1948, "New Methods in Valve Cam Design", *SAE Quarterly Transactions*, 2(7), 19–33.

Dudley, D. W., 1962, *Gear Handbook*, MacMillan, New York.

Erdman, A. G. and Gustafson, J. E., 1977, "LINCAGES: Linkage INteractive Computer Analysis and Graphically Enhanced Synthesis Package," *ASME Design Engineering Technical Conference*, Chicago, Illinois, Paper No. 77–DET–5.

Fenton, R., G., 1965, "Dynamic Analysis of Geneva Mechanisms," *Machine Design*, Vol. 37, No. 2, pp. 177–182.

Fenton, R., G., 1966*a*, "Determining Minimum Cam Size," *Machine Design*, Vol. 38, No. 2, pp. 155–158.

Fenton, R., G., 1966*b*, "Reducing Noise in Cams," *Machine Design*, Vol. 38, No. 8, pp. 187–190.

Fenton, R., G., 1975*a*, "Geneva Mechanisms Connected in Series," *Journal of Engineering for Industry, Trans. ASME*, Vol. 97, pp. 603–608.

Fenton, R., G., 1975*b*, "Dynamic Analysis of Geneva Mechanisms Connected in Series," *Proc. 4th World Congress on Theory of Machines and Mechanisms*, Newcastle-upon-Tyne, 8–12 September, Vol. 4, pp. 885–888.

Fenton, R., G., 1975*c*, "Optimum Design of Disc Cams," *Proc. 4th World Congress on Theory of Machines and Mechanisms*, Newcastle-upon-Tyne, Vol. 4, pp. 781–784.

Gibson, C., G., and Hunt, K., H., 1990*a*, "Geometry of Screw Systems-1. Screws: Genesis and Geometry," *Mechanism and Machine Theory*, Vol. 25, No. 1 pp. 1–10.

Gibson, C., G., and Hunt, K., H., 1990*b*, "Geometry of Screw Systems-2. Classification of Screw Systems," *Mechanism and Machine Theory*, Vol. 25, No. 1 pp. 11–27.

González-Palacios, M. A. and Angeles, J., 1990, "The Generation of Contact Surfaces of Indexing Cam Mechanisms. A Unified Approach," *Proc. 1990 ASME Design Automation Conference, Advances in Design Automation*, Vol. 2, pp. 359–364. Also: *ASME Journal of Mechanical Design*, in press.

González-Palacios, M. A. and Angeles, J., 1991*a*, "Synthesis of Contact Surfaces of Spherical Cam-Oscillating Roller-Follower Mechanisms. A General Approach", *Proc. 1991 ASME Design Automation Conference, Advances in Design Automation*, Vol. 2, pp. 255–260. Also: *ASME Journal of Mechanical Design*, in press.

González-Palacios, M. A. and Angeles, J., 1991*b*, "SIXPAQ: A Comprehensive Software Package for Analysis and Synthesis of Six-Bar Dwell Linkages", *Proc. 1991 ASME Computers in Engineering*, Santa Clara, Cal., Vol. 1, pp. 309–314.

González-Palacios, M. A., 1992, *A Unified Theoretical Framework for the Synthesis of Cam Mechanisms*, Ph.D. Thesis, Department of Mechanical Engineering, McGill University, Montreal.

González-Palacios, M. A. and Angeles, J., 1992*a*, "The Synthesis of Cam-Oscillating Roller-Follower Mechanisms. A Unified Approach", *Proc. 1992 ASME 22nd Biennial Mechanisms Conference*, Scottsdale, Vol. 46, pp. 329–335.

González-Palacios, M. A. and Angeles, J., 1992*b*, "On the Design of Planar and Spherical Pure-Rolling Indexing Cam Mechanisms", *Proc. 1992 ASME 22nd Biennial Mechanisms Conference*, Scottsdale, Vol. 46, pp. 323–328.

González-Palacios, M. A., Angeles, J. and Ranjbaran, F., 1993, "The Kinematic Synthesis of Serial Manipulators with a Prescribed Jacobian", *Proc. IEEE International Conference on Robotics and Automation*, Atlanta, Vol. 1, pp. 450–455.

Gouxun, P., Zhengyang, X. and Huimin, T., 1988, "Unified Optimal Design of External and Internal Parallel Indexing Cam Mechanisms," *Mechanism and Machine Theory*, Vol. 23, pp. 313–318.

Grewal, P., S. and Newcombe, W., R., 1988, "Dynamic Performance of High-Speed Semi-Rigid Follower Cam Systems—Effects of Cam Profile Errors," *Mechanism and Machine Theory*, Vol. 23, pp. 121–133.

Gürsoy, O., 1990, "The Dual Angle of Pitch of a Closed Ruled Surface," *Mechanism and Machine Theory*, Vol. 25, No. 2 pp. 131–140.

Hain, K., 1970, "Challenge: To Design Better Cams," *J. Mechanisms*, Vol. 5, pp. 283–286.

Hain, K., 1971, "Optimization of a Cam Mechanism—To Give Good Transmissibility, Maximal Output Angle of Swing and Minimal Acceleration," *J. Mechanisms*, Vol. 6, pp. 419–434.

Hain, K., 1978, "Optimization of an Intermittent Motion-Mechanism with Fixed Cam," in J. Rees Jones, (editor), *Cams and Cam Mechanisms*, The Institution of Mechanical Engineers, London,, pp. 111–115.

Hartenberg, R.S. and Denavit, J., 1964, *Kinematic Synthesis of Linkages*, McGraw-Hill, New York.

Hon-Cheung, Y., 1987, "Four Co-reciprocal Screws and Their Kinematic Significance," *Mechanism and Machine Theory*, Vol. 22, pp. 190–203.

Hunt, K., H., 1967*a*, "Prismatic Pairs in Spatial Linkages," *J. Mechanisms*, Vol. 2, pp. 213–230.

Hunt, K., H., 1967*b*, "Screw Axes and Mobility in Spatial Mechanisms Via the Linear Complex," *J. Mechanisms*, Vol. 3, pp. 307–327.

Hunt, K. H., 1973, "Profiled-Follower Mechanisms," *Mechanism and Machine Theory*, Vol. 8, pp. 371–395.

IFToMM Commission A, 1991, "Terminology for the Theory of Machines and Mechanisms", *Mechanism and Machine Theory*, Vol. 26, No. 5, pp. 435–539.

Jackowski, C. S., and Dubil, J. F., 1967, "Single-Disk Cams with Positively Controlled Oscillating Followers," *J. Mechanisms*, Vol. 2, pp. 157–184.

Jacobs, R. J., 1949, "Indexing with Concave Barrel Cams", *Machine Design*, Vol. 21, No. 3, February, pp. 93–96.

Jensen, P. W., 1965, *Cam Design and Manufacture*, Industrial Press, New York.

Johnson, R., C., 1955, "Cam Design," *Machine Design*, Vol. 27, No. 11, November, pp. 195–204.

Johnson, R., C., 1956*a*, "Cam Mechanisms," *Machine Design*, Vol. 28, No. 2, January, pp. 105–108.

Johnson, R., C., 1956*b*, "Minimizing Cam Vibrations," *Machine Design*, Vol. 28, No. 16, August, pp. 103–104.

Johnson, R., C., 1956*c*, "Cam Profiles," *Machine Design*, Vol. 28, No. 25, December, pp. 129–132.

Johnson, R. C., 1958, "Development of a High-Speed Indexing Mechanism," *Machine Design*, Vol. 30, Sept., pp. 134–138.

Jones, J., R., 1978*a*, "Mechanisms. Cam Cutting Co-ordinates," *Engineering*, Vol. 218, March, pp. 220–224.

Jones, J., R., 1978*b*, "Mechanisms. Pressure Angles and Forces in Cams," *Engineering*, Vol. 218, July, pp. 703–706.

Jones, J. R., and Tsang, K. S., 1987, "Optimal Configurations for Parallel Shaft Indexing Mechanisms," *Proc. 7th World Congress on Theory of Machines and Mechanisms*, Sevilla, 17–22 September, Vol. 3, pp. 1687–1690.

Kass, R., C., and Chace, M. A., 1975, "An Approach to the Simulation of Two-Dimensional Higher Pair Contacts," *Proc. 4th World Congress on Theory of Machines and Mechanisms*, Newcastle-upon-Tyne, Vol. 4, pp. 1057–1063.

Kennedy, A. B. W., 1886, *Mechanics of Machinery*, 1st Ed., Macmillan, London.

Kerr, D. R., and Sanger, D., J., 1989, "The Inner Product in the Evaluation of Reciprocal Screws," *Mechanism and Machine Theory*, Vol. 24, pp. 87–92.

Koloc, V., and Václavíc, M., 1988, *Vačkové Mechanismy*, SNTL-Nakladatelství Techniké Literatury, Prague.

Köse, Ö., 1982*a*, "On the Dual Spherical Motions-I," *Mechanism and Machine Theory*, Vol. 17, pp. 185–190.

Köse, Ö., 1982*b*, "On the Dual Spherical Motions-II," *Mechanism and Machine Theory*, Vol. 17, pp. 185–190.

Koster, M., P., 1975, "Digital Simulation of the Dynamics of Cam Followers and Camshafts," *Proc. 4th World Congress on Theory of Machines and Mechanisms*, Newcastle-upon-Tyne, Vol. 4, pp. 969–974.

Kota, S., Erdman, A. G. and Riley, D. R., 1987, "MINN-DWELL – Computer Aided Design and Analysis of Linkage-Type Dwell Mechanisms", *Proc. ASME Internation-all Computers in Engineering Conference*, Vol. 2, New York, pp. 351–358.

Loeff, L., and Soni, H., 1975, "Optimum Sizing of Planar Cams," *Proc. 4th World Congress on Theory of Machines and Mechanisms*, Newcastle-upon-Tyne, Vol. 4, pp. 777–780.

Makino, H., 1979, "Basic Analysis and Optimal Design of In-Line Transfer Indexing Cam," *Proc. 5th World Congress on Theory of Machines and Mechanisms*, Montreal, 8–13 July, Vol. 2, pp. 875–878.

McCarthy, J., M., and Roth, B., 1981, "The Curvature Theory of Line Trajectories in Spatial Kinematics," *ASME Journal of Mechanical Design*, Vol. 103, pp. 718–724.

McCarthy, J. M., 1987, "On the Scalar and Dual Formulation of the Curvature Theory of Line Trajectories", *ASME J. Mechanisms, Transmissions, and Automation Design*, Vol. 109, pp. 101–106.

Meriam, J. L., 1993, *Engineering Mechanics*, 3rd ed., Wiley, New York.

Müller, J., 1987, "The history of Cams and Cam Mechanisms", *Proc. 7th World Congress on Theory of Machines and Mechanisms*, Seville, pp. 1649–1652.

Müller, J. and Mauersberger, K., 1988, "Zur Entwicklungsgeschichte der Kurvengetriebe", *Wissenschaftliche Zeitung der Wilhelm-Pieck Universität Rostock*, W.-P. Universität Rostock, Rostock, Vol. 37, No. 7, pp. 66-88.

Neklutin, C. N., 1959, "Tring-Type Cam Profiles", *Machinery Design*, Oct. 15, pp. 175–187.

Norton, R., L., 1988, "Effect of Manufacturing Method on Dynamic Performance of Cams—An Experimental Study. Part I—Eccentric Cams," *Mechanism and Machine Theory*, Vol. 23, pp. 191–199.

Norton, R., L., Levasseur, D., Pettit, A., and Alamsyah, C., 1988, "Analysis of the Effect of Manufacturing Methods and Heat Treatment on the Performance of Double Dwell Cams," *Mechanism and Machine Theory*, Vol. 23, pp. 461–473.

Ohwovoriole, M., S., and Roth, B., 1981, "An Extension of Screw Theory," *Journal of Engineering for Industry, Trans. ASME*, Vol 103, pp. 725–735.

Olędzki, A., and Szydlowski, W., 1975, "Modelling of the Geneva Mechanisms," *Proc. 4th World Congress on Theory of Machines and Mechanisms*, Newcastle-upon-Tyne, 8–12 September, Vol. 4, pp. 889–893.

Osman, M., O., M., Bahgat, B., M., and Osman M., 1987, "Dynamic Analysis of a Cam Mechanism with Bearing Clearances," *Mechanism and Machine Theory*, Vol. 22, pp. 303–314.

Pandrea, N., and Voiculescu, D., 1975, "Analysis of Screw Displacements of the Right Line and the Plane by Means of the Screw Theory and of the Dual Matrices," *Proc. 4th World Congress on Theory of Machines and Mechanisms*, Newcastle-upon-Tyne, Vol. 4, pp. 285–288.

Phillips, J. R., and Hunt, K., H., 1964, "On The Theorem of Three Axes in the Spatial Motion of Three Bodies," *Aust. J. Appl. Sci.*, Vol. 15, pp. 267–287.

Rao, J., S., and Raghavacharyulu, E., 1975, "Experimental Determination of Jump Characteristics in Cam-Follower Systems," *Proc. 4th World Congress on Theory of Machines and Mechanisms*, Newcastle-upon-Tyne, Vol. 4, pp. 951–956.

Raven, F., H., 1959, "Analytical Design of Cams and Three-Dimensional Cams by Independent Position Equations," *J. Appl. Mech. Trans. ASME*, pp. 18–24.

Rooney, J., 1975*a*, "On Obtaining the Velocity Motor of Any link in a General N-Link Spatial Manipulator," *Proc. 4th World Congress on Theory of Machines and Mechanisms*, Newcastle-upon-Tyne, Vol. 4, pp. 1083–1087.

Rooney, J., 1975*b*, "On the Principle of Transference," *Proc. 4th World Congress on Theory of Machines and Mechanisms*, Newcastle-upon-Tyne, Vol. 4, pp. 1089–1094.

Rosenberg J., and Křen, K., 1987, "Kinematic Analysis of Mechanism Links Connected by a Higher Kinematic Pair and Its Application," *Mechanism and Machine Theory*, Vol. 22, pp. 393–398.

Rothbart, H. A., 1956, *Cams. Design, Dynamics, and Accuracy*, John Wiley, New York.

Rubel, A. J. and Kaufman, R. E., 1977, "KINSYN III: A New Human-Engineered System for Interactive Computer-Aided Design of Planar Linkages", *Journal of Engineering for Industry, Trans. ASME*, Vol. 99, pp. 440–448.

Sadler, J., P., and Yang, Z., 1990, "Optimal Design of Cam-Linkage Mechanisms for Dynamic-Force Characteristics," *Mechanism and Machine Theory*, Vol. 25, pp. 41–57.

Shadek, K., S., H., Lloyd, J., L., and Smith, M., R., 1990, "A New Design of Geneva Drive to Reduce Shock Loading," *Mechanism and Machine Theory*, Vol. 25, pp. 589–595

Sheth, P. N. and Uicker, J. J., 1972, "IMP (Integrated Mechanism Program), A Computed-Aided Design Analysis System for Mechanisms and Linkages", *Journal of Engineering for Industry, Trans. ASME*, Vol. 94, pp. 454–464.

Stoddart, D. S., 1953*a*, "Polydyne Cam Design," *Machine Design*, Vol. 25, January, pp. 121–135.

Stoddart, D. S., 1953*b*, "Polydyne Cam Design-II," *Machine Design*, Vol. 25, February, pp. 146–154.

Struik, D. J., 1961, *Lectures on Classical Differential Geometry*, Addison-Wesley Company, Inc.

Sugimoto, K., and Duffy, J., 1982, "Application of Linear Algebra to Screw Systems," *Mechanism and Machine Theory*, Vol. 17, pp. 73–83.

Takano, M., and Toyama, S., 1979, "Dynamics of Indexing Cam Mechanism and Speed-Up of Its Motion," *Proc. 5th World Congress on Theory of Machines and Mechanisms*, Montreal, Vol. 2, pp. 1408–1411.

Tesar, D. and Matthew, G. K., 1976, *The Dynamic Synthesis, Analysis, and Design of Modeled Cam Systems*, Lexington Books, Lexington (Massachusetts).

Vadasz, A., F., and Soni, A., H., 1979, "Analytical Determination of the Space Burmester Lines for Cylindric-Cylindric Cranks," *Proc. 5th World Congress on Theory of Machines and Mechanisms*, Montreal, Vol. 2, pp. 1036–1039.

Veldkamp, G., R., 1967*a*, "Canonical Systems and Instantaneous Invariants in Spatial Kinematics," *J. Mechanisms*, Vol. 3, pp. 329–388.

Veldkamp, G. R., 1976, "On the Use of Dual Numbers, Vectors and Matrices in Instantaneous, Spatial Kinematics," *Mechanism and Machine Theory*, Vol. 11, pp. 141–156.

Waldron, K. J., 1972, "A Method of Studying Joint Geometry," *Mechanism and Machine Theory*, Vol. 7, pp. 347–353.

Wiederrich, J., L., and Roth, B., 1975, "Dynamic Synthesis of Cams Using Finite Trigonometric Series," *Journal of Engineering for Industry, Trans. ASME*, Vol. 97, pp. 287–293.

Willis, R., 1841, *Principles of Mechanism*, London.

Wilson, C. E., Sadler, J. P. and Michels, W., J., 1983, *Kinematics and Dynamics of Machinery*, Harper & Row, Publishers, New York.

Wolfram, S., 1992, *Mathematica*, Wolfram Research, Inc., Version 2.1, Champaign, Illinois.

Wunderlich, W., 1971, "Contributions to the Geometry of Cam Mechanisms with Oscillating Followers", *J. Mechanisms*, Vol. 6, pp. 1–20.

Xiao, D. Z.,, and Yang, A. T., 1989, "Kinematics of Three Dimensional Gearing," *Mechanism and Machine Theory*, Vol. 24, No. 4 pp. 245–255.

Yang, A., T., 1963, *Application of Quaternion Algebra and Dual Numbers to the Analysis of Spatial Mechanisms*, Doctoral Dissertation, Columbia University, New York City, No. 64-2803 (University Microfilm, Ann Arbor, Michigan).

Yang, A. T., Freudenstein, F., 1964, "Application of Dual-Number Quaternion Algebra to the Analysis of Spatial Mechanisms", *J. Appl. Mech. Trans. ASME*, 31, pp. 300–308.

Yang, A., T., 1974, "Calculus of Screws," *Basic Questions of Design Theory*, in W. R. Spillers (editor), American Elsevier Publishing Co., Inc., New York, pp. 265–281.

Yang, A., T., Kirson, Y., and Roth, B., 1975, "On a Kinematic Curvature Theory for Ruled Surfaces," *Proc. 4th World Congress on Theory of Machines and Mechanisms*, Newcastle-upon-Tyne, Vol. 4, pp. 737–742.

Zhong-Tang, R., Jing-Ping, H., 1989, "Synthesis of Planar Cam Mechanisms Based on Theory of Conjugate Surfaces," *Vortragssammelband II IFToMM-Symposium Kurvengetriebe*, Berlin, pp. 73–83.

Zigo, M., 1967, "A General Numerical Procedure for the Calculation of Cam Profiles from Arbitrarily Specified Acceleration Curves," *J. Mechanisms*, Vol. 2, pp. 407–414.

Appendix B

DISPLACEMENT PROGRAMS

The functions describing the rise or return in the displacement program of the synthesis of cam mechanisms have been studied extensively in the literature (Rothbart, 1956; Jensen, 1965; Tesar and Matthew, 1976; Chen, 1982; Angeles and López-Cajún, 1991). Because of the scope of this book, only some of the functions of those types having the property of zero velocity and acceleration at the ends of the rise (or return) phase are included here. Moreover, these functions are defined in normal form, i.e., if x and τ denote normalized input and output variables, then

$$\tau = \tau(x), \qquad 0 \leq \tau \leq 1, \quad 0 \leq x \leq 1 \tag{B.1}$$

B.1 Generalized Input-Output Function

In the theory presented in Chapters 3 and 4, the dimensions and symbols of the variables of the input-output functions change according to the type of kinematic pair of the mechanism to be considered. Two kinds of pairs have been assumed either for the input or the output motions, namely, revolute and prismatic. Thus, a total of four combinations are achieved, namely, RR, RP, PR and PP, which are applicable to both three- and four-link cam mechanisms. However, all of them can be regarded as one generalized input-output function, namely,

$$\varphi(x) = h\tau(x) \tag{B.2a}$$

h being the rise of the follower and τ the normal input-output function, as defined in eq.(B.1) and shown in Fig. B.1. Definitions of φ, h, and x, for each of the four types of the mechanisms mentioned above are shown in Table B.1. The derivatives of $\varphi(x)$ are taken with respect to ψ or z_2, depending on the type of input motion at hand. Thus, the chain rule is applied to φ to obtain its first and second derivatives with respect to the input variable, and denoted by φ' and φ'', thereby obtaining

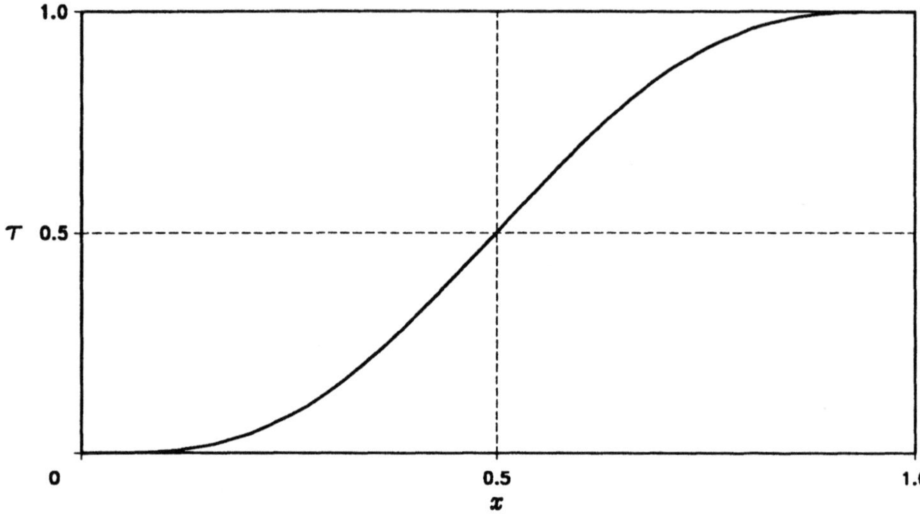

Figure B.1 Normalized input-output function

Table B.1 Generalized Input-Output Function

Type	$\varphi \equiv$	$h \equiv$	$x \equiv$	$x' \equiv$
R-R	$\phi(\psi)$	$\Delta\phi$	$\psi/\Delta\psi$	$dx/d\psi = 1/\Delta\psi$
R-P	$z_3(\psi)$	Δz_3	$\psi/\Delta\psi$	$dx/d\psi = 1/\Delta\psi$
P-R	$\phi(z_2)$	$\Delta\phi$	$z_2/\Delta z_2$	$dx/dz_2 = 1/\Delta z_2$
P-P	$z_3(z_2)$	Δz_3	$z_2/\Delta z_2$	$dx/dz_2 = 1/\Delta z_2$

$$\varphi' = hx'\frac{d\tau}{dx} \tag{B.2b}$$

$$\varphi'' = hx'^2\frac{d^2\tau}{dx^2} \tag{B.2c}$$

where x' is defined as in Table B.1

B.2 Cycloidal Function

The cycloidal function satisfies the condition of zero velocity and zero acceleration at the ends. This function and its first and second derivatives are displayed below:

$$\tau = x - \frac{1}{2\pi}\sin 2\pi x \quad \text{(B.3a)}$$

$$\frac{d\tau}{dx} = (1 - \cos 2\pi x) \qquad 0 \leq x \leq 1 \quad \text{(B.3b)}$$

$$\frac{d^2\tau}{dx^2} = 2\pi \sin 2\pi x \quad \text{(B.3c)}$$

B.3 Polynomial Functions

If the rise is represented by a polynomial, then its coefficients are determined from the conditions to be satisfied. The methodology to determine the polynomial coefficients can be found in (Dudley, 1948; Angeles and López-Cajún, 1991). Some of the solutions are presented below:

B.3.1 3-4-5 Polynomial

$$\tau = 10x^3 - 15x^4 + 6x^5 \quad \text{(B.4a)}$$

$$\frac{d\tau}{dx} = 30x^2 - 60x^3 + 30x^4 \qquad 0 \leq x \leq 1 \quad \text{(B.4b)}$$

$$\frac{d^2\tau}{dx^2} = 60x - 180x^2 + 120x^3 \quad \text{(B.4c)}$$

B.3.2 4-5-6-7 Polynomial

$$\tau = 35x^4 - 84x^5 + 70x^6 - 20x^7 \quad \text{(B.5a)}$$

$$\frac{d\tau}{dx} = 140x^3 - 420x^4 + 420x^5 - 140x^6 \qquad 0 \leq x \leq 1 \quad \text{(B.5b)}$$

$$\frac{d^2\tau}{dx^2} = 420x^2 - 1680x^3 + 2100x^4 - 840x^5 \quad \text{(B.5c)}$$

B.4 Combined Functions

In order to improve the performance of the basic curves as defined above, designers have tried combinations of them. The aim has been to produce a follower motion with bounded jerk. One of these combinations is the called the *trapezoid* function, which is a combination of cubic and parabolic curves. This type, from the point of view of the maximum value of $d^2\tau/dx^2$, is slightly better than the cycloidal curve (Chen, 1982). From the same point of view, an even better function was proposed, the *modified trapezoidal* function (Neklutin, 1959), which replaces the cubic curves by cycloidal curves. This function is described below, a detailed derivation being found in (Tesar and Matthew, 1976; Chen, 1982).

B.4.1 Modified Trapezoidal Acceleration

$$\tau = 0.09724612(4x - \frac{1}{\pi}\sin 4\pi x) \tag{B.6a}$$
$$\frac{d\tau}{dx} = 0.3889845(1 - \cos 4\pi x) \qquad 0 \le x < \frac{1}{8} \tag{B.6b}$$
$$\frac{d^2\tau}{dx^2} = 4.888124\sin 4\pi x \tag{B.6c}$$

$$\tau = 2.444406184x^2 - 0.22203097x + 0.00723407 \tag{B.6d}$$
$$\frac{d\tau}{dx} = 4.888124x - 0.22203097 \qquad \frac{1}{8} \le x < \frac{3}{8} \tag{B.6e}$$
$$\frac{d^2\tau}{dx^2} = 4.888124 \tag{B.6f}$$

$$\tau = 1.6110154x - 0.0309544\sin(4\pi x - \pi) - 0.3055077 \tag{B.6g}$$
$$\frac{d\tau}{dx} = 1.6110154 - 0.3889845\cos(4\pi x - \pi) \qquad \frac{3}{8} \le x < \frac{1}{2} \tag{B.6h}$$
$$\frac{d^2\tau}{dx^2} = 4.888124\sin(4\pi x - \pi) \tag{B.6i}$$

$$\tau = 1.6110154x + 0.03009544\sin(4\pi x - 2\pi) - 0.3055077 \tag{B.6j}$$
$$\frac{d\tau}{dx} = 1.6110154 + 0.3889845\cos(4\pi x - 2\pi) \qquad \frac{1}{2} \le x < \frac{5}{8} \tag{B.6k}$$
$$\frac{d^2\tau}{dx^2} = -4.888124\sin(4\pi x - 2\pi) \tag{B.6l}$$

$$\tau = 4.6660917x - 2.44406184x^2 - 1.2292648 \tag{B.6m}$$
$$\frac{d\tau}{dx} = 4.6660917 - 4.888124x \qquad \frac{5}{8} \le x < \frac{7}{8} \tag{B.6n}$$
$$\frac{d^2\tau}{dx^2} = -4.888124 \tag{B.6o}$$

$$\tau = 0.6110154 + 0.3889845x + 0.0309544\sin(4\pi x - 3\pi) \tag{B.6p}$$

$$\frac{d\tau}{dx} = 0.3889845[1 + \cos(4\pi x - 3\pi)] \qquad \frac{7}{8} \leq x \leq 1 \tag{B.6q}$$

$$\frac{d^2\tau}{dx^2} = -4.888124\sin(4\pi x - 3\pi) \tag{B.6r}$$

Appendix C

SYMBOLIC DUAL ALGEBRA

In Chapter 2 we discussed dual numbers as the basic tool for the formulation introduced in Chapters 3 and 4. Some derivations of these chapters are rather difficult to obtain by hand. However, with the aid of *Mathematica*, a software package for symbolic computations (Wolfram, 1992), we succeeded in obtaining and simplifying results. Since operations with dual numbers are not available in *Mathematica*, we wrote our own functions, called *Dualfunctions*. These functions are presented below in alphabetical order.

General definitions like the dual unit ϵ, identified in the code as **ee**, as well as the sine and the cosine of dual numbers, should be written in the first line of the program, namely,

```
Unprotect[Power]; Unprotect[Sin]; Unprotect[Cos];
ee^n_ := 0 /; n >= 2;
dtrigrule = {Sin[Dual[x_ + ee y_]] :> Sin[x] + ee y Cos[x],
             Cos[Dual[x_ + ee y_]] :> Cos[x] - ee y Sin[x],
             Sin[Dual[ee y_]] :> ee y ,
             Cos[Dual[ee y_]] :> 1,
             Sin[Dual[x_]] :> Sin[x],
             Cos[Dual[x_]] :> Cos[x]};
Protect[Power]; Protect[Sin]; Protect[Cos];
```

Moreover, we define de cross product of two vectors as:

```
cross[v_, u_]:= {v[[2]] u[[3]] - u[[2]] v[[3]],
                 u[[1]] v[[3]] - v[[1]] u[[3]],
                 v[[1]] u[[2]] - u[[1]] v[[2]]} ;
```

To simplify expressions in Dualfunctions we apply the following rules:

```
Unprotect[Sqrt]
Sqrt[a_^2] := a;
Protect[Sqrt];

trigrule = {Sin[xx_]^nn_Integer :>
           Sin[xx]^(nn-2) - Cos[xx]^2 Sin[xx]^(nn-2) /; nn > 1,
           xx_ (yy_ + zz_) :> xx yy + xx zz
           };

trigrule2 = {Cos[xx_]^nn_Integer :>
            Cos[xx]^(nn-2) - Sin[xx]^2 Cos[xx]^(nn-2) /; nn > 1,
            xx_ (yy_ + zz_) :> xx yy + xx zz
            };
```

C.1 CrossUnitDual

```
BeginPackage["CrossUnitDual`","DecompDual`"]

CrossUnitDual::usage = "Evaluates the dual cross product of
                                two dual unit vectors"
Begin["`private`"]

   << rules

CrossUnitDual[u_, v_] :=
   Block[{e1x,e1y,e1z,e1,m1x,m1y,m1z,m1,p1,
              e2x,e2y,e2z,e2,m2x,m2y,m2z,m2,p2,b},
      {e1x,m1x}=DecompDual[u[[1]]];
      {e1y,m1y}=DecompDual[u[[2]]];
      {e1z,m1z}=DecompDual[u[[3]]];
      e1 = {e1x,e1y,e1z} //. trigrule;
      m1 = {m1x,m1y,m1z};

      {e2x,m2x}=DecompDual[v[[1]]];
      {e2y,m2y}=DecompDual[v[[2]]];
      {e2z,m2z}=DecompDual[v[[3]]];
      e2 = {e2x,e2y,e2z} //. trigrule;
      m2 = {m2x,m2y,m2z};

      If[e1===e2,
      p1=cross[e1,m1];
      p2=cross[e2,m2];
      b = Sqrt[(p2[[1]]-p1[[1]])^2+(p2[[2]]-p1[[2]])^2+
               (p2[[3]]-p1[[3]])^2];
      b = b //. trigrule;
      PowerExpand[(p2-p1 + Global`ee cross[p1,p2])/b]
      ,
      cross[u,v]
      ]
   ]

End[]

EndPackage[]
```

C.2 DecompDual

```
BeginPackage["DecompDual`"]

DecompDual::usage = "Returns the primal and dual parts of a dual numbe

Begin["`private`"]

DecompDual[ex_] :=
   Block[{ex2,dual,real},
      ex2 = Collect[ex,Global`ee];
      dual = Coefficient[ex2,Global`ee];
      If[FreeQ[ex2,Global`ee],Return[{ex2,0}]];
      real = If[Head[ex2] === Plus,
                 Drop[ex2,{Position[ex2,Global`ee][[1,1]]}],
                 0
             ];
      {real,dual} (* real+Global`ee dual*)
   ]

End[]

EndPackage[]
```

C.3 DivDual

```
BeginPackage["DivDual`"]

DivDual::usage = "Returns a dual number from division of
                            two dual numbers "

Begin["`private`"]

DivDual[x_,y_] :=
   Block[{x1,realx,dualx,y1,realy,dualy},
      x1 = Collect[x,Global`ee];
      dualx = Coefficient[x1,Global`ee];
      If[FreeQ[x1,Global`ee], realx = x1,
      realx = If[Head[x1] === Plus,
                  Drop[x1,{Position[x1,Global`ee][[1,1]]}],
                  0
               ]
      ];
      y1 = Collect[y,Global`ee];
      dualy = Coefficient[y1,Global`ee];
      If[FreeQ[y1,Global`ee], realy = y1,
      realy = If[Head[y1] === Plus,
                  Drop[y1,{Position[y1,Global`ee][[1,1]]}],
                  0
               ]
      ];
      realx/realy + Global`ee (dualx/realy - realx dualy/realy^2)
   ]

End[]

EndPackage[]
```

C.4 SqrtDual

```
BeginPackage["SqrtDual`","DecompDual`","DivDual`"]

SqrtDual::usage = "Evaluates Sqrt[pa + ee da]"

Begin["`private`"]

   << rules

SqrtDual[n_] :=
   Block[{pn,dn},
      {pn,dn} = DecompDual[n];
      pn = pn //. trigrule2;
      dn = dn //. trigrule2;
      If[pn === 0,
      (* "Sqrt of a dual number undefined"*)
      Global`undefinedsqrt
      ,
      Sqrt[pn] + Global`ee dn/2/Sqrt[pn]
      ]
   ]

End[]

EndPackage[]
```

C.5 TanDual

```
BeginPackage["TanDual`","DecompDual`","DivDual`"]

TanDual::usage = "Evaluates a dual angle tan(da)=tan(a)+
                  ee h (1+tan^2(a). Returns: num., den. and offset"

Begin["`private`"]

   << rules

TanDual[n_, d_] :=
   Block[{pn,dn,pd,dd,a,h},
      {pn,dn} = DecompDual[n];
      {pd,dd} = DecompDual[d];
      pn = pn //. trigrule2;
      dn = dn //. trigrule2;
      pd = pd //. trigrule2;
      dd = dd //. trigrule2;
      If[pd === 0,
         h = (dn pd - pn dd)/pn^2;
         {pn,pd,h}
         ,
         h=(dn pd - pn dd) / (pn^2 + pd^2);
         {pn,pd,h}
      ]
   ]

End[]

EndPackage[]
```

C.6 UnitDual

```
BeginPackage["UnitDual`","DecompDual`"]

UnitDual::usage = "Evaluates a dual unit vector from an arbitrary dual vecto

Begin["`private`"]

UnitDual[ex_] :=
   Block[{x1,x2,a1,a2,a3,b1,b2,b3,nx1,x3},
      {a1,b1} = DecompDual[ex[[1]]];
      {a2,b2} = DecompDual[ex[[2]]];
      {a3,b3} = DecompDual[ex[[3]]];
      x1={a1,a2,a3};
      x2={b1,b2,b3};
      nx1=Sqrt[x1 .  x1];
      x3=cross[cross[x1,x2],x1];
      x1/nx1+Global`ee x3/nx1^3
   ]

End[]

EndPackage[]
```

INDEX

A

accurate calculattion, 6.1
active viewport, 7.3
Aronhold-Kennedy Theorem, 1.4, 2.2, 3.1, 4.2
asymptotic normal, 2.3
AutoCAD^R, 6.1
  AME^{TM}, 6.1, 6.6
auxiliary roller, 4.1, 4.4
axial piston pump, 1.2.1

B

backlash, 1.2.1
balancing
  dynamic, 6.1
  static, 6.1
bearing clearance, 1.2.1
bottom surface, 4.2.2, 6.5

C

cam
  kinematics, 7.1
  mechanism, 1.1
cam-follower offset, 1.2.1
canonical systems, 1.2.2
central
closest principal axis of inertia, 6.1
  normal, 2.3
  point, 2.3
composite solid, 6.6
computer graphics, 7.1
conical
  face follower, 4.4
  surface, 4.4
conjugate surfaces, 1.2.1
connectivity, 6.5
constant
  breadth cam mechanism, 5.6
  pressure angle, 4.4
constructive solid geometry (CSG), 6.1, 6.6
contact surface, 4.2, 4.2.1, 4.2.2, 4.2.3, 4.2.4, 5.1, 6.5
cusp, 5.5.2
cycloidal motion, 4.2.2, 5.2, 5.6
cylindrical
  cam mechanism, 3.2.2, 4.3.1
  pair, 1.1, 4.2, 4.2.3
  surface, 3.2.2, 4.1
cylindroid, 1.2.2

D

dextrous, 2.1
differential geometry, 1.2.2
direct
  contact, 5.1
  operation, 3.3, 3.3.1
directrix, 2.3
displacement program, 4.2.1
Divergence Theorem, 6.1, 6.2
driven element, 3.1, 3.3
driving element, 3.1, 3.3
dual
  algebra, 1.4
  arc, 4.1
  angle, 2.1, 4.2.1
  angular velocity, 2.2, 3.2
  curve, 1.2.2, 4.2

matrix, 1.2.2, 2.1, 4.2
number, 2.1
part, 2.1, 3.2
pitch curve, 4.2
quaternion algebra, 1.2.2
radius of curvature, 2.3, 4.2
rotation, 2.1
scalar, 2.1
screw matrix, 2.1
space, 3.1
unit vector, 2.1, 3.2, 4.1
unit sphere, 2.1, 2.2, 3.1
unity, 2.1, 3.2
vector, 2.1
dwell, 5.1, 5.6
dynamic
synthesis, 1.2.1
unbalance, 6.1
dynamics, 1.2.1

E

envelope theory, 1.2.1
error, 1.2.1
assembly, 1.2.1
manufacturing, 1.2.1
Euclidean
norm, 2.1
space, 2.1, 2.2, 6.2
external
RHR-ICM, 5.3
secondary mechanism (SM), 5.5.2

F

fast calculation, 6.1
Ferguson indexing mechanism, 5.1
finite
differences, 1.2.1
-element analysis (FEA), 7.1
trigonometric series, 1.2.1
first-class solid, 6.5
five-link mechanism, 1.2.1
flat-face follower, 3.1, 4.4.1
follower, 1.1
roller, 1.1
frame, 1.1, 3.1
Frenet equations, 2.3
fundamental dimensions, 8.3

G

Gauss Divergence Theorem *see Divergence Theorem*
generatrix, 2.3, 5.5
geodesic
curvature, 2.3
Frenet equations, 2.3
geometric properties, evaluation, 6.1, 6.[illegible]
Geneva mechanism, 1.2.1, 5.1
gyroscopic moment, 6.1
globoidal cam, 4.3.1
great circle, 2.2, 4.1, 4.2

H

helical springs, 5.1
homogeneous
function, 6.2
object, 6.1
hyperboloid of one sheet, 2.3, 4.2.1
hypoid gears, 2.1

I

indexing cam mechanism, 1.1, 3.1, 5.1, 5.1
indexing step, 5.2, 5.5.2
inertia matrix, 6.1
input-output function, 3.1, 3.2.1, 3.2.2, 3.2.3, 3.2.4, 4.3, 5.3, 5.6, 7.3
instan screw axis (ISA), 1.2.2
instantaneous invariants, 1.2.2
Integrated Mechanisms Program (IMP), 7.1
interactive synthesis, 7.1
interactivity, 7.2
intermittent motion, 1.2.1

internal
  RHR-ICM, 5.3
  secondary mechanism (SM), 5.5.2
inverse operation, 3.3, 3.3.1
ISA, 1.2.2

J

jerk, 1.2.1

K

kinematic 1.1
  pair, 1.1
    higher-pair, 1.1, 3.1, 5.5
    lower-pair, 1.1, 3.1
  synthesis, 1.2.1
kinematics, 1.2.1
KINSIN III, 7.1

L

line coordinates, 2.1
LINCAGES, 7.1

M

Mathematica, Appendix C
mean-value theorem, 6.1
mechanical considerations, 8.3
mechanism, 1.1
mesh, 6.5
minimization of cam size, 1.2.1
minimum sliding, 3.1
  velocity, 3.2, 3.2.3
MINN-DWELL, 7.1
MOMENT, 6.1
moment, 6.1
  of a region, 6.2
  of inertia, 6.1, 6.2
MOMENTDB, 6.5
motor, 1.2.2
multi-piston pump, 1.2.1

N

natural trihedron, 2.3
negative action, 5.6
node, 6.5
noise in cams, 1.1.1
norm of a dual vector, 2.1
normalized function, 5.2

O

off-line synthesis, 7.2
on-line synthesis, 7.2
optimization of cam mechanisms, 1.2.1
oscillating follower, 1.2.1, 7.4

P

parallel
  -axes theorem, 6.1, 6.4
  dual vectors, 2.1, 4.2.2, 4.2.3
parallelepiped, 6.5
passive viewport, 7.3
pitch, 2.1
  curve, 5.5.2
  ruled surface, 4.2, 4.2.1, 4.2.2, 4.2.3, 4.2.4, 5.5.2
planar cam mechanism, 1.2.1, 3.1, 3.2.2
Plücker coordinates, 2.1
point coordinates, 2.1, 4.2
pole of the motion, 2.2, 3.1
polode, 2.2, 4.1
polynomial equation of motion, 1.2.1
polydyne, 1.2.1
positive
  action, 5.6
  -definite matrix, 6.1
  motion, 1.2.1, 5.1, 5.6
power losses, 5.1
prescribed motion, 3.1
pressure angle, 1.2.1, 3.1, 3.3, 4.1, 4.3, 5.1, 5.5, 5.6
primal part, 2.1, 3.2
primary mechanism, 5.5, 5.6
primitive solid, 6.1, 6.6
principal axis of inertia, 6.1

prismatic-higher-cylindric-revolute (PHCR) mechanism, 4.2.3
prismatic-higher-cylindric-prismatic (PHCP) mechanism, 4.2.4
prismatic-higher-revolute (PHR) mechanism, 3.2.3
prismatic-higher-revolute-prismatic (PHRP) mechanism, 4.2.4
prismatic pair, 1.1, 3.1, 4.2.3
prismatic-prismatic-prismatic (PPP) mechanism, 3.2.4
proper dual number, 2.1
profile, 1.2.1, 3.1
profiled-follower mechanism, 1.2.1
pure
    dual number, 2.1
    -rolling, 3.2.1, 3.2.3, 4.2.2, 4.2.3, 4.2.4, 4.4, 5.1
        indexing cam mechanism (PRICAM), 5.1, 6.6

R

radial cam mechanism, 1.2.1
revolute-higher-cylindrical-prismatic (RHCP) mechanism, 4.2.2
revolute-higher-cylindrical-revolute (RHCR) mechanism, 4.2.1
revolute-higher-prismatic (RHP) mechanism, 1.4, 3.2.2
revolute-higher-revolute (RHR) mechanism, 1.4, 3.2.1
revolute-higher-revolute-revolute (RHRR) mechanism, 4.2.1
revolute pair, 1,1, 3.1, 4.2, 4.2.2, 4.2.3,
residual vibration, 1.2.1
rotating follower, 3.1
ruled surface, 1.2.2, 2.3, 3.1, 3.2, 3.3, 5.5

S

screw
    axis, 2.2
    pair, 1.1, 3.1
screws
    calculus of, 1.2.2
    contrary, 1.2.2
    reciprocal, 1.2.2
    repelling, 1.2.2
    system of, 1.2.2
    theory of, 1.2.2
second-class solid, 6.5
secondary mechanism, 5.5, 5.6
Silicon Graphics Inc. IRIS workstation, 7.3
single disk planar cam, 1.2.1
SIXPAQ, 7.1
slider-crank mechanism, 1.2.1
sliding velocity, 3.2
solid
    model, 4.2.2, 5.3, 5.4, 5.6
    modeling, 6.1
spatial
    cam mechanism, 1.2.1, 3.1, 3.2.2
    linkage, 1.2.2
spherical
    cam mechanism, 1.2.1, 3.1
    indicatrix, 2.3
    linkage, 1.2.2
    pair, 1.1
    surface, 4.2.2
static unbalance, 6.1
striction curve, 2.3, 3.3
strip, 6.5
symbolic algebra, 5.5.2
symmetric matrix, 6.1
synthesis procedure, 7.4.1

T

tensor, 6.2
    product, 6.2.2
theorem of three axes, 1.2.2
top surface, 4.2.2, 6.5

translating follower, 3.1, 3.2.2

U

undercuting, 4.2, 5.5.2, 7.1
unit normal, 2.3
UNIX workstation, 7.3
USYCAMS, 5.4, 7.1

V

vector basis, 2.1
vertex, 6.5
viewport, 7.3

W

weighted volume, 6.1

Mechanics

SOLID MECHANICS AND ITS APPLICATIONS

Series Editor: G.M.L. Gladwell

Aims and Scope of the Series

The fundamental questions arising in mechanics are: *Why?*, *How?*, and *How much?* The aim of this series is to provide lucid accounts written by authoritative researchers giving vision and insight in answering these questions on the subject of mechanics as it relates to solids. The scope of the series covers the entire spectrum of solid mechanics. Thus it includes the foundation of mechanics; variational formulations; computational mechanics; statics, kinematics and dynamics of rigid and elastic bodies; vibrations of solids and structures; dynamical systems and chaos; the theories of elasticity, plasticity and viscoelasticity; composite materials; rods, beams, shells and membranes; structural control and stability; soils, rocks and geomechanics; fracture; tribology; experimental mechanics; biomechanics and machine design.

1. R.T. Haftka, Z. Gürdal and M.P. Kamat: *Elements of Structural Optimization.* 2nd rev.ed., 1990 ISBN 0-7923-0608-2
2. J.J. Kalker: *Three-Dimensional Elastic Bodies in Rolling Contact.* 1990 ISBN 0-7923-0712-7
3. P. Karasudhi: *Foundations of Solid Mechanics.* 1991 ISBN 0-7923-0772-0
4. N. Kikuchi: *Computational Methods in Contact Mechanics.* (forthcoming) ISBN 0-7923-0773-9
5. *Not published.*
6. J.F. Doyle: *Static and Dynamic Analysis of Structures.* With an Emphasis on Mechanics and Computer Matrix Methods. 1991 ISBN 0-7923-1124-8; Pb 0-7923-1208-2
7. O.O. Ochoa and J.N. Reddy: *Finite Element Analysis of Composite Laminates.* ISBN 0-7923-1125-6
8. M.H. Aliabadi and D.P. Rooke: *Numerical Fracture Mechanics.* ISBN 0-7923-1175-2
9. J. Angeles and C.S. López-Cajún: *Optimization of Cam Mechanisms.* 1991 ISBN 0-7923-1355-0
10. D.E. Grierson, A. Franchi and P. Riva: *Progress in Structural Engineering.* 1991 ISBN 0-7923-1396-8
11. R.T. Haftka and Z. Gürdal: *Elements of Structural Optimization.* 3rd rev. and exp. ed. 1992 ISBN 0-7923-1504-9; Pb 0-7923-1505-7
12. J.R. Barber: *Elasticity.* 1992 ISBN 0-7923-1609-6; Pb 0-7923-1610-X
13. H.S. Tzou and G.L. Anderson (eds.): *Intelligent Structural Systems.* 1992 ISBN 0-7923-1920-6
14. E.E. Gdoutos: *Fracture Mechanics.* An Introduction. 1993 ISBN 0-7923-1932-X
15. J.P. Ward: *Solid Mechanics.* An Introduction. 1992 ISBN 0-7923-1949-4
16. M. Farshad: *Design and Analysis of Shell Structures.* 1992 ISBN 0-7923-1950-8
17. H.S. Tzou and T. Fukuda (eds.): *Precision Sensors, Actuators and Systems.* 1992 ISBN 0-7923-2015-8
18. J.R. Vinson: *The Behavior of Shells Composed of Isotropic and Composite Materials.* 1993 ISBN 0-7923-2113-8

Kluwer Academic Publishers – Dordrecht / Boston / London

Mechanics

SOLID MECHANICS AND ITS APPLICATIONS

Series Editor: G.M.L. Gladwell

19. H.S. Tzou: *Piezoelectric Shells.* Distributed Sensing and Control of Continua. 1993 ISBN 0-7923-2186-3
20. W. Schiehlen: *Advanced Multibody System Dynamics.* Simulation and Software Tools. 1993 ISBN 0-7923-2192-8
21. C.-W. Lee: *Vibration Analysis of Rotors.* 1993 ISBN 0-7923-2300-9
22. D.R. Smith: *An Introduction to Continuum Mechanics.* 1993 ISBN 0-7923-2454-4
23. G.M.L. Gladwell: *Inverse Problems in Scattering.* An Introduction. 1993 ISBN 0-7923-2478-1
24. G. Prathap: *The Finite Element Method in Structural Engineering.* 1993 ISBN 0-7923-2492-7
25. J. Herskovits (ed.): *Structural Optimization '93.* 1993 ISBN 0-7923-2510-9
26. M.A. González-Palacios and J. Angeles: *Cam Synthesis.* 1993 ISBN 0-7923-2536-2

Kluwer Academic Publishers – Dordrecht / Boston / London

GPSR Compliance

The European Union's (EU) General Product Safety Regulation (GPSR) is a set of rules that requires consumer products to be safe and our obligations to ensure this.

If you have any concerns about our products, you can contact us on ProductSafety@springernature.com

In case Publisher is established outside the EU, the EU authorized representative is:

Springer Nature Customer Service Center GmbH
Europaplatz 3
69115 Heidelberg, Germany

Batch number: 10584568

Printed by Printforce, the Netherlands